Applications of Electromagnetic Waves

Applications of Electromagnetic Waves

Special Issue Editor

Reza K. Amineh

MDPI • Basel • Beijing • Wuhan • Barcelona • Belgrade • Manchester • Tokyo • Cluj • Tianjin

Special Issue Editor
Reza K. Amineh
Department of Electrical and Computer Engineering
New York Institute of Technology (NYIT)
USA

Editorial Office
MDPI
St. Alban-Anlage 66
4052 Basel, Switzerland

This is a reprint of articles from the Special Issue published online in the open access journal *Electronics* (ISSN 2079-9292) (available at: https://www.mdpi.com/journal/electronics/special_issues/electromagnetic_waves).

For citation purposes, cite each article independently as indicated on the article page online and as indicated below:

LastName, A.A.; LastName, B.B.; LastName, C.C. Article Title. *Journal Name* **Year**, *Article Number*, Page Range.

ISBN 978-3-03936-300-1 (Hbk)
ISBN 978-3-03936-301-8 (PDF)

Contents

About the Special Issue Editor

Reza K. Amineh (Assistant Professor) is currently with the Department of Electrical and Computer Engineering, New York Institute of Technology. Prior to that, he was a Principal Scientist in the Department of Sensor Physics at Halliburton Co. He received his Ph.D. degree in electrical engineering from McMaster University, Canada, in 2010. He was a postdoctoral fellow at University of Toronto and McMaster University, from 2012 to 2013 and from 2010 to 2012, respectively. He was a Ph.D. intern with the Advanced Technology Group, BlackBerry, in 2009. He has authored/co-authored over 75 journal and conference papers, two book chapters, and a book titled Real-Time Three-Dimensional Imaging of Dielectric Bodies Using Microwave/Millimeter Wave Holography published by Wiley & IEEE Press. He contributed in more than 40 patent disclosures in applied electromagnetics while working at Halliburton Co. and received several industrial awards. His research interests include applied electromagnetics with applications in imaging and sensing. Amineh was a recipient of the prestigious Banting Postdoctoral Fellowship from the Government of Canada in 2012 and the Ontario Ministry of Research and Innovation (OMRI) Postdoctoral Fellowship in 2010. During his Ph.D. program, he was awarded the McMaster Internal Prestige Scholarship Clifton W. Sherman for two consecutive years. He co-authored a paper that was selected as a finalist in the student paper competition at IEEE Wireless and Microwave Technology Conference in 2019, an Honorable Mention Paper presented at the IEEE Symposium on Antennas and Propagation in 2008, and a paper selected among the journal Inverse Problems' "Highlights Collection of 2010". Amineh is a senior member of IEEE.

Preface to "Applications of Electromagnetic Waves"

Electromagnetic (EM) waves carry energy through propagation in space. This radiation associates with entangled electric and magnetic fields which must exist simultaneously. Although all EM waves travel at the speed of light in vacuum, they cover a wide range of frequencies. This full range is called the EM spectrum. The various portions of the EM spectrum are referred to by various names based on their different attributes in the emission, transmission, and absorption of the corresponding waves and also based on their different practical applications. There are no certain boundaries separating these various portions, and the ranges tend to overlap. Overall, the EM spectrum, from the lowest to the highest frequency (longest to shortest wavelength) contains the following waves: radio frequency (RF), microwaves, millimeter waves, terahertz, infrared, visible light, ultraviolet, X-rays, and gamma rays.

This Special Issue consists of sixteen papers covering a broad range of topics related to the applications of EM waves, from the design of filters and antennas for wireless communications to biomedical imaging and sensing and beyond.

I am grateful to the Multidisciplinary Digital Publishing Institute (MDPI) for enabling the creation of this Special Issue and the production of this book.

As a final note, I hope that the reader of this book has a pleasant reading experience. I also hope that she/he will be inspired to download additional articles from the Special Issue that are freely available at https://www.mdpi.com/journal/electronics/special_issues/electromagnetic_waves."

Reza K. Amineh
Special Issue Editor

 *electronics*

Editorial

Applications of Electromagnetic Waves: Present and Future

Reza K. Amineh

Department of Electrical and Computer Engineering, New York Institute of Technology, New York, NY 10023, USA; rkhalaja@nyit.edu

Received: 5 May 2020; Accepted: 11 May 2020; Published: 15 May 2020

1. Introduction

Electromagnetic (EM) waves carry energy through propagation in space. This radiation associates with entangled electric and magnetic fields which must exist simultaneously. Although all EM waves travel at the speed of light in vacuum, i.e., 3×10^8 m/s, they cover a wide range of frequencies called the EM spectrum. The various portions of the EM spectrum are referred to by various names based on their different attributes in the emission, transmission, and absorption of the corresponding waves, and also based on their different practical applications. There are no certain boundaries separating these various portions and the ranges tend to overlap. Overall, the EM spectrum, from the lowest to the highest frequency (longest to shortest wavelength) contains the following waves: radio frequency (RF), microwaves, millimeter waves, terahertz, infrared, visible light, ultraviolet, X-rays, and gamma rays.

In general, the applications of EM waves significantly depend on their corresponding frequency (wavelength). Harnessing the capabilities of EM waves has led to great impacts on various fields such as wireless communication (e.g., see [1]), industrial sensing/imaging (e.g., see [2,3]), biomedical sensing/imaging (e.g., see [4,5]) and treatment (e.g., see [6]), remote sensing (e.g., see [7]), radar (e.g., see [8]), security screening (e.g., see [9]), wireless power transfer (e.g., see [10]), and so on.

2. The Present Issue

This Special Issue consists of sixteen papers covering a broad range of topics related to the applications of EM waves, from the design of filters and antennas for wireless communications to biomedical imaging and sensing and beyond. The contents of these papers are briefly introduced here.

Regarding imaging efforts with EM waves, in [11] a compact and cost-effective three-dimensional (3D) microwave imaging system is proposed based on a fast and robust holographic technique. Unlike the previous 3D holographic imaging techniques which are based on wideband data collection, here, narrow-band microwave data are employed along with an array of receiver antennas. To achieve a low cost and compact size, off-the-shelf components have been employed to build a data acquisition system replacing the costly and bulky vector network analyzers (VNAs). In [12], the feasibility study of a microwave imaging technique is studied based on the Huygens principle for bone lesion detection. An artificial multilayered bone phantom comprised of cortical bone and bone marrow layers has been constructed and the imaging has been implemented based on the measurements in the frequency range of 1–3 GHz. In [13], a non-invasive and repeatable blood glucose monitoring technique is proposed at microwave frequencies by eliminating the leaky modes through the use of surface EM waves from a curved Goubau line. In [14], reverse time migration (RTM) technique is employed to process the permafrost ground penetration radar (GPR) data of the Tibetan highway. The RTM profiles clearly reflect the internal fine structure of permafrost and the thawing state.

Regarding high frequency component design, in [15] the design of a fifth order bandpass waveguide filter with Chebyshev response is proposed, which operates in the X-band at a center frequency of 10 GHz. The structure is based on complementary split ring resonators (CSRRs) and reduces the overall physical length by 31% while enhancing the bandwidth up to 37.5% compared to the conventional designs. In [16], an ultra-wideband bandpass filter (UWB-BPF) with a notch band and a wide upper stopband is proposed. Two pairs of half-wavelength high-impedance line resonators tightly and strongly coupled to the input/output lines are used to provide the wideband responses. In [17], an ultra-broadband terahertz bilayer graphene-based absorption structure is proposed which has high absorption and independence of polarization property. It has two stacking graphene layers sandwiched by an Au cylinders array, backed by a metallic ground plane. The structure shows a bandwidth of 7.1 THz with the absorption exceeding 80%. In [18], a technique is proposed to enhance the bandwidth and gain of an endfire radiating open-ended waveguide using a thin slow-wave surface plasmon structure. Mounted on the E-plane of the stated waveguide, a thin corrugated slow-wave structure has been used in conjunction with a waveguide transition to generate an endfire electromagnetic beam. For the proposed structure, an impedance bandwidth from 8 to 18 GHz has been achieved along with a gain enhancement from 7 to 14.8 dBi. In [19], single-ended and balanced bandpass filters are proposed for multi-channel applications. The proposed U-shaped stepped impedance resonator (USIR) can achieve size miniaturization. Moreover, by using the source–load coupling scheme, two transmission zeros (TZs) are respectively generated at the lower and upper sides of the passbands, which is useful for improvement of the selectivity performance. In addition, spurlines are introduced at the input and output ports to produce another TZ to further enhance the stopband performance. In [20], first applications of metamaterials to microwave antennas are reviewed over the past decade. Then, the manufacturing of microwave antennas using graphene-containing carbon composite materials has been developed and prototypes of dipole and horn antennas made from such materials have been created and studied.

Among another set of diverse applications, in [21] a novel methodology is proposed for material identification based on the use of a microwave sensor array with the elements of the array resonating at various frequencies within a wide range and applying machine learning algorithms on the collected data. The performance of the proposed methodology is tested via the use of easily available materials such as woods, cardboards, and plastics. In [22], the Fourier series expansion method (FSEM) is employed to calculate the complex propagation constants of plasma structures consisting of infinitely long, silver nanorod arrays in the range of 180–1900 nm, and the characteristics of the complex propagation constant are analyzed in depth. In [23], a technical analysis of LED arrays used in monochrome computer printers is presented along with their contribution to unintentional EM emanations. Analyses are based on realistic type sizes and distribution of glyphs. Usable pictures are reconstructed from intercepted RF emanations. In [24], the analysis of levels of EM disturbances from different types of electronic devices is studied. Obtained results are connected with possibilities of existence of sensitive emissions correlating with processed data. The devices of a given type are measured in similar conditions. In [25], an energy verification method for the nozzle of the SC200 proton therapy facility is proposed to ensure safe redundancy of treatment. In [26], electrical characteristic analysis and corresponding experimental tests on gold bonding wire are presented. Firstly, according to EIA (Electronic Industries Association)/JEDEC97 standards, this paper establishes the electromagnetic structure model of gold bonding wire. The parameters, including flat length ratio, diameter, span, and bonding height, are analyzed. In addition, the influence of three kinds of loops of bonding wire is discussed in relation to the S parameters.

3. Future

While some applications of EM waves, such as communication systems and radar, can be considered more traditional, others, such as biomedical imaging and treatment, wireless power transfer, and security screening, are more recent and rapidly growing. This is in part due to the introduction of new concepts such as metamaterials (e.g., see [27]), holographic processing (e.g., see [28]), wireless power transfer methods, radio-frequency identification (RFID) (e.g., see [29]), and so on, which has resonated well with the rapid and significant progress in the field of RF electronics, leading to new commercial products. For instance, for biomedical imaging, microwave imaging systems have been developed and have recently been commercialized (e.g., see [30,31]). As the implementation cost of the EM systems reduces due to the emergence of cost-effective hardware components, it is expected that such systems will grow significantly in the near future and their applications will be expanded in various unexplored directions.

Acknowledgments: First of all, I would like to thank all researchers who submitted articles to this special issue for their excellent contributions. I am also grateful to all reviewers who helped in the evaluation of the manuscripts and made very valuable suggestions to improve the quality of the contributions. I would like to acknowledge the editorial board of *Electronics*, who invited me to guest edit this special issue. I am also grateful to the *Electronics* Editorial Office staff who worked thoroughly to maintain the rigorous peer-review schedule and timely publication. R.K.A. is supported by US National Science Foundation (NSF Award No. 1920098) and New York Institute of Technology's ISRC grants during the course of editing this special issue.

Conflicts of Interest: The author declares no conflicts of interests.

References

1. Tse, D.; Viswanath, P. *Fundamentals of Wireless Communication*; Cambridge University Press: Cambridge, UK, 2005.
2. Zoughi, R. *Microwave Non-Destructive Testing and Evaluation*; Kluwer Academic Publishers: Norwell, MA, USA, 2000.
3. Amineh, R.K.; Martin, L.E.S.; Donderici, B. Holographic Techniques for Corrosion Evaluation of Wellbore Pipes. US Patent No. 9488749, 8 November 2016.
4. Nikolova, N.K. Microwave Imaging for Breast Cancer. *IEEE Microw. Mag.* **2011**, *12*, 78–94. [CrossRef]
5. Sun, Q.; He, Y.; Liu, K.; Fan, S.; Parrott, E.P.G.; Pickwell-MacPherson, E. Recent Advances in Terahertz Technology for Biomedical Applications. *Quant. Imaging Med. Surg.* **2017**, *7*, 345–355. [CrossRef] [PubMed]
6. Yerushalmi, A. Localized, non-invasive deep microwave hyperthermia for the treatment of prostatic tumors: The first 5 years. In *Application of Hyperthermia in the Treatment of Cancer*; Recent Results in Cancer Research; Issels, R.D., Wilmanns, W., Eds.; Springer: Amsterdam, The Netherlands, 1988; p. 107.
7. Campbell, J.B.; Wynne, R.H. *Introduction to Remote Sensing*, 5th ed.; The Guilford Press: New York, NY, USA, 2011.
8. Maio, A.D.; Greco, M.S. *Modern Radar Detection Theory*; SciTech Publishing: Raleigh, NC, USA, 2016.
9. Sheen, D.M.; McMakin, D.L.; Hall, T.E. Three-Dimensional Millimeter-Wave Imaging for Concealed Weapon Detection. *IEEE Trans. Microw. Theory Tech.* **2001**, *49*, 1581–1592. [CrossRef]
10. Artan, N.S.; Amineh, R.K. Wireless Power Transfer to Medical Implants with Multi-Layer Planar Coils. Chapter 9; In *Emerging Capabilities and Applications of Wireless Power Transfer*; IGI Global: Hershey, PA, USA, 2018.
11. Wu, H.; Amineh, R.K. A Low-Cost and Compact Three-Dimensional Microwave Holographic Imaging System. *Electronics* **2019**, *8*, 1036. [CrossRef]
12. Khalesi, B.; Sohani, B.; Ghavami, N.; Ghavami, M.; Dudley, S.; Tiberi, G. A Phantom Investigation to Quantify Huygens Principle Based Microwave Imaging for Bone Lesion Detection. *Electronics* **2019**, *8*, 1505. [CrossRef]
13. Liu, L.W.; Kandwal, A.; Cheng, Q.; Shi, H.; Tobore, I.; Nie, Z. Non-Invasive Blood Glucose Monitoring Using a Curved Goubau Line. *Electronics* **2019**, *8*, 662. [CrossRef]
14. Wang, Y.; Fu, Z.; Lu, X.; Qin, S.; Wang, H.; Wang, X. Imaging of the Internal Structure of Permafrost in the Tibetan Plateau Using Ground Penetrating Radar. *Electronics* **2020**, *9*, 56. [CrossRef]
15. AbuHussain, M.; Hasar, U.C. Design of X-Bandpass Waveguide Chebyshev Filter Based on CSRR Metamaterial for Telecommunication Systems. *Electronics* **2020**, *9*, 101. [CrossRef]
16. Weng, M.-H.; Hsu, C.-W.; Lan, S.-W.; Yang, R.-Y. An Ultra-Wideband Bandpass Filter with a Notch Band and Wide Upper Bandstop Performances. *Electronics* **2019**, *8*, 1316. [CrossRef]
17. Liu, S.; Deng, L.; Qu, M.; Li, S. Polarization-Independent Tunable Ultra-Wideband Meta-Absorber in Terahertz Regime. *Electronics* **2019**, *8*, 831. [CrossRef]
18. Kandwal, A.; Nie, Z.; Li, J.; Liu, Y.; WY Liu, L.; Das, R. Bandwidth and Gain Enhancement of Endfire Radiating Open-Ended Waveguide Using Thin Surface Plasmon Structure. *Electronics* **2019**, *8*, 504. [CrossRef]

19. Yan, F.; Huang, Y.M.; Huang, T.; Ding, S.; Wang, K.; Bozzi, M. Transversely Compact Single-Ended and Balanced Bandpass Filters with Source–Load-Coupled Spurlines. *Electronics* **2019**, *8*, 416. [CrossRef]

20. Dugin, N.A.; Zaboronkova, T.M.; Krafft, C.; Belyaev, G.R. Carbon-Based Composite Microwave Antennas. *Electronics* **2020**, *9*, 590. [CrossRef]

21. Harrison, L.; Ravan, M.; Tandel, D.; Zhang, K.; Patel, T.; K Amineh, R. Material Identification Using a Microwave Sensor Array and Machine Learning. *Electronics* **2020**, *9*, 288. [CrossRef]

22. Zong, C.; Zhang, D. Analysis of Propagation Characteristics along an Array of Silver Nanorods Using Dielectric Constants from Experimental Data and the Drude-Lorentz Model. *Electronics* **2019**, *8*, 1280. [CrossRef]

23. Kubiak, I.; Loughry, J. LED Arrays of Laser Printers as Valuable Sources of Electromagnetic Waves for Acquisition of Graphic Data. *Electronics* **2019**, *8*, 1078. [CrossRef]

24. Kubiak, I. Impact of IT Devices Production Quality on the Level of Protection of Processed Information against the Electromagnetic Infiltration Process. *Electronics* **2019**, *8*, 1054. [CrossRef]

25. Wang, P.; Zheng, J.; Song, Y.; Zhang, W.; Wang, M. Analysis and Design of an Energy Verification System for SC200 Proton Therapy Facility. *Electronics* **2019**, *8*, 541. [CrossRef]

26. Tian, W.; Cui, H.; Yu, W. Analysis and Experimental Test of Electrical Characteristics on Bonding Wire. *Electronics* **2019**, *8*, 365. [CrossRef]

27. Eleftheriades, G.V.; Balmain, K.G. *Negative-Refraction Metamaterials: Fundamental Principles and Applications*; Wiley: Hoboken, NJ, USA; IEEE Press: Piscataway, NJ, USA, 2005.

28. Amineh, R.K.; Nikolova, N.K.; Ravan, M. *Real-Time Three-Dimensional Imaging of Dielectric Bodies Using Microwave/Millimeter Wave Holography*; Wiley: Hoboken, NJ, USA; IEEE Press: Piscataway, NJ, USA, 2019.

29. Hunt, V.D.; Puglia, A.; Puglia, M. *RFID: A Guide to Radio Frequency Identification*; Wiley: Hoboken, NJ, USA, 2007.

30. Micrima Limited. Available online: https://micrima.com/ (accessed on 31 May 2002).

31. EMTensor GmbH. Available online: https://www.emtensor.com/ (accessed on 13 May 2002).

 *electronics*

Article

A Low-Cost and Compact Three-Dimensional Microwave Holographic Imaging System

Hailun Wu and Reza K. Amineh *

Department of Electrical and Computer Engineering, New York Institute of Technology, New York, NY 10023, USA; hwu28@nyit.edu
* Correspondence: rkhalaja@nyit.edu; Tel.: +1-646-273-6204

Received: 17 August 2019; Accepted: 12 September 2019; Published: 15 September 2019

Abstract: With the significant growth in the use of non-metallic composite materials, the demands for new and robust non-destructive testing methodologies is high. Microwave imaging has attracted a lot of attention recently for such applications. This is in addition to the biomedical imaging applications of microwave that are also being pursued actively. Among these efforts, in this paper, we propose a compact and cost-effective three-dimensional microwave imaging system based on a fast and robust holographic technique. For this purpose, we employ narrow-band microwave data, instead of wideband data used in previous three-dimensional cylindrical holographic imaging systems. Three-dimensional imaging is accomplished by using an array of receiver antennas surrounding the object and scanning that along with a transmitter antenna over a cylindrical aperture. To achieve low cost and compact size, we employ off-the-shelf components to build a data acquisition system replacing the costly and bulky vector network analyzers. The simulation and experimental results demonstrate the satisfactory performance of the proposed imaging system. We also show the effect of number of frequencies and size of the objects on the quality of reconstructed images.

Keywords: holography; microwave imaging; microwave measurement system; nondestructive testing

---

1. Introduction

Recently, microwave imaging (MWI) is gaining significant attention, and its applications are growing fast due to the penetration of microwave inside many optically opaque materials. Nowadays, MWI is widely employed to do nondestructive testing (NDT) [1], through-the-wall imaging [2], biomedical imaging [3], etc. One of the most successful applications is the use of MWI in security screening [4,5]. There, direct holographic MWI is employed to measure magnitude and phase of the back-scattered fields over a wide band. Then, fast Fourier-based reconstruction is employed to provide three-dimensional (3D) images. In ref. [4,5], far-field approximations have been employed to derive the 3D image reconstruction process. However, different from concealed weapon detection, microwave imaging techniques for nondestructive testing (NDT) and biomedical applications are mainly applications in near-field regions.

Plastic or newly-developed non-metallic composite materials are widely used in the industrial field these days due to concerns associated with the corrosion of metallic parts. Traditional detection methods such as eddy current testing [6], magnetic flux leakage [7], and magnetic particle testing [8] cannot be applied to detect defects on nonmetallic materials. Aside from NDT for imaging of nonmetallic materials, microwave imaging has been also widely developed for biomedical applications [9–12] which are also considered as near-field applications. This is due to the non-ionizing nature of microwave radiation and its ability to differentiate normal and malignant tissues with different dielectric properties in the human body. Example applications that are being pursued actively include early stage breast cancer detection [3] and brain stroke detection [13].

To address the above-mentioned needs for fast and robust near-field microwave imaging, holographic imaging techniques have been adapted for such applications. In near-filed holographic microwave imaging, back-scattered signals are collected over rectangular [14–16] or cylindrical [17,18] apertures and reconstruction can be performed to volumetrically image the dielectric bodies. A summary of near-field microwave holographic imaging techniques can be found in ref. [19]. In ref. [17], it has been shown that using a cylindrical setup leads to higher quality of images due to the fact that scattered data is collected over all possible angles around the object. To deal with the periodicity of functions along the azimuthal direction in a cylindrical setup, circular convolution theory has been employed along with Fourier transform (FT), solution to linear systems of equations, and inverse Fourier transform (IFT) to reconstruct images. Two-dimensional (2D) images are reconstructed over cylindrical surfaces at multiple radii distances. The stack of these 2D images provides a 3D image.

In ref. [18], wideband data is required to perform 3D imaging in a cylindrical setup. However, a wideband system suffers multiple drawbacks in certain applications including: (1) Data acquisition hardware including antennas and circuitry becomes complex, costly, and bulky. (2) Compact and low-cost data acquisition techniques such as modulated scatterer technique (MST) [20] cannot be implemented easily and efficiently for wideband systems. (3) Additional errors may occur due to dispersive properties of media which may not be modeled accurately in a wideband system. (4) Sweeping scattering (S) parameters over a wideband takes time and this may hinder imaging in applications, in which imaging time is critical such as object tracking or medical imaging (patient movement during data acquisition may generate artifacts). Due to these drawbacks, in ref. [21], near-field holographic 3D MWI has been proposed using single frequency microwave data and an array of receiver antennas in a rectangular scanning setup. Only simulations results were presented in ref. [21].

Here, for the first time, we extend the narrow-band near-field holographic 3D MWI to a cylindrical setup while we employ an array of receiver antennas to collect the scattered data. This allows for benefitting from the advantages of a cylindrical system in providing high quality images while mitigating drawbacks of a wideband system numerated above. Besides, employing narrow-band data in the proposed imaging system allows for building a cost-effective data acquisition circuitry replacing the commonly used vector network analyzer (VNA). In other words, instead of using VNA which is bulky and costly, in this paper, a data acquisition system composed of commercial off-the-shelf microwave components is proposed for near-field 3D holographic MWI. Recently, low-cost microwave measurement systems have been proposed mainly to be used with time-domain microwave imaging systems such as delay and sum (confocal) [22], and multiple signal classification (MUSIC) [23] techniques. Here, we propose the construction of a cost-effective system used with frequency-domain near-field holographic MWI. To allow for collection of sufficient data, a microwave switch is employed along with an array of receiver antennas moving together with a transmitter antenna to scan over a cylindrical aperture.

The validity of the proposed imaging system is first demonstrated via simulation data. We also show the effect of number of frequencies and size of the objects on the quality of images. Then, the construction of a compact and cost-effective imaging system will be explained followed by showing some experimental results.

2. Theory

Figure 1 illustrates the proposed microwave imaging setup including a transmitter antenna to illuminate objects and an array of N_A receiver antennas that scans the scattered fields. The transmitter antenna and the array of receiver antennas scan a cylindrical aperture with radius of r_A and height of z_A. The scattered field is recorded at N_ϕ angles along the azimuthal direction ϕ (within $[0, 2\pi]$) and at N_z positions along the longitudinal direction z. The complex-valued scattered field $E^{sc}(\phi, z)$ is measured, at each sampling position, at N_ω frequencies within the narrow band of ω_1 to ω_{N_ω}, by each receiver. The image reconstruction process then provides images over cylindrical surfaces with radii r_i, where $i = 1, \ldots, N_r$ and r_i is within $(0, r_A)$. It is worth noting that the imaging system is assumed to be linear and space-invariant (LSI). The use of Born approximation for the scattering integral leads to the linear property of the imaging system [15].

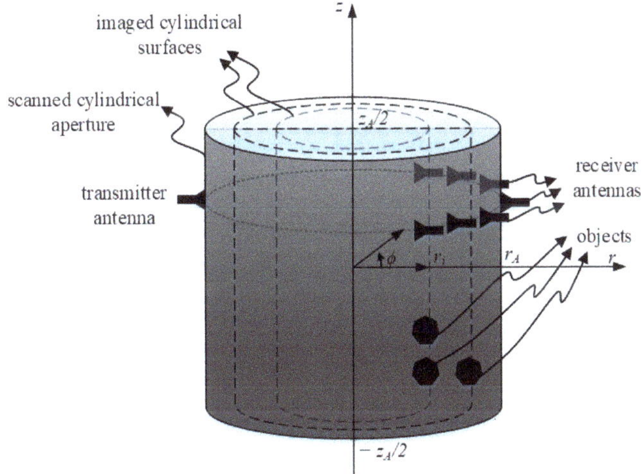

Figure 1. The proposed microwave imaging setup in which a transmitter antenna scans a cylindrical aperture together with an array of receiver antennas. The images are then reconstructed over cylindrical surfaces with radii $r = r_i$.

For implementation of the holographic imaging, first, the responses $E^{sc,co}$ due to small objects called calibration objects (COs) placed at $(r_i,0,0)$, $i = 1, \ldots, N_r$, are recorded. CO is the smallest object with the largest possible contrast with respect to the background medium that can be measured by the system. It approximates an impulse function (Dirac delta function) as an input for the imaging system. The scattered response recorded for a CO placed at $(r_i,0,0)$ is denoted by $E_i^{sc,co}(\phi,z)$ which approximately represents the point-spread function (PSF) of the imaging system. PSF is the impulse response of the system, i.e., the response collected for a point-wise object (here, named CO) which approximates an impulse function as an input for the imaging system. Then, the response due to objects under test (OUT) $E^{sc}(\phi,z)$ can be written as the sum of responses due to objects at cylindrical surfaces $r = r_i$, $i = 1, \ldots, N_r$. The object response at each cylindrical surface, in turn, can be written, according to the convolution theory, as the convolution of the collected PSF for that cylindrical surface $E_i^{sc,co}(\phi,z)$ with the contrast function of the object over that surface $f_i(\phi,z)$. This is written as:

$$E^{sc}(\phi,z) = \sum_{i=1}^{N_r} E_i^{sc}(\phi,z) = \sum_{i=1}^{N_r} E_i^{sc,co}(\phi,z) *_\phi *_z f_i(\phi,z) \tag{1}$$

In Equation (1), PSF functions $E_i^{sc,co}(\phi,z)$ are known due to the measurement or simulation of the CO responses. This indicates that a database of PSFs is built *a priori* for the relevant background medium and imaged surfaces inside them by placing a CO at on that surface and recording the responses over the aperture. Such a database can be created either through measurements or simulations. Then, the recorded PSFs will be employed in the imaging of unknown objects. Besides, $E^{sc}(\phi,z)$ is known due to the recording of the response for the OUT. The goal is then to estimate the contrast functions of objects $f_i(\phi,z)$. To provide more data for image reconstruction, measurements can be implemented at multiple frequencies (over a narrow-band), ω_n, $n = 1, \ldots, N_\omega$ and multiple receivers, a_m, $m = 1, \ldots, N_A$. Thus, for each receiver a_m, Equation (1) can be re-written at all the frequencies to provide the following system of Equations:

$$
\begin{cases}
E_{a_m}^{sc}(\phi, z, \omega_1) = \sum_{i=1}^{N_r} E_{i,a_m}^{sc,co}(\phi, z, \omega_1) *_\phi *_z f_i(\phi, z) \\
\qquad\qquad\vdots \\
E_{a_m}^{sc}(\phi, z, \omega_{N_\omega}) = \sum_{i=1}^{N_r} E_{i,a_m}^{sc,co}(\phi, z, \omega_{N_\omega}) *_\phi *_z f_i(\phi, z)
\end{cases}
\tag{2}
$$

We can get such systems of equations for each receiver a_m, $m = 1, \ldots, N_A$, and then combine all these systems of equations since they share the same unknown parameters $f_i(\phi, z)$, $i = 1, \ldots, N_r$. In order to solve the system of equations we transform the equations to the spatial frequency domain. In ref. [14], doing such transformation is straight-forward along x and y directions. However, here, the functions are periodic along ϕ direction. This necessitates modification of the processing.

Let us first consider the spatially-sampled versions of $E_{a_m}^{sc}(\phi, z, \omega_n)$, $E_{i,a_m}^{sc,co}(\phi, z, \omega_n)$, and $f_i(\phi, z)$ denoted by $E_{a_m}^{sc}(n_\phi, n_z, \omega_n)$, $E_{i,a_m}^{sc,co}(n_\phi, n_z, \omega_n)$, and $f_i(n_\phi, n_z)$, $n_\phi = 1, \ldots, N_\phi$ and $n_z = 1, \ldots, N_Z$, with spatial and angular intervals denoted by Δz and $\Delta\phi$, respectively. Thus, the convolutions in Equation (1) can be written in spectral domain as [24]:

$$
\text{DTFT}_{z,\phi}\{E_{a_m}^{sc}(n_\phi, n_z, \omega_n)\} = \sum_{i=1}^{N_r} \text{DTFT}_{z,\phi}\{E_{i,a_m}^{sc,co}(n_\phi, n_z, \omega_n)\}\text{DTFT}_{z,\phi}\{f_i(n_\phi, n_z)\}
\tag{3}
$$

where $\text{DTFT}_{z,\phi}$ denotes discrete time FT (DTFT) along azimuthal and longitudinal directions, respectively. Sequences $E_{a_m}^{sc}(n_\phi, n_z, \omega_n)$, $E_{i,a_m}^{sc,co}(n_\phi, n_z, \omega_n)$, and $f_i(n_\phi, n_z)$ are aperiodic along the longitudinal direction z. The number of samples along z, namely N_z, is taken sufficiently large such that the values outside the sampled window are negligible. Their DTFT is, however, a periodic function versus the spatial frequency variable k_z (corresponding to z), with period of $1/\Delta z$. Besides, these DTFTs are periodic sums of the FT of their corresponding continuous functions. Thus, the value of the continuous FT of these functions (with negligible aliasing from the adjacent terms) can be obtained from DTFT values within the range $[-1/(2\Delta z), +1/(2\Delta z)]$, provided that Δz is sufficiently small. The DTFTs with respect to z are denoted by $\tilde{E}_{a_m}^{sc}(n_\phi, k_z, \omega_n)$, $\tilde{E}_{i,a_m}^{sc,co}(n_\phi, k_z, \omega_n)$, and $\tilde{f}_i(n_\phi, k_z)$. Since these functions are periodic along ϕ, the convolution along that direction can be considered as a circular convolution [24]. Then, the DTFTs for the N_ϕ-periodic sequences along ϕ are computationally reduced to discrete Fourier transforms (DFT) of these sequences [24]. The DFTs with respect to the ϕ variable for sequences $\tilde{E}_{a_m}^{sc}(n_\phi, k_z, \omega_n)$, $\tilde{E}_{i,a_m}^{sc,co}(n_\phi, k_z, \omega_n)$, and $\tilde{f}_i(n_\phi, k_z)$ are denoted by $\tilde{\tilde{E}}_{a_m}^{sc}(k_\phi, k_z, \omega_n)$, $\tilde{\tilde{E}}_{i,a_m}^{sc,co}(k_\phi, k_z, \omega_n)$, and $\tilde{\tilde{f}}_i(k_\phi, k_z)$, where k_ϕ is an integer from 0 to $N_\phi - 1$.

Using the transformations discussed above at all the frequencies for each receiver a_m leads to the following system of equations at each spatial frequency pair $\kappa = (k_\phi, k_z)$:

$$
\begin{cases}
\tilde{\tilde{E}}_{a_m}^{sc}(\kappa, \omega_1) = \sum_{i=1}^{N_r} \tilde{\tilde{E}}_{i,a_m}^{sc,co}(\kappa, \omega_1)\tilde{\tilde{f}}_i(\kappa) \\
\qquad\qquad\vdots \\
\tilde{\tilde{E}}_{a_m}^{sc}(\kappa, \omega_{N_\omega}) = \sum_{i=1}^{N_r} \tilde{\tilde{E}}_{i,a_m}^{sc,co}(\kappa, \omega_{N_\omega})\tilde{\tilde{f}}_i(\kappa)
\end{cases}
\tag{4}
$$

After combining the systems of equations for all the N_A receivers, the following system of equations is obtained at each spatial frequency pair $\kappa = (k_\phi, k_z)$:

$$
\tilde{\tilde{\underline{E}}}^{sc} = \tilde{\tilde{\underline{\underline{D}}}}\, \underline{F}
\tag{5}
$$

where

$$\tilde{\bar{E}}^{sc} = \begin{bmatrix} \tilde{\bar{E}}_1^{sc} \\ \vdots \\ \tilde{\bar{E}}_{N_A}^{sc} \end{bmatrix} \quad \tilde{\bar{\bar{D}}} = \begin{bmatrix} \tilde{\bar{D}}_1 \\ \vdots \\ \tilde{\bar{D}}_{N_A} \end{bmatrix} \quad \tilde{\bar{f}} = \begin{bmatrix} \tilde{f}_1(\kappa) \\ \vdots \\ \tilde{f}_{N_r}(\kappa) \end{bmatrix} \tag{6}$$

And

$$\tilde{\bar{E}}_{a_m}^{sc} = \begin{bmatrix} \tilde{\bar{E}}_{a_m}^{sc}(\kappa,\omega_1) \\ \vdots \\ \tilde{\bar{E}}_{a_m}^{sc}(\kappa,\omega_{N_\omega}) \end{bmatrix}, \tilde{\bar{\bar{D}}}_{a_m} = \begin{bmatrix} \tilde{\bar{E}}_{1,a_m}^{sc,co}(\kappa,\omega_1) & \cdots & \tilde{\bar{E}}_{N_r,a_m}^{sc,co}(\kappa,\omega_1) \\ \vdots & \ddots & \vdots \\ \tilde{\bar{E}}_{1,a_m}^{sc,co}(\kappa,\omega_{N_\omega}) & \cdots & \tilde{\bar{E}}_{N_r,a_m}^{sc,co}(\kappa,\omega_{N_\omega}) \end{bmatrix} \tag{7}$$

These systems of equations are solved at each spatial frequency pair $\kappa = (k_\phi, k_z)$ to obtain the values for $\tilde{f}_i(\kappa)$, $i = 1,\ldots,N_r$. Then, inverse DTFT along longitudinal direction z and inverse DFT along azimuthal direction ϕ can be applied to reconstruct images $f_i(n_\phi, n_z)$ over all cylindrical surfaces $r = r_i$, $i = 1,\ldots,N_r$. At the end, the normalized modulus of $f_i(n_\phi, n_z)$, $|f_i(n_\phi, n_z)|/M$, where M is the maximum of $|f_i(n_\phi, n_z)|$ for all r_i, is plotted versus ϕ and z to obtain 2D images of the objects at all N_r cylindrical surfaces. By putting together all 2D images, a 3D image of the objects is obtained. We call this process normalization of the images.

3. Simulation Results

In this section, we present the imaging results obtained from applying the proposed holographic imaging technique on the data simulated in FEKO which is a high frequency electromagnetic simulation software [25]. For 2D imaging, we conduct the simulations to collected data when the transmitter and receivers antennas rotate $360°$ around the objects. For 3D imaging, in addition to the azimuthal rotation as mentioned above, the transmitter and receivers antennas scan together along z direction as well. Further details are mentioned in the following subsections. First, we show the performance of the technique in 2D imaging and study the effect of number of frequencies and object size. Then, we demonstrate the satisfactory performance of the technique by 3D imaging examples.

3.1. 2D Imaging with Single-Frequency Data

Figure 2 shows the FEKO simulation setup consisting of one transmitter antenna and eight receiver antennas rotating together on a circle of radius $R = 60$ mm. The antennas are resonant dipoles. The azimuthal angle between the receiver antennas is $\Delta\phi_a = 20°$ Properties of the background medium are $\varepsilon_r = 22$ and $\sigma = 1.25$ S/m. We perform imaging over three circles with radii of $r_1 = 24$ mm, $r_2 = 36$ mm, and $r_3 = 48$ mm. The objects are cubes of size $D = 3$ mm. There are two objects at r_1 with angular separation of $\Delta\phi_1 = 40°$, one object at r_2, and two objects at r_3 with angular separation of $\Delta\phi_3 = 60°$. The properties of objects are $\varepsilon_r = 55$ and $\sigma = 4$ S/m. Data is collected at 181 samples along the azimuthal direction (every $2°$) and at 1.7 GHz.

The reconstructed images over three circles with radii of 24 mm, 36 mm and 48 mm are shown in Figure 3. On the circle with radius $r_3 = 48$ mm, two high-level peaks are observed at $\pm20°$ which means that two cubes on the outer circle can be reconstructed well. On the middle circle with radius of $r_2 = 36$ mm, one high-level peak is observed at 0 degree correctly representing an object at that position. However, many high-level artifacts are present compared to the reconstructed image on the outer circle. The reconstructed image on the inner circle with radius of $r_1 = 24$ mm shows no distinct high-level peaks representing the objects.

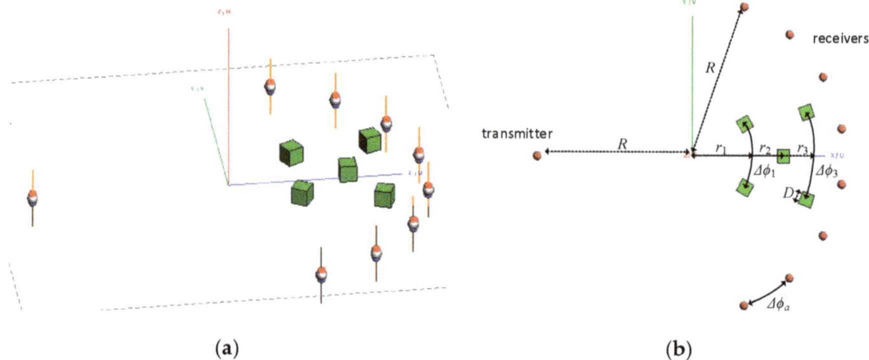

(a)

(b)

Figure 2. (**a**) Angled view of the FEKO simulation setup, and (**b**) Top-view of the setup consisting of one transmitter antenna and eight receiver antennas rotating together on a circle of radius $R = 60$ mm. The properties of the background medium are $\varepsilon_r = 22$ and $\sigma = 1.25$ S/m. The properties of the objects are $\varepsilon_r = 55$ and $\sigma = 4$ S/m.

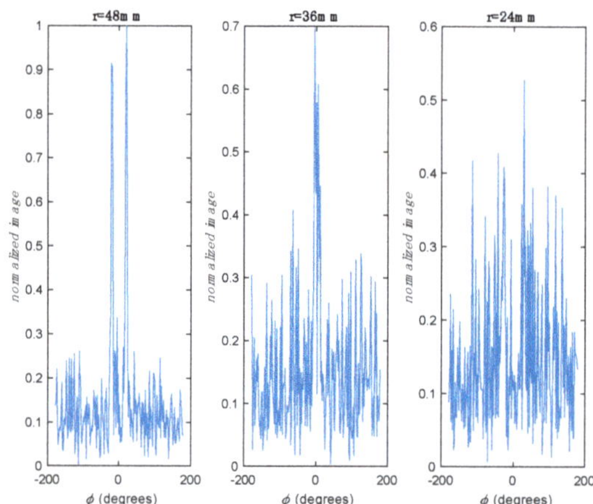

Figure 3. Normalized 2D reconstructed image of the objects with side of $D = 3$ mm on three imaging circles with radii of 48, 36, and 24 mm using single frequency data collected by eight receiver antennas.

3.2. 2D Imaging with Double Frequency Data

In order to improve the quality of the reconstructed images, double frequency data is collected by eight receiver antennas. With more frequency information, better reconstructed images are expected.

Figure 4 shows the reconstructed images when using double frequency data at 1.5 GHz and 1.9 GHz. Compared to the images shown in Figure 3, it is observed that when using double frequency data, the reconstructed image on the circle with radius of $r_3 = 48$ mm has better quality showing two distinct peaks representing the presence of the two objects on that circle and lower level of artifacts. The reconstructed image on the middle circle also shows lower level of artifacts compared to those in Figure 3 and the high-level peak representing the object on that circle has better quality.

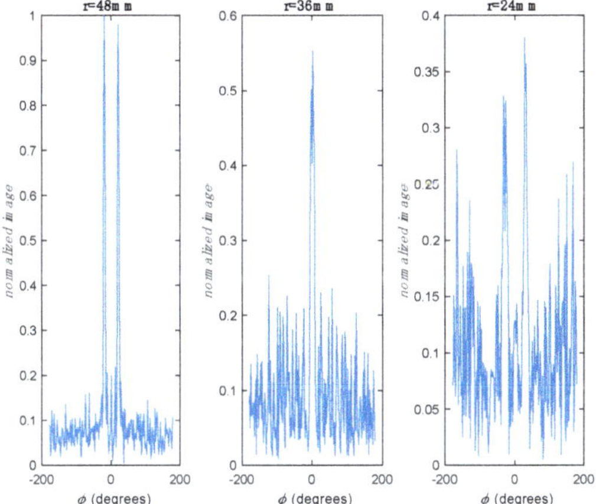

Figure 4. Normalized 2D reconstructed image of objects with side of $D = 3$ mm on three imaging circles with radii of 48, 36 and 24 mm using double frequency data collected by eight receiver antennas.

On the inner circle with a radius of $r_1 = 24$ mm, two peaks are observed at $\pm30^\circ$. This indicates that the two cubes on the inner circle can be reconstructed when double frequency data is collected although we still observe some high level of artifacts on this circle.

3.3. 2D Imaging of Larger Objects

In order to see the effect of size of the objects on the quality of the reconstructed images, we increase the size of objects in Figure 2 to $D = 5$ mm and 10 mm. The images in this section are reconstructed with data collected at 1.5 GHz and 1.9 GHz.

The reconstructed images for object sizes of $D = 5$ mm are shown in Figure 5. By comparing these results with those in Figure 4, we conclude that the quality of the reconstructed images is better, in particular, for the imaged circle with radius of $r_1 = 24$ mm. The levels of the artifacts on the imaged circles with radii of $r_1 = 24$ mm and $r_2 = 36$ mm get much lower than those in Figure 4.

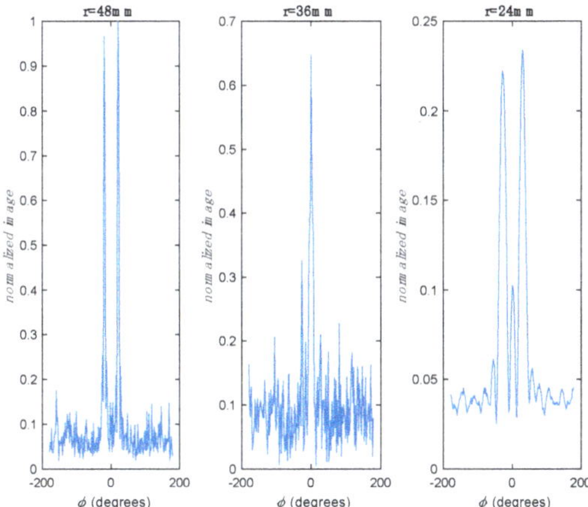

Figure 5. Normalized 2D reconstructed image of objects with a side of $D = 5$ mm on three imaged circles with radii of 48, 36 and 24 mm using double frequency data collected by eight receiver antennas.

Next, we increase the size of objects to $D = 10$ mm. Figure 6 shows the reconstructed images for this case. Compared to the reconstructed images of objects with $D = 5$ mm, the quality of the reconstructed images gets worse. On the imaged circle with radius of $r_2 = 36$ mm, two peaks are observed which might be the shadow of objects on the imaged circle with radius of $r_3 = 48$ mm. On the imaged circle with radius of $r_1 = 24$ mm, the objects at $\pm 30^{\circ}$ are not reconstructed well and the image includes many spurious peaks. The degradation of the imaging quality is mainly due to the use of Born approximation in holographic imaging which indicates that the image reconstruction quality deteriorates for larger or higher contrast objects [15].

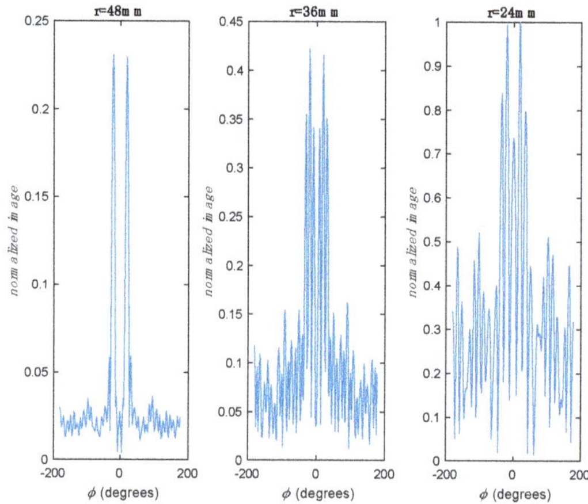

Figure 6. Normalized 2D reconstructed image of objects with a side of $D = 10$ mm on three imaged circles with radii of 48, 36 and 24 mm using double frequency data collected by eight receiver antennas.

3.4. Study of the Imaging Quality

In this section, we study the quality of the imaging process when using single frequency data, double frequency data, and when a single object with sizes of 3 mm, 5 mm, and 10 mm is placed on circles with radii of $r_1 = 24$ mm, $r_2 = 36$ mm, and $r_3 = 48$ mm. To evaluate the quality of the reconstructed images, we define a reconstruction error E_T parameter as:

$$E_T = \sum_{i=1}^{N_r} \left\| |f_i(n_\phi, n_z)|/M - f_{i,ideal}(n_\phi, n_z) \right\| \tag{8}$$

where $f_{i,ideal}(n_\phi, n_z)$ is the ideal image for which the values are all 0 except being 1 at the true positions of the objects.

Form Tables 1 and 2, it is observed that the quality of the imaging degrades for inner circles. This has been notified in the previous works as well (e.g., see ref. [18]). Also, it is observed that the quality of the imaging improves when using double frequency data compared to single frequency data.

Table 1. Reconstruction error when using single frequency data.

Size	$r = 48$ mm	$r = 36$ mm	$r = 24$ mm
$D = 3$ mm	7.62	9.12	10.73
$D = 5$ mm	9.52	9.64	11.83
$D = 10$ mm	10.55	13.15	13.52

Table 2. Reconstruction error when using double frequency data.

Size	$r = 48$ mm	$r = 36$ mm	$r = 24$ mm
$D = 3$ mm	6.41	6.91	8.86
$D = 5$ mm	6.44	8.52	8.90
$D = 10$ mm	8.08	8.85	9.23

3.5. 3D Imaging with Double Frequency Data and Eight Receivers

Figure 7 shows the FEKO simulation model for the first 3D imaging example. The transmitter-receivers configuration, number of antennas, properties of the background and objects are similar to those in Figure 2. The objects are two cuboids with square cross-section of size $S = 4$ mm and height of $L = 67.5$ mm. The angular separation between the cuboids is $\Delta\phi = 40°$ and they are placed at radius of $r = 36$ mm. In order to have a realistic study, we add White Gaussian noise with signal-to-noise ratio (SNR) of 30 dB to the simulated data. Scanning step along the azimuthal direction is the same as in the 2D imaging simulations. Along z axis scanning is performed at 21 steps over -5λ to 5λ, where λ is the wavelength at center frequency 1.7 GHz. Data is collected at 1.5 GHz and 1.9 GHz. Image reconstruction is implemented over three cylindrical surfaces at $r_1 = 24$ mm, $r_2 = 36$ mm, and $r_3 = 48$ mm. Sample raw responses for the 4th receiver (in the middle), at frequency of 1.5 GHz, and for PSFs for $r_1 = 24$ mm, $r_2 = 36$ mm, and $r_3 = 48$ mm for object response are shown in Figure 8. Figure 9 shows the reconstructed images. Two cuboids on the middle surface are reconstructed well. Two bright lines are observed at $\pm20°$ on the middle surface and the extent of them along the z axis is consistent with the actual height of the objects. This clearly shows the capability of the proposed imaging technique in reconstruction of 3D images using narrow-band microwave data.

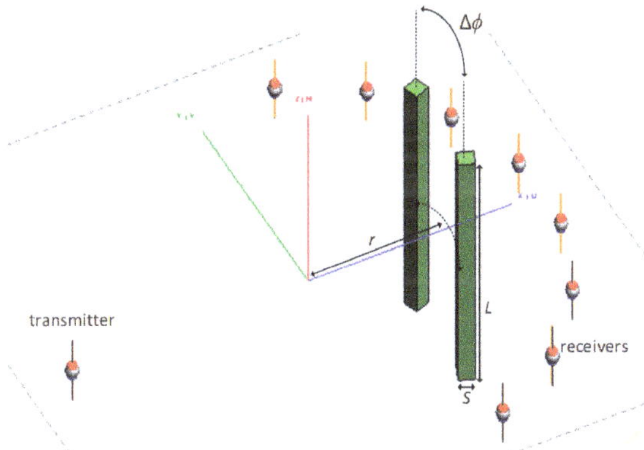

Figure 7. FEKO simulation model for the first 3D imaging example. The transmitter-receivers configuration, number of antennas, properties of the background and objects are similar to those in Figure 2. The objects are two cuboids with square cross-section of size $S = 4$ mm and height of $L = 67.5$ mm. The angular separation between the cuboids is $\Delta\phi = 40°$ and they are placed at radius of $r = 36$ mm.

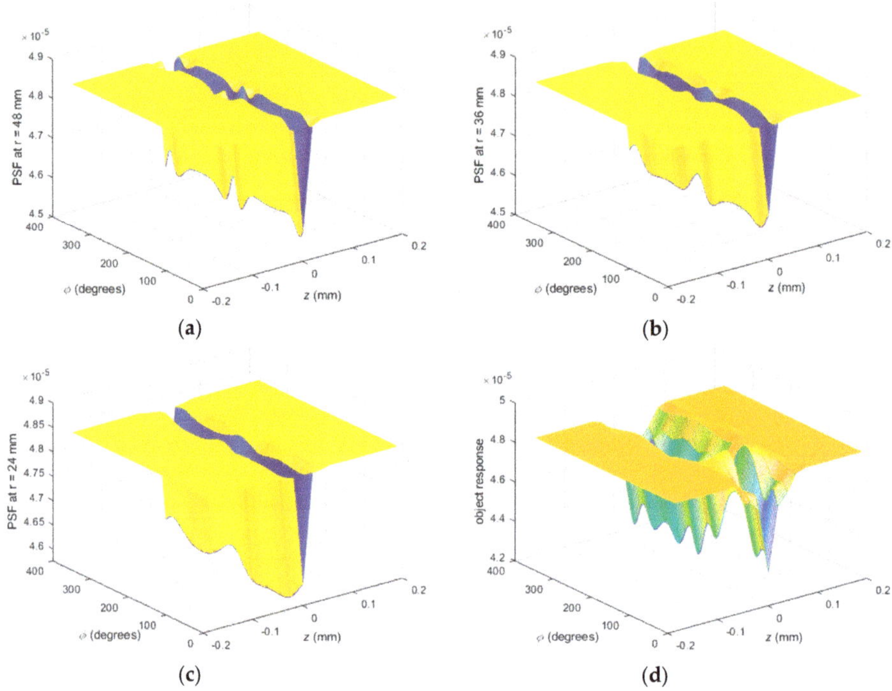

Figure 8. Sample raw responses for the 4th receiver (in the middle) and at frequency of 1.5 GHz: (a) point-spread function (PSF) for r = 48 mm, (b) PSF for r = 36 mm, (c) PSF for r = 24 mm, and (d) objects response.

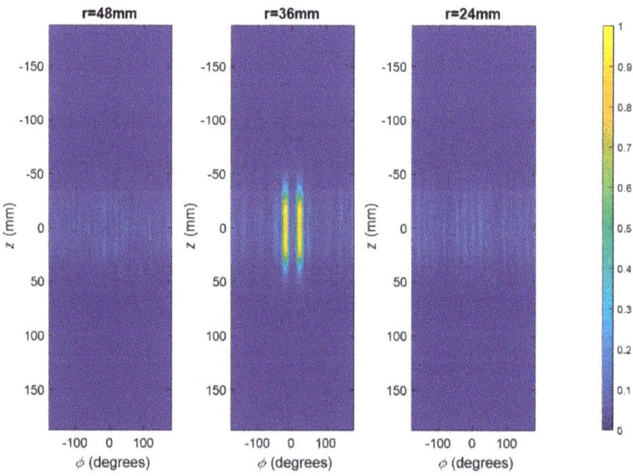

Figure 9. Normalized 3D reconstructed image of the objects shown in Figure 7 on three imaged surfaces at radii of 48, 36, and 24 mm using double frequency data collected by eight receiver antennas.

Figure 10 shows the FEKO simulation model for the second 3D imaging example. The transmitter-receivers configuration, number of antennas, properties of the background medium and objects are similar to those in Figure 2. There is an X-shaped object with square cross-section of size S = 4 mm and length of arms L = 56.25 mm. It is placed at radius of r = 36 mm with its arms rotated

45° with respect to the x and z axes. Similar to the previous example, in order to have a realistic study, we add White Gaussian noise with signal-to-noise ratio (SNR) of 30 dB to the simulated data. Scanning step along the azimuthal direction is the same as in the 2D imaging simulations. Along z axis scanning is performed at 21 steps over -5λ to 5λ, where λ is the wavelength at center frequency 1.7 GHz. Data is collected at 1.5 GHz and 1.9 GHz. Image reconstruction is implemented over three cylindrical surfaces at $r_1 = 24$ mm, $r_2 = 36$ mm, and $r_3 = 48$ mm. Figure 11 shows the reconstructed images. The X-shaped object is reconstructed well at the middle-imaged surface confirming the satisfactory performance of the proposed imaging technique.

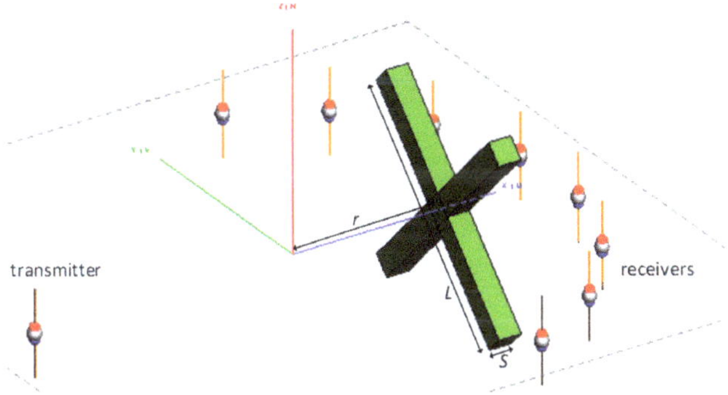

Figure 10. FEKO simulation model for the second 3D imaging example. The transmitter-receivers configuration, number of antennas, properties of the background and objects are similar to those in Figure 2. There is an X-shape object with square cross-section of size $S = 4$ mm and length of each arm of $L = 56.25$ mm. It is placed at radius of $r = 36$ mm with its arms rotated 45° with respect to the x and z axes.

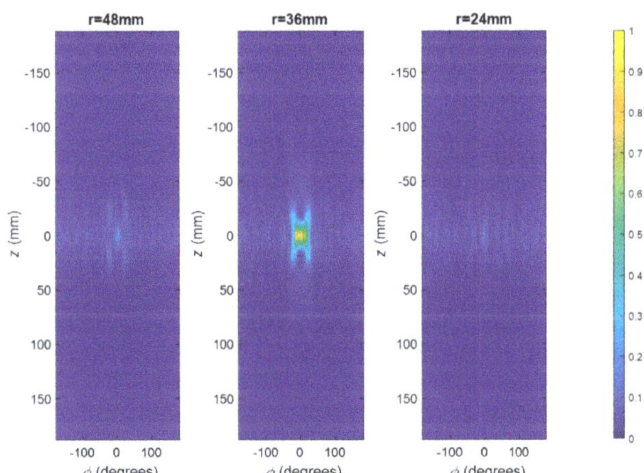

Figure 11. Normalized 3D reconstructed image of the object in Figure 10 on three imaged surfaces at radii of 48, 36, and 24 mm using double frequency data collected by eight receiver antennas.

3.6. Resolution along Azimuthal and Longitudinal Directions

Figure 12 shows the normalized 1D slices along ϕ and z directions for the 3D reconstructed images of a single object placed at $r_3 = 48$ mm, $r_2 = 36$ mm, and $r_1 = 24$ mm (obtained from three separate image reconstruction processes). The resolution is evaluated by computing the distance bounded by two points on the image on opposite sides of the peak and marked by 0.7 times the peak value.

This corresponds to approximately 6 mm and 14 mm along the azimuthal and longitudinal directions, respectively, and it is approximately similar for all the radii (only slight degradation in the order of 1 mm or 2 mm is observed for the smaller radii compared to larger ones). The resolution along azimuthal direction has been evaluated by $\Delta\phi$ in radian multiplied by the radius for the corresponding surface, where $\Delta\phi$ is the angular width of the 0.7 level discussed above. That is why although the angular widths of the 0.7 levels look different in Figure 12a–c, they all lead to approximately similar azimuthal resolutions. Please note that to have a realistic evaluation of the resolution, the data has been corrupted with noise of SNR = 30 dB.

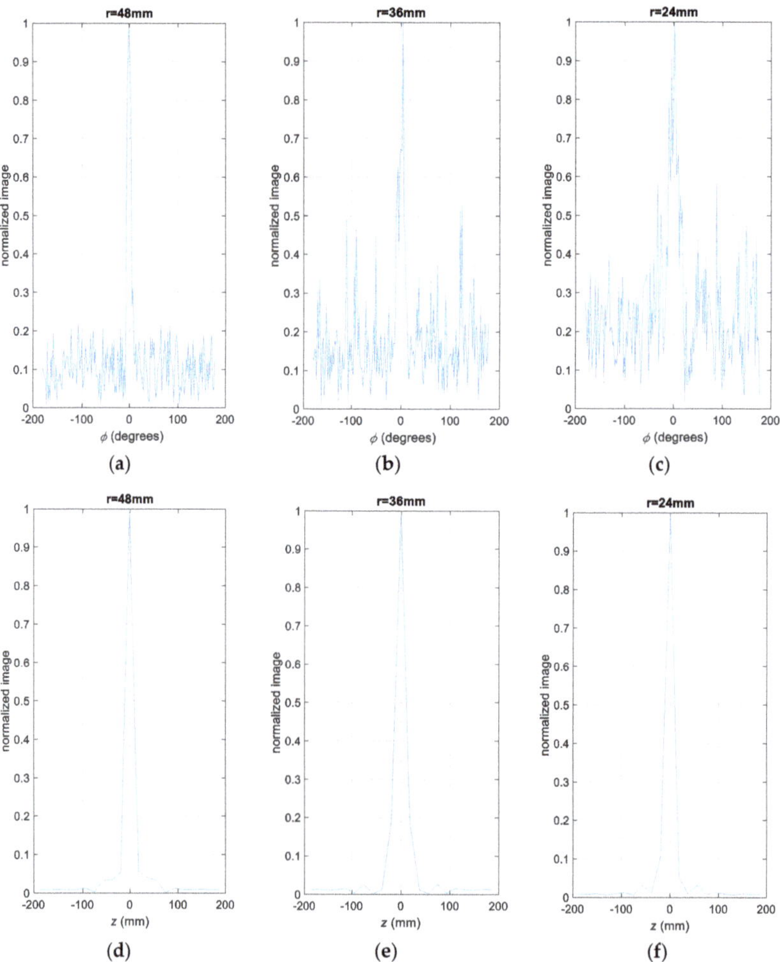

Figure 12. Depiction of 1D slices of 3D reconstructed images of a single object placed: (**a**) at 48 mm, slice along ϕ direction, (**b**) at 36 mm, slice along ϕ direction, (**c**) at 24 mm, slice along ϕ direction, (**d**) at 48 mm, slice along z direction, (**e**) at 36 mm, slice along z direction, (**f**) at 24 mm, slice along z direction.

3.7. Study the Effect of Noise

In order to study the effect of noise on the reconstructed images, in this section we present the results for the 3D imaging example in Figure 10 with the results shown with noise of SNR = 30 dB in

Figure 11. Here, we decrease SNR to 20 dB and 10 dB. Figure 13 shows the reconstructed images. It is observed that with SNR = 20 dB, the results are still satisfactory but with SNR = 10 dB the quality of the images deteriorates significantly.

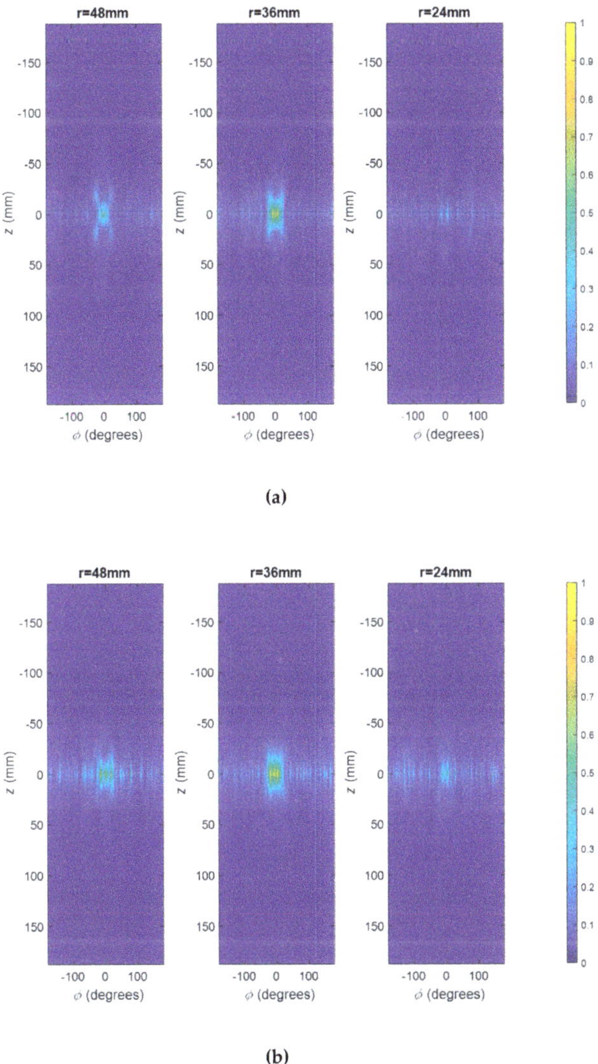

(a)

(b)

Figure 13. Reconstructed 3D images of the object shown in Figure 11 when the data is corrupted by noise of: (**a**) SNR = 20 dB and (**b**) SNR = 10 dB.

4. Experimental Results

In this section, we present the construction of a low-cost microwave data acquisition system including a transmitter unit, an in-phase quadrature (I/Q) receiver unit, a cylindrical scanning system, antennas, and computer for controlling and processing. Then, we present the 3D imaging results demonstrating the satisfactory performance of the system.

4.1. Microwave Measurement System

Figure 14 shows the block diagram of the constructed low-cost and compact 3D microwave holographic imaging system. For our measurements, we use a Plexiglas container including a mixture of water (20%) and glycerin (80%). According to ref. [26], this mixture has properties of approximately $\varepsilon_r = 22$ and $\sigma = 1.25$ S/m within the frequency range of 1.5 GHz to 1.8 GHz which is the targeted operation range in our experimental study. We confirmed this by dielectric property measurements using a Keysight Dielectric probe kit (performance probe N1501A) together with the relevant measurement Software N1500A and a VNA (E5063A from Keysight). The liquid mixture container has diameter of 120 mm and height of 200 mm. The objects to be imaged are plastic cylinders with diameter of 18 mm and height of 50 mm covered by thin copper sheets. These objects are held inside the liquid at the desired positions with thin wooden sticks that are clipped to a cylindrical foam placed at the top of the liquid container. Imaging will be performed over cylindrical surfaces (3D imaging) at radii of $r_1 = 20$ mm, $r_2 = 35$ mm, and $r_3 = 50$ mm. Figure 15 shows the imaging system along with the zoomed view for the main components that will be described in the following. To reduce electromagnetic interferences (EMI), the data acquisition circuitry and the scanning setup are placed inside boxes covered by microwave absorbing sheets.

In the data acquisition system, a transmitter module, DC1705C from Analog Devices, with frequency range from 700 MHz to 6.39 GHz is connected to a 10 MHz precision pocket reference oscillator, PPRO30–10.000 from Crystek Corporation, which provides a reference frequency. A USB serial controller, DC590B from Analog Devices, is connected to DC1705C so that the transmitter unit can be controlled by PC via MATLAB software [27]. The output of the transmitter unit is connected to a variable gain amplifier (VGA), ADL5330 from Analog Devices, operating from 10 MHz to 3 GHz which is then connected to a transmitter antenna. In this way, a microwave signal can be transmitted with variable frequencies and powers to illuminate the imaged medium.

For transmitting and receiving the microwave power, we employ commercial monopole antennas, Mini GSM/Cellular Quad-Band Antenna-2 dBi SMA Plug from Adafruit Co., covering frequency bands of 850/900/1800/1900/2100 MHz. Figure 16 shows the measured $|S_{11}|$ (by VNA) for the nine antennas used as transmitter and receivers. These measurements are performed while the antennas are placed in a 3D-printed holder around the liquid container. The values of $|S_{11}|$ are mostly below -10 dB over the targeted operation band of 1.5 GHz to 1.8 GHz.

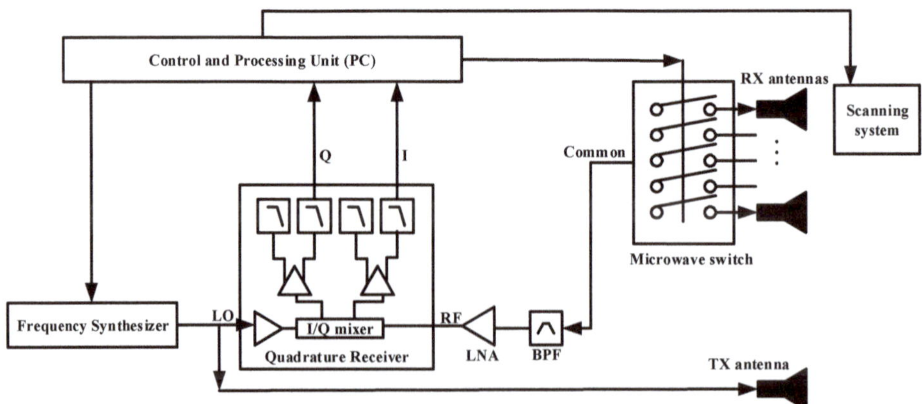

Figure 14. Block diagram of the imaging system.

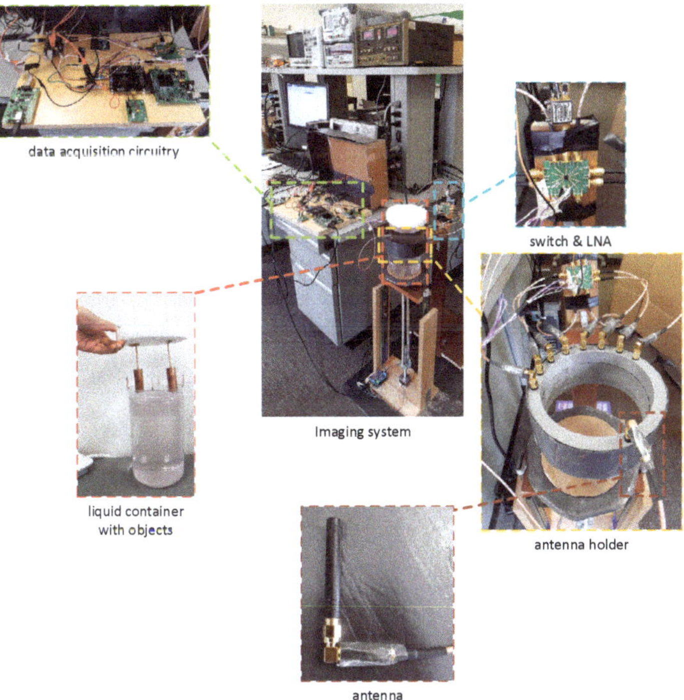

Figure 15. Imaging system with its main components.

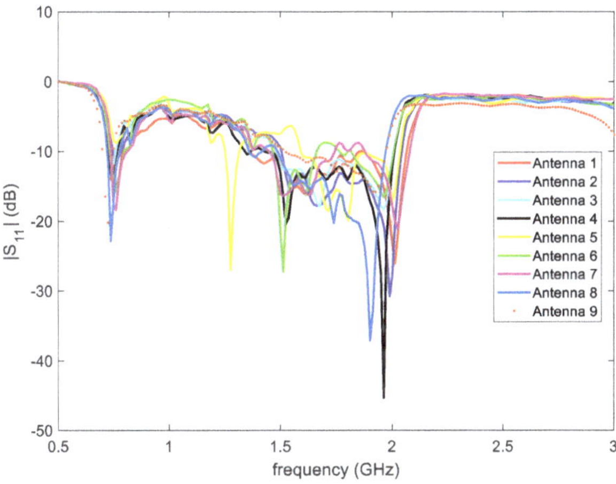

Figure 16. Measured $|S_{11}|$ for the nine antennas used as transmitter and receivers.

The receiver unit of the data acquisition system includes a 14-Bit, 125 Msps Direct Conversion Receiver, DC1513B-AB from Analog Devices, which has a frequency range from 0.7 to 2.7 GHz. To control this receiver unit with PC, this unit is connected to a USB data acquisition controller, DC890B from Analog Devices. The clock source for the receiver board DC1513B-AB is provided by High Speed ADC Clock Source, DC1216A-C from Analog Devices, with clock speed of 80 MHz.

One output of the transmitter is connected to a 6 dB attenuator, CATTEN-06R0 form Crystek Corporation, covering 0 GHz to 3 GHz, and used as referenced signal for the receiver unit. Eight receiver antennas are connected to an RF SP8T switch, EV1HMC321ALP4E from Analog Devices, that operates from 0 to 8 GHz. This switch is controlled by an Arduino Uno demo board controlled by MATLAB. Received signal after the switch is fed to the receiver unit through a wideband low-noise amplifier (LNA), ZX60-33LN-S+ from Mini-Circuits, operating from 50 MHz to 3 GHz, as well as a bandpass filter (BPF), VBFZ-1690-S+ from Mini-Circuits, covering 1455 MHz to 1925 MHz The transmitter and receiver units are powered separately via high-precision lab power supplies. All ground pins are connected to a common ground pin. The transmitter antenna and the receiver antenna arrays are placed on opposite sides of the container similar to the simulation setup in Figure 2. The angular separation between the receiver antennas is also similar to the simulation study, i.e., $\Delta\phi_a = 20^\circ$.

The cylindrical scanning system contains two stepper motors, NEMA 17 from Adafruit, connected, via an Arduino stepper motor shield board and an Arduino Uno board, to computer to be controlled via MATLAB. One of the motors moves the container along the longitudinal direction and the other one moves that along the azimuthal direction from 0° to 360°. The antennas are stationary and placed on an antenna holder which has been custom-designed and 3D-printed in our lab.

To control the whole system using computer, a MATLAB code has been developed that performs the following tasks: (1) control the transmitter, (2) control the switch, (3) control the receiver unit and acquire data, (4) control motors, and (5) implement holographic imaging. In the following we briefly describe each part.

Transmitter code controls the transmitter unit and generates the signal at two frequencies, 1.5 GHz and 1.8 GHz. To set two frequencies, the Serial Port Register Contents in the DC1705C board are written through serial peripheral interface (SPI) bus.

Switch code is used to select which of the eight receiver antennas is connected to the receiver unit to collect data. Three of the output pins on the Arduino Uno demo board are connected to three control pins on the switch. By setting the voltage level for each pin according to the truth table of the switch, one of the antennas can be chosen at a time to collect data.

The main body of the receiver unit code is used to collect data from two channels of the I/Q receiver unit, one channel providing the real part and another the imaginary part. These two parts are combined in MATLAB to form a complex number. Thus, at each sampling position, we collect two complex numbers corresponding to two frequencies of 1.5 GHz and 1.8 GHz for each antenna. Also, to make the measurements more robust to noise, we collect 4096 samples for each channel (real and imaginary channels), per position, per frequency, and per antenna. Since the local oscillator (LO) and the radio frequency (RF) inputs of the receiver unit have the same frequency, the I/Q output signals, i.e., the intermediate frequency (IF) outputs of the receiver unit are DC signals.

Motor code controls the motors to move along the longitudinal axis and the azimuth axis from 0° to 360° in a desired speed and direction. The number of sampling positions can be changed.

The imaging code is used to implement holographic imaging. Imaging is performed over cylindrical surfaces (3D imaging) at radii of $r_1 = 20$ mm, $r_2 = 35$ mm, and $r_3 = 50$ mm.

4.2. Experimental 3D Imaging Results

The container including the objects is scanned by the transmitter and receiver antennas over a cylindrical aperture. At each longitudinal position z, scanning is performed along the azimuthal direction ϕ in 100 steps to cover 360°. Scanning along the longitudinal direction is performed over one half of a cylindrical aperture with length of 80 mm and in 10 steps. Then, due to approximate symmetry of the structure along the longitudinal direction, the other half is acquired by flipping and combining that with the first half. The complex-valued data collected by the eight receiver antennas are then processed using the 3D holographic imaging technique. In the first experiment, we place two objects on the outer surface $r_3 = 50$ mm with, approximately, one object at $\phi = 0^\circ$ and another one at $\phi = 180^\circ$. Figure 17 shows the reconstructed images. It is observed that the two objects can be reconstructed well at $r_3 = 50$ mm with the images at the other surfaces showing small artifacts. We use the reconstruction error parameter defined in Equation (8) to evaluate the quality of image reconstruction. The computed reconstruction error for this experiment is 17.92.

We then repeat this experiment but, this time, putting the two objects on the middle surface $r_2 = 35$ mm. Figure 18 shows the reconstructed images over the three cylindrical surfaces for this case. Again, it is observed that the two objects on the middle surface can be reconstructed well at their true positions of $\phi = 0°$ and $\phi = 180°$. We use the reconstruction error parameter defined in Equation (8) to evaluate the quality of image reconstruction. The computed reconstruction error for this experiment is 20.84. Comparing the reconstruction error parameter with the previous example, we observe the degradation of the image quality. This is mainly due to the fact that the background medium is lossy, and the responses of the objects are weaker for the objects on the surface $r_2 = 35$ mm compared to those at $r_3 = 50$ mm.

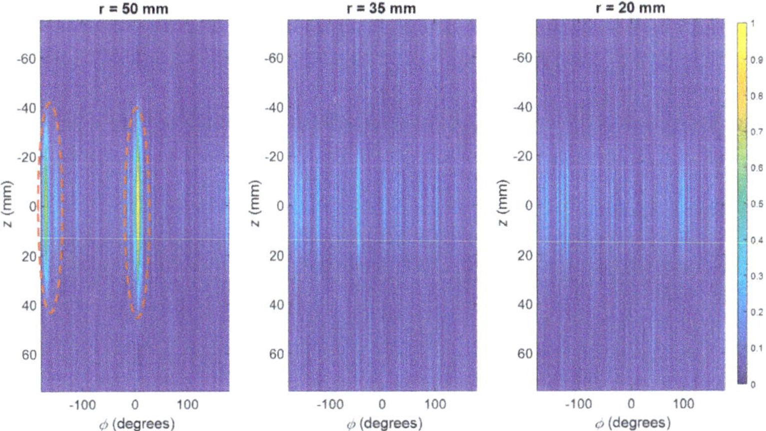

Figure 17. Normalized 3D reconstructed image of the objects on the outer surface for three imaged surfaces at radii of 50, 35, and 20 mm using double frequency data collected by eight receiver antennas. The red dashed lines show the reconstructed objects.

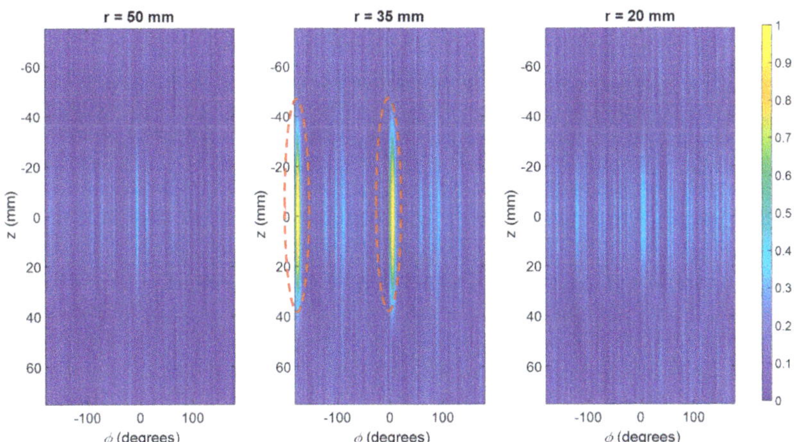

Figure 18. Normalized 3D reconstructed image of the objects on the middle surface for three imaged surfaces at radii of 50, 35, and 20 mm using double frequency data collected by eight receiver antennas. The red dashed lines show the reconstructed objects.

5. Conclusions

In this paper, we proposed a microwave imaging system capable of reconstructing 3D images of objects in the near field of the antennas. The applications are in non-destructive testing of non-metallic composite materials and biomedical imaging. To achieve low-cost, compactness, and reduce the component count for a data acquisition system, narrow-band data is acquired as opposed to the previously proposed 3D cylindrical holographic imaging techniques. An I/Q detection data acquisition system is built based on the cost-effective off-the-shelf components to replace the expensive and bulky VNAs. The system is tailored for imaging over a narrow band and this makes it much less expensive, affordable, and compact.

In this work, we used a mixture of water and glycerin which is very lossy. We selected this mixture considering both biomedical applications and non-destructive testing of the pipes which may carry mixtures of water and other substances. For microwave imaging of such media, the optimal frequency band to provide sufficient penetration while having acceptable resolution is within the range of 1 GHz to 10 GHz (e.g., see ref. [3,22,23]). This justifies the chosen frequencies in this work.

The reconstructed images using this data acquisition system confirms the satisfactory performance of this narrow-band system in 3D imaging. To evaluate the quality of the image reconstruction process, we defined a reconstruction error to compare the images with an ideal image (true image). We observed that, in general, the reconstruction error increases for reconstructing objects on surfaces farther away from the antennas. Besides, the reconstruction error decreases when using double frequency data compared to single frequency data. Furthermore, we observed that increasing the object size improves the quality of the images but there is a limit for that due to the use of Born approximation in the holographic imaging.

Although scanning of the objects over a cylindrical aperture takes a few hours, the holographic imaging technique itself is fast. Here, we provide an estimate of the computational complexity of our 3D image reconstruction process. The number of samples along ϕ and z are N_ϕ and N_z, respectively. The number of receiver antennas and measurement frequencies are N_A and N_ω, respectively. The number of imaged surfaces is N_r. We denote the number of samples of k_z by N_{k_z}. The number of samples along k_ϕ is N_ϕ. Table 3 summarizes the computational complexity of our approach. The computational complexity of solving the systems of equations has been provided with the assumption that they are solved with QR factorization. The total number of flops for the image reconstruction process is the sum of all the flops in Table 3. The 3D image reconstruction process takes 10 s on a regular PC with Intel Core i5 CPU at 3.2 GHz and 8 GB RAM.

Table 3. Details of computational complexity of the proposed image reconstruction process.

Operation	Number of Flops
FT of the scattered fields	$N_\omega N_A N_\phi^2 N_z \log(N_z)$
FT of PSF functions	$N_r N_\omega N_A N_\phi^2 N_z \log(N_z)$
Solving the systems of equations for all combinations of k_ϕ and k_z	$N_\phi N_{k_z}(2 N_A N_\omega N_r^2 - (2/3)N_r^3)$
Inverse FT of the contrast function	$N_r N_\phi^2 N_{k_z} \log(N_{k_z})$

To expedite the imaging process and move toward real-time or quasi real-time imaging, an array of antennas can be employed similar to the setups used for security screening in the airports [4].

As a final note, in this work, the goal was to propose an imaging system which is: (a) more cost-effective, (b) accessible and affordable outside microwave laboratories in various industrial or biomedical applications, and (c) compact and portable due to the fact it is tailored for measurement over a very narrow band and for a specific purpose. VNAs normally are general-purpose equipment that are capable of performing measurements over a very wideband. This makes them expensive and bulky which is not suitable for widespread use in various industrial settings. We should emphasize that the proposed system has been built with a cost of less than $1000 USD and can be made very compact by arranging the boards in a small box. The drawback of the proposed system compared to a regular VNA is lower dynamic range. For a regular modern VNA, the dynamic range at frequencies around 1 GHz to 2 GHz is around 100 dB while the dynamic range for the receiver used in our proposed system is 63.5 dB. This limits the sensitivity of the proposed system indicating that the imaged objects need to be larger to provide measurable signatures in the proposed system compared to a VNA-based measurement system.

Author Contributions: Conceptualization, R.K.A.; methodology, R.K.A.; software, H.W., R.K.A.; validation, H.W., R.K.A.; formal analysis, H.W., R.K.A.; investigation, H.W., R.K.A.; resources, R.K.A.; data curation, H.W., R.K.A.; writing—original draft preparation, H.W., R.K.A.; writing—review and editing, H.W., R.K.A.; visualization, H.W., R.K.A.; supervision, R.K.A.; project administration, R.K.A.; funding acquisition, R.K.A.

Funding: This project has been supported by US national science foundation (NSF), award No. 1920098, and New York Institute of Technology's (NYIT) Institutional Support for Research and Creativity (ISRC) Grants.

Conflicts of Interest: The authors declare no conflict of interest.

References

1. Zoughi, R. *Microwave Non-Destructive Testing and Evaluation*; Kluwer: Dordrecht, The Netherlands, 2000.
2. Amin, M.G. *Through-the-Wall Radar Imaging*; CRC Press: Boca Raton, FL, USA, 2016.
3. Nikolova, N.K. Microwave imaging for breast cancer. *IEEE Microw. Mag.* **2011**, *12*, 78–94. [CrossRef]
4. Sheen, D.M.; McMakin, D.L.; Hall, T.E. Three-dimensional millimeter-wave imaging for concealed weapon detection. *IEEE Trans. Microw. Theory Tech.* **2001**, *49*, 1581–1592. [CrossRef]
5. Sheen, D.M.; McMakin, D.; Hall, T.E. Near-field three-dimensional radar imaging techniques and applications. *Appl. Opt.* **2010**, *49*, E83–E93. [CrossRef] [PubMed]
6. Hagemaier, D.J. *Fundamentals of Eddy Current Testing*; Amer Society for Nondestructive: Columbus, OH, USA, 1990.
7. Amineh, R.K.; Koziel, S.; Nikolova, N.K.; Bandler, J.W.; Reilly, J.P. A space mapping methodology for defect characterization from magnetic flux leakage measurement. *IEEE Trans. Mag.* **2018**, *44*, 2058–2065. [CrossRef]
8. Betz, C.E. *Principles of Magnetic Particle Testing*, 1st ed.; Magnaflux Corporation: Chicago, IL, USA, 1967.
9. Meaney, P.M.; Goodwin, D.; Golnabi, A.H.; Zhou, T.; Pallone, M.; Geimer, S.D.; Burke, G.; Paulsen, K.D. Clinical microwave tomographic imaging of the calcaneus: A first-in-human case study of two subjects. *IEEE Trans. Biomed. Eng.* **2012**, *59*, 3304–3313. [CrossRef] [PubMed]
10. Fear, E.C.; Hagness, S.C.; Meaney, P.M.; Okoniewski, M.; Stuchly, M.A. Enhancing breast tumor detection with near-field imaging. *IEEE Microw. Mag.* **2002**, *3*, 48–56. [CrossRef]
11. Xie, Y.; Guo, B.; Xu, L.; Li, J.; Stoica, P. Multistatic adaptive microwave imaging for early breast cancer detection. *IEEE Trans. Biomed. Eng.* **2016**, *53*, 1647–1657. [CrossRef] [PubMed]
12. Tobon, J.A.V.; Attardo, E.A.; Dassano, G.; Vipiana, F.; Casu, M.R.; Vacca, M.; Vecchi, G. Design and modeling of a microwave imaging system for breast cancer detection. In Proceedings of the 2015 9th European Conference on Antennas and Propagation, Lisbon, Portugal, 13–17 April 2015.
13. Semenov, S.Y.; Corfield, D.R. Microwave tomography for brain imaging: Feasibility assessment for stroke detection. *Int. J. Antennas Propag.* **2008**, *2008*, 254830. [CrossRef]
14. Amineh, R.K.; McCombe, J.; Khalatpour, A.; Nikolova, N.K. Microwave holography using point-spread functions measured with calibration objects. *IEEE Trans. Instrum. Meas.* **2015**, *64*, 403–417. [CrossRef]
15. Amineh, R.K.; Ravan, M.; Khalatpour, A.; Nikolova, N.K. Three-dimensional near-field microwave holography using reflected and transmitted signals. *IEEE Trans. Antennas Propag.* **2011**, *59*, 4777–4789. [CrossRef]
16. Amineh, R.K.; Ravan, M.; McCombe, J.; Nikolova, N.K. Three-dimensional microwave holographic imaging employing forward- scattered waves only. *Int. J. Antennas Propag.* **2013**, *2013*, 897287. [CrossRef]
17. Wu, H.; Amineh, R.K.; Ravan, M. Near-field holographic microwave imaging using data collected over cylindrical apertures. In Proceedings of the 18th Int. Symp. on Antenna Technology and Applied Electromagnetics (ANTEM), Waterloo, ON, Canada, 19–22 August 2018.
18. Amineh, R.K.; Ravan, M.; Wu, H.; Kasturi, A. Three-dimensional holographic imaging using data collected over cylindrical apertures. *Microwave Opt. Technol. Lett.* **2019**, *61*, 907–911. [CrossRef]

19. Amineh, R.K.; Nikolova, N.K.; Ravan, M. *Real-Time Three-Dimensional Imaging of Dielectric Bodies Using Microwave/Millimeter Wave Holography*; Wiley-IEEE Press: Hoboken, NJ, USA, 2019.

20. Bolomey, J.C.; Gardiol, F.E. *Engineering Applications of the Modulated Scattering Technique*; Artech House Publishers: Norwood, MA, USA, 2001.

21. Amineh, R.K.; Ravan, M.; Sharma, R.; Baua, S. Three-dimensional holographic imaging using single frequency microwave data. *Int. J. Antennas Propag.* **2018**, *2018*, 6542518. [CrossRef]

22. Marimuthu, J.; Bialkowski, K.S.; Abbosh, A.M. Software-defined radar for medical imaging. *IEEE Trans. Microw. Theory Tech.* **2016**, *64*, 643–652. [CrossRef]

23. Pagliari, D.J.; Pulimeno, A.; Vacca, M.; Tobon, J.A.; Vipiana, F.; Casu, M.R.; Solimene, R.; Carloni, L.P. A Low-cost, fast, and accurate microwave imaging system for breast cancer detection. In Proceedings of the IEEE Biomedical Circuits and Systems Conference (BioCAS), Atlanta, GA, USA, 22–24 October 2015.

24. Oppenheim, A.V.; Schafer, R.W.; Buck, J.R. *Discrete-Time Signal Processing*, 2nd ed.; Prentice Hall: Upper Saddle River, NJ, USA, 1999.

25. Altair-FEKO Software. Available online: https://altairhyperworks.com/product/FEKO (accessed on 14 September 2019).

26. Meaney, P.M.; Fox, C.J.; Geimer, S.D.; Paulsen, K.D. Electrical characterization of glycerin: water mixtures: Implications for use as a coupling medium in microwave tomography. *IEEE Trans. Microw. Theory Tech.* **2017**, *65*, 1471–1478. [CrossRef] [PubMed]

27. MATLAB Software, MathWorks. Available online: https://www.mathworks.com (accessed on 14 September 2019).

Article

A Phantom Investigation to Quantify Huygens Principle Based Microwave Imaging for Bone Lesion Detection

Banafsheh Khalesi [1,*], Behnaz Sohani [1], Navid Ghavami [2], Mohammad Ghavami [1], Sandra Dudley [1] and Gianluigi Tiberi [1]

[1] School of Engineering, London South Bank University, London SE1 0AA, UK; sohanib@lsbu.ac.uk (B.S.); ghavamim@lsbu.ac.uk (M.G.); dudleyms@lsbu.ac.uk (S.D.); g.tiberi@iet.unipi.it (G.T.)

[2] UBT-Umbria Bioengineering Technologies, Spin off of University of Perugia, 06081 Assisi, Italy; navgh1988@yahoo.co.uk

* Correspondence: khalesib@lsbu.ac.uk

Received: 4 November 2019; Accepted: 3 December 2019; Published: 9 December 2019

Abstract: This paper demonstrates the outcomes of a feasibility study of a microwave imaging procedure based on the Huygens principle for bone lesion detection. This study has been performed using a dedicated phantom and validated through measurements in the frequency range of 1–3 GHz using one receiving and one transmitting antenna in free space. Specifically, a multilayered bone phantom, which is comprised of cortical bone and bone marrow layers, was fabricated. The identification of the lesion's presence in different bone layers was performed on images that were derived after processing through Huygens' principle, the S21 signals measured inside an anechoic chamber in multi-bistatic fashion. The quantification of the obtained images was carried out by introducing parameters such as the resolution and signal-to-clutter ratio (SCR). The impact of different frequencies and bandwidths (in the 1–3 GHz range) in lesion detection was investigated. The findings showed that the frequency range of 1.5–2.5 GHz offered the best resolution (1.1 cm) and SCR (2.22 on a linear scale). Subtraction between S21 obtained using two slightly displaced transmitting positions was employed to remove the artefacts; the best artefact removal was obtained when the spatial displacement was approximately of the same magnitude as the dimension of the lesion.

Keywords: microwave imaging; phantom measurement system; bone lesion detection

---

1. Introduction

Bone fracture can be caused as a result of high force impact, a simple accident, stress or certain medical conditions that weaken the bones. The structure of the bones includes two principle parts: (i) cortical (compact) bone, which is a hard outer layer and is dense, strong, durable and surrounded by the cancellous tissue, and (ii) bone marrow, which is the inner layer, less dense and with a lighter content. There are many types of bone fractures [1]. Depending on the fracture severity, the injuries can lead to a reduction in the mobility of the patient [2]. X-rays, computed tomography (CT) and magnetic resonance imaging (MRI) are used as essential tools in the diagnosis and monitoring of bone conditions, including fractures, and joint abnormalities [3]. However, each technique suffers from its own negative aspects. For instance, fractures can be commonly detected by X-rays [4], which is the fastest and easiest way to assess bone injuries, including fractures. However, since this technique involves radiation and can potentially cause damage, it raises major concerns especially in the cases of infants and stages of pregnancy. In addition, X-rays provide limited information about muscles, tendons or joints [5]. Nevertheless, CT is very effective for imaging and gives better quality images for body organs, such as an image of complicated fractures, subtle fractures or dislocations. However, similar to X-rays, ionizing radiation is the major problem of using this technique, which leads to limits in its application [6]. There is no ionizing radiation in the MRI technique, and it may be more useful in identifying bone and joint injuries. MRI can also detect occult fractures or bone bruises that are not visible on X-ray images, but the high cost of purchasing and maintaining such systems and their long time duration cause financial restrictions. Moreover, none of these devices are portable and cannot be used at the accident site. Thus, a fast and portable imaging system could be particularly useful locally for rapid diagnosis of bone injuries. A wide range of research concentrates on the development of new medical imaging techniques to achieve a portable, low cost and safe imaging alternative. Among these, using microwave imaging techniques has attracted the attention of researchers due to its various benefits such as the use of non-ionizing signals, low cost, low complexity and its ability to penetrate through different mediums (air, skin, bones and tissues [7]). The dielectric properties of human tissues can be used as an effective and accurate indicator for diagnostic purposes [8]. The significant difference between the dielectric properties of tissues with lesion and healthy tissues of the human body at microwave frequencies is the basis of microwave medical imaging techniques. Microwave tomography techniques, which give the maps of dielectric properties [9–11], and the UWB radar techniques, which aim to find and locate the significant scatterers [11–14], are recognized as the two main branches of microwave imaging techniques [14]. Nevertheless, microwave tomography has its drawbacks such as low signal-to-clutter and complex mathematical formulation. Both microwave tomography and UWB radar techniques have been increasingly well investigated for stroke detection [15], breast cancer detection [16], bone imaging [17,18], and skin cancer detection [19] through using different approaches at different frequency ranges.

UWB imaging for bone lesion detection has been studied through using matching liquids [17,18]. We propose imaging execution using two antennas operating in free space. Imaging was performed via a Huygens principle (HP) based algorithm, which was initiated originally for breast imaging. Explicitly, S21 signals in the frequency range 1–3 GHz [18] were collected in a multi-bistatic fashion in an anechoic chamber setting using a multilayer phantom, mimicking bones. A realistic multilayer bone mimicking phantom comprised of cortical bone mimicking and bone marrow mimicking layers was constructed. Subsequently, a large inclusion was placed within the marrow layer to represent bone marrow lesion, and afterwards, a small inclusion was placed in the cortical layer to represent the bone lesion or fracture. Angular subtraction rotation was used for artefact removal, allowing lesion detection.

Additionally, quantification of the images obtained through HP microwave imaging was performed by introducing dedicated parameters such as the resolution and signal-to-clutter ratio. Furthermore, investigating the impact of different frequencies and bandwidths, lesion size and angular rotation subtraction in the detection procedure were also addressed.

This paper is organized as follows. The phantom construction procedure, the experimental setup used for phantom measurements, the imaging procedure and image quantification are described in Section 2. Section 3 represents the corresponding experimental results and discussions. Finally, conclusions are presented in Section 4.

2. Materials and Methods

2.1. Phantom Construction

This section presents the design and fabrication of a multilayered cylindrical phantom mimicking the human bone by considering the relative permittivity and conductivity with the aim of performing microwave imaging experiments in the frequency range of 1 to 3 GHz. Hence, our proposed multilayered bone phantom was comprised of two layers, which included: (i) an external layer representing the cortical bone tissue (radius = 5.5 cm) and (ii) an internal layer representing the bone marrow tissue (radius = 3.5 cm). A small size inclusion (radius = 0.3 cm) was placed in the cortical bone layer to represent the bone fracture, and a larger sized inclusion (radius = 0.7 cm) was placed in the bone marrow layer to represent the internal bone lesion. In this paper, the lesion was assumed to have the dielectric properties of blood.

The phantom fabrication procedure for each layer of phantom was performed by considering the following factors: (i) dielectric property (permittivity and conductivity) similarity of the layers with the tissues to be mimicked, (ii) an easy construction process, (iii) the stability of the materials and (iv) the geometric dimension similarity between each layer and the realistic scenario. The upper half of Table 1 shows the dielectric properties of each tissue to be mimicked, where the values were derived from [20], while the lower half of the Table indicates the dielectric properties of the tissue mimicking materials used.

Table 1. Relative permittivity and conductivity at a frequency of 2 GHz.

	Relative Permittivity	Conductivity (S/m)
Bone marrow	5.35	0.07
Bone cortical	11.7	0.31
Lesion (assumed here as blood)	59	2.19
Bone marrow tissue equivalent material (ZMT Zurich MedTech Company, TLec24 oil)	5	0.2
Bone cortical tissue equivalent material (ZMT Zurich MedTech Company, TLe11.5c.045 oil))	7	0.3
Blood tissue equivalent material (40% glycerol and 60% water)	60	2

To construct the multilayered bone phantom appropriately, different volumes of cylindrically shaped plastic containers and tubes were used and are shown in Figure 1.

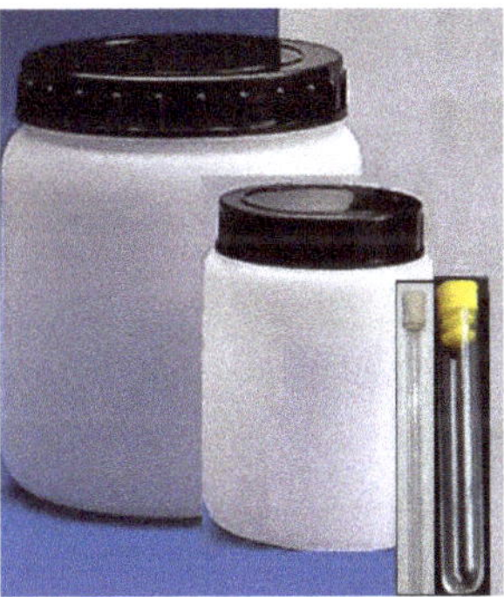

Figure 1. Design of the different layers of the phantom.

As shown in Table 2, the phantom fabrication was performed using a large cylindrically shaped plastic container with a radius of 5.5 cm and a height of 13 cm filled with cortical bone equivalent material to represent the cortical bone layer. Then, a medium sized cylindrically shaped plastic container with a radius of 3.5 cm and a height of 9 cm was placed inside the large container after filling it up with bone marrow equivalent material representing the bone marrow layer. Subsequently, the small cylindrically shaped plastic tube with radius = 0.3 cm and a height of 13 cm filled up with lesion equivalent material was placed inside the cortical bone layer to represent bone fracture (see Figure 2a). In the next scenario, the larger cylindrically shaped plastic tube having a radius equal to 0.7 cm and a height equal to 11 cm, again filled up with lesion equivalent material, was placed inside the bone marrow layer to represent bone marrow lesion (see Figure 2b).

Table 2. Phantom layers' design height and size.

Different Layers of the Phantom	Radius (cm)	Height (cm)
Bone marrow (internal layer)	3.5	9
Bone cortical (external layer)	5.5	13
Small lesion	0.3	13
Large lesion	0.7	11

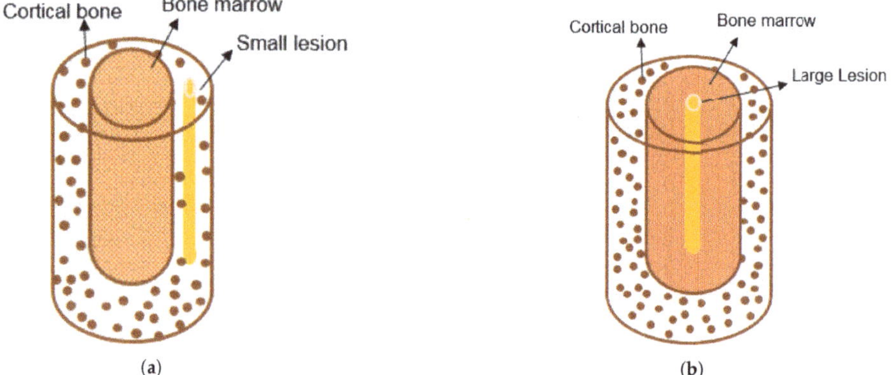

Figure 2. Design of the proposed bone fracture (**a**) and bone marrow lesion (**b**).

Different recipes for each layer of the phantom were tested to select those showing dielectric properties similar to those given in the upper half of Table 1. In this context, dedicated liquids were purchased from the ZMT Zurich MedTech Company [21]. As shown in the lower half of Table 1, the TLe11.5C.045 oil displayed (at 2 GHz) a dielectric permittivity of 7 and a conductivity of 0.3 S/m; the Tle5C24 displayed (at 2 GHz) a dielectric permittivity of 5 and a conductivity of 0.2 S/m. Thus, TLe11.5C.045 was selected as a cortical bone tissue equivalent material and Tle5C24 as a bone marrow tissue equivalent material. In addition, a mixture of glycerol and water with a ratio of 40% and 60%, respectively, was chosen as the recipe mimicking the lesion (blood), giving (at 2 GHz) a permittivity value equal to 60 and conductivity of 2 S/m [22]. Figure 3a,b shows the fabricated multilayered bone fracture and bone marrow lesion, respectively.

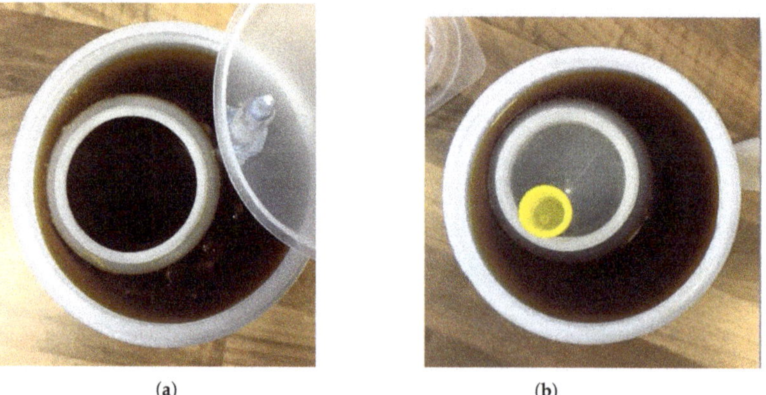

Figure 3. Fabricated bone fracture phantom (**a**) and bone marrow lesion phantom (**b**).

2.2. Experimental Configurations in an Anechoic Chamber

All microwave images presented in this paper were obtained by processing the frequency domain measurements obtained in the band of 1 to 3 GHz. Measurements were performed inside an anechoic chamber using a vector network analyser (VNA) (model MS2028C, Anritzu) and PulsON P200 antennas. Specifically, the measurement setup was comprised of one transmitting antenna and one receiving antenna connected to the VNA device. The phantom was placed at the centre of a rotatable table. Transmitting antenna was located 12 cm away from the centre of the table, while the receiving antenna was placed nearer to the object (i.e., 8.5 cm from the centre of the table). Both receiving and transmitting antennas were vertically polarized and omni-directional in the azimuth plane and were calibrated and operated in free space.

The receiving antenna was configured to rotate azimuthally around the phantom to collect the reflected signals in the different directions. For each receiving position, the complex S21 was recorded over a wide frequency range of 1 to 3 GHz with a frequency step of 10 MHz [23] in order to exploit the variation of the signal over the different frequencies. To allow artefact removal, the measurement procedure was repeated using M = 3 transmitting positions displaced 5° from each other (considered as a transmitting triplet displaced at positions 0°, 5°, and 10°). It should be pointed out that the 3 transmitting positions were synthesized by appropriately rotating the phantom instead of rotating the transmitting antenna. For each transmitting position, the receiving antenna rotated to measure the receiving signal every 6°, which led to a total of NPT = 60 receiving points. Figure 4 shows the measurement setup of the multilayered bone phantom inside the anechoic chamber.

Figure 4. Position of the bone marrow lesion phantom inside the anechoic chamber. The phantom was placed in the centre of a rotatable table. The external PulsON P200 antenna is the transmitter, and the internal PulsON P200 antenna is the receiver.

The positions of the phantoms with respect to the transmitting antenna are shown in Figure 5a,b, which represents the pictorial views of the measurement setup.

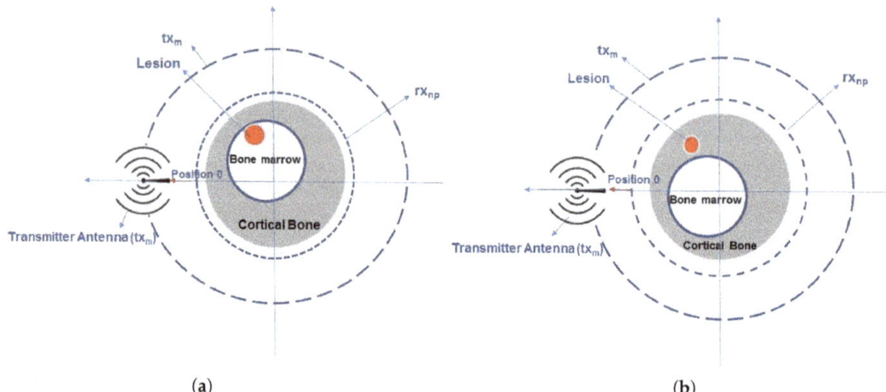

(a)   (b)

Figure 5. Pictorial view of bone marrow lesion (**a**) and bone fracture lesion (**b**) measurement setups.

2.3. Imaging Procedure and Image Quantification

In order to reconstruct the image, the complex measured S21 was processed through the Huygens principle [24]. Specifically, the cylindrically shaped object (the phantom) was placed in free space and illuminated by a transmitting antenna located at the position tx_m. The receiving antenna rotated around it to measure the signals at the point $rx_{np} \equiv (a_0, \phi_{np}) \equiv \overrightarrow{\rho}_{np}$ displaced along a circular surface having radius a_0 (see Figure 5):

$$S21_{tx_m}^{known}\,|_{rx_{np}} = S21_{np,tx_m}^{known} \qquad \text{with} \qquad np = 1, \ldots, N_{PT} \qquad (np = 1, 2, \ldots 60) \tag{1}$$

where rx_{np} is the position of the receiving antenna, np is the number of receiving point, which varies from 1 to $N_{PT} = 60$, and m represents the transmitting position with $m = 1, 2, 3$.

According to HP: "Each locus of a wave excites the local matter which re-radiates secondary wavelets, and all wavelets superpose to a new, resulting wave (the envelope of those wavelets), and so on" [25]; the field inside the object can be calculated as the superposition of the fields radiated by the N_{PT} receiving points of Equation (1):

$$E_{HP,2D}^{rcstr}(\rho, \phi; tx_m; f) = \Delta s \sum_{np=1}^{N_{PT}} S21_{np,tx_m}^{known} G(k_1|\overrightarrow{\rho_{np}} - \overrightarrow{\rho}|) \tag{2}$$

In Equation (2), $(\rho, \phi) \equiv \overrightarrow{\rho}$ is the observation point, f is the frequency, k_1 indicates the wave number and Δs is the spatial sampling. The "reconstructed" internal field is indicated by the string "rcstr", while the string HP indicates that the Huygens based procedure will be employed in Equation (2). In order to propagate the field (since we are dealing with a 3D problem), Green's function G for homogeneous problems was used [24].

It was shown in [24,26] that Equation (2) can capture the contrast, i.e., mismatch boundaries, and locate an inclusion within the volume.

In Equation (2), k_1 can be set to represent the wave number for the media constituting the external layer (if known). However, it has been shown that detection can be achieved also if setting k_1 as the free space wave number [27].

By the assumption of using N_F frequencies f_i, the intensity of the final image I was obtained through Equation (3), i.e., by summing incoherently all the solutions.

$$I(\rho, \phi; tx_m) = \sum_{i=1}^{N_F} |E_{HP,2D}^{rcstr}(\rho, \phi; tx_m; f_i)|^2 \tag{3}$$

As Equation (4) shows, the subtraction between S21 obtained using two slightly displaced transmitting position was employed in order to remove the artefacts, i.e., the image of the transmitter and the reflection of the first layers [26]:

$$E_{\text{HP,2D}}^{\text{rcstr}}(\rho, \phi; tx_m - tx_{m'}; f) = \Delta s \sum_{np=1}^{N_{PT}} \left((S21_{np,tx_m}^{\text{known}} - S21_{np,tx_{m'}}^{\text{known}}) G(k_1 | \overrightarrow{\rho_{np}} - \overrightarrow{\rho} |) \right) \tag{4}$$

with tx_m and $tx_{m'}$ belonging to the transmitting triplet. This procedure will be referred to as rotation subtraction artefact removal.

Image Quantification

Images may contain some clutter even following artefact removal procedures. Thus, it is appropriate to introduce some parameters in order to compare and quantify the performance of microwave imaging. The parameters that will be introduced are the resolution and signal-to-clutter ratio (SCR). Specifically, here, the resolution is defined as a dimension of the region whose normalized intensity is above 0.5 [28]; SCR was defined as the ratio between maximum intensity evaluated in the region of the lesion divided by the maximum intensity outside the region of the lesion [29]. The evaluation of these parameters was performed for both external and internal lesion placement in the two following scenarios: (i) calculating Equation (3) maintaining the same bandwidth of 0.5 GHz and varying the central frequency and (ii) calculating Equation (3) increasing the bandwidth. Then, to evaluate the impact of the rotation angle of the transmitter for artefact removal, the procedure was repeated using two transmitting positions displaced 5° (i.e., transmitting $\Delta\phi = 5°$) and using two transmitting positions displaced 10° (i.e., transmitting $\Delta\phi = 10°$).

3. Results and Discussions

The authors in [18] performed a study for bone imaging, collecting the signals in multi-monostatic fashion and using antennas immersed in a coupling liquid. In [18], imaging was performed using a beamforming procedure named non-coherent migration, after applying an average trace subtraction strategy to remove the artefact. Instead, here, we collect the signals in multi-bistatic fashion, using antennas in free space; imaging was performed using an HP based algorithm, which operated in the frequency domain, after applying a rotation subtraction strategy to remove the artefact.

It may happen that artefact removal is not effective to cancel the artefact fully, partially or completely masking the inclusion, i.e., the lesion. This may be due to imperfect cancellation of the transmitting antenna or appropriate cancellation of the first layers' reflection or even due to multiple reflections occurring inside the phantom that cannot cancel completely. Figures 6 and 7 show all the images of the experimental phantom investigated in various frequencies and various bandwidths for bone marrow lesion and bone fracture scenarios, respectively. The correct position of the lesion is indicated by the arrow. The left columns refer to artefact removal performed using a transmitting step $\Delta\phi = 5°$, while the right columns refer to artefact removal performed using a transmitting step $\Delta\phi = 10°$.

It is important to point out that all the images shown here were normalized and adjusted, forcing the intensity values below 0.5 to zero. However, SCR was calculated before performing the image adjustment.

As Figures 6a,b and 7a,b show, in the frequency range from 1 GHz to 1.5 GHz, the artefact masked the inclusion. When using the frequency range from 1.5 GHz to 2 GHz (see Figures 6c,d and 7c,d) and from 2 GHz to 2.5 GHz (see Figures 6e,f and 7e,f), only the lesion was visible, without any residual clutter. Images corresponding to a frequency range of 2.5 GHz to 3 GHz depicted that, although the lesion was detectable, the presence of residual clutters could not be ignored (see Figures 6g,h and 7g,h).

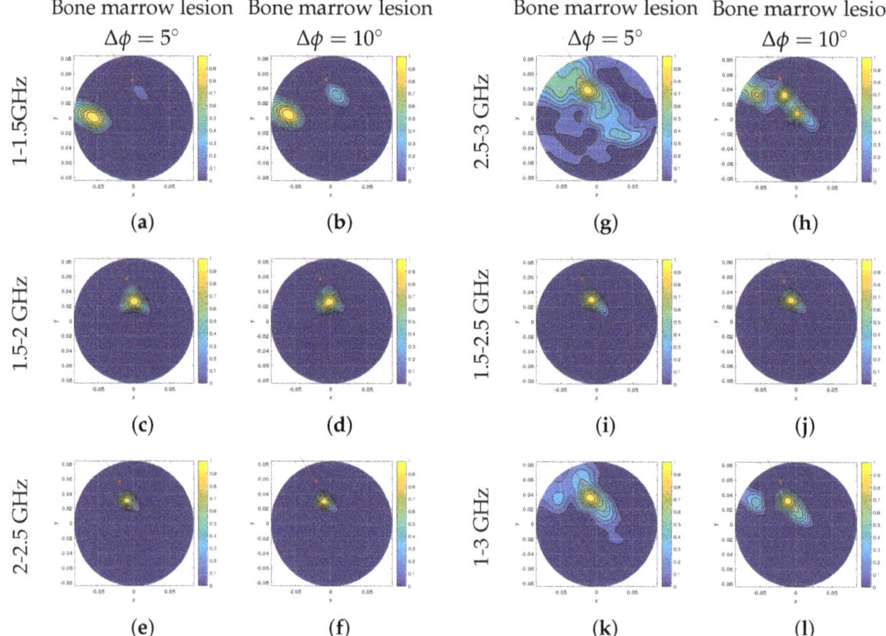

Figure 6. Microwave images of the bone marrow lesion employing various frequencies and bandwidths; (**a,b**), (**c,d**), (**e,f**) and (**g,h**) represent the resulting images when employing frequency ranges 1–1.5 GHz, 1.5–2 GHz, 2–2.5 GHz and 2.5–3 GHz, respectively, while (**i,j**) and (**k,l**) represent the images when considering bandwidth equal to 1 GHz and 2 GHz, respectively. Images are obtained following normalization to their correspondent maximum values and forcing to zero the intensity values below 0.5 (X and Y are given in meters).

After visual inspection, image quantification was performed calculating the resolution and SCR. Table 3 summarises such parameters.

Table 3. Resolution (m) and SCR (linear) for various frequency ranges.

Frequency	Bone Marrow Lesion				Bone Fracture			
GHz	$\Delta\phi = 5°$		$\Delta\phi = 10°$		$\Delta\phi = 5°$		$\Delta\phi = 10°$	
	Resolution, m	SCR	Resolution, m	SCR	Resolution, m	SCR	Resolution, m	SCR
1–1.5	N/A	<1	N/A	<1	N/A	<1	N/A	<1
1.5–2	0.015	2.06	0.015	2.13	0.016	1.85	0.016	1.51
2–2.5	0.012	2.13	0.012	1.88	0.011	2.09	0.011	1.92
2.5–3	0.017	1.38	0.015	1.52	N/A	<1	N/A	<1

Concerning the bone marrow lesion, as Table 3 shows, the highest SCR values were related to the frequency ranges from 1.5 GHz to 2 GHz and from 2 GHz to 2.5 GHz. The same holds for the bone fracture scenarios. Concerning resolution, resolutions of 1.2 cm and of 1.1 cm were achieved for the bone marrow lesion and bone fracture, respectively, when employing a frequency range from 2 GHz to 2.5 GHz. The result of operating frequency range from 2.5 GHz to 3 GHz showed that a further increase of the central frequency did not imply high SCR and/or better resolution, since residual clutter may be enhanced.

Turning now to investigate the impact of increasing the bandwidth, SCR and resolution values were calculated employing a frequency from 1.5 GHz to 2.5 GHz (i.e., a bandwidth of 1 GHz) and a frequency from 1 GHz to 3 GHz (i.e., a bandwidth of 2 GHz). As Table 4 shows, the highest SCR (2.22) corresponded to the bandwidth of 1 GHz compared to employing a bandwidth of 2 GHz (SCR = 1.49). It was evident that, although increasing the bandwidth of operation might be beneficial for SCR [28], such a bandwidth increase should be performed carefully to avoid including a region of frequency where residual clutter may be enhanced.

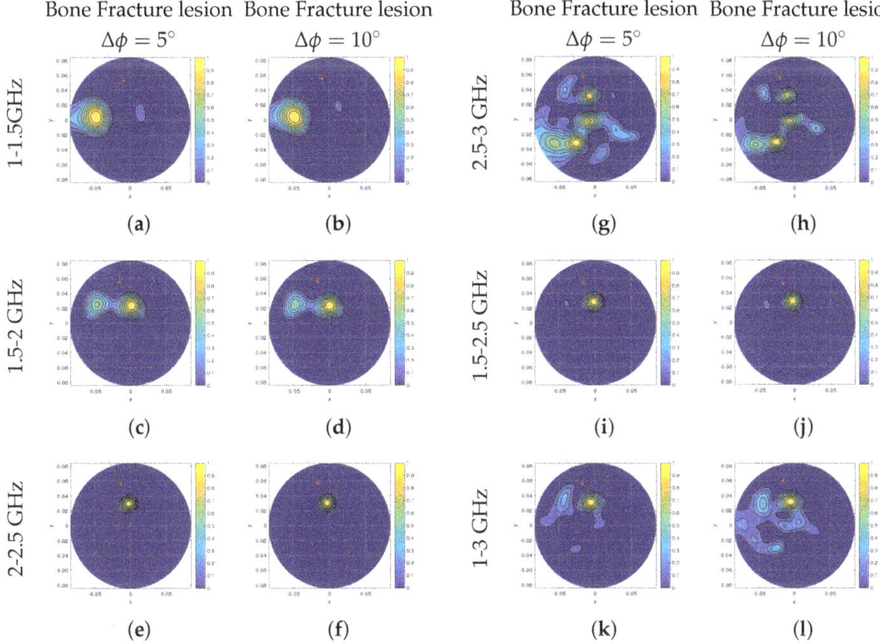

Figure 7. Microwave images of the bone fracture lesion employing various frequencies and bandwidths; (**a,b**), (**c,d**), (**e,f**) and (**g,h**) represent the resulting images when employing frequency ranges 1–1.5 GHz, 1.5–2 GHz, 2–2.5 GHz and 2.5–3 GHz, respectively, while (**i,j**) and (**k,l**) represent the images when considering bandwidth equal to 1 GHz and 2 GHz, respectively. Images are obtained following normalization to their correspondent maximum values and forcing to zero the intensity values below 0.5 (X and Y are given in meters).

Table 4. Resolution (m) and SCR (linear) for various bandwidths.

Bandwidth	Bone Marrow Lesion				Bone Fracture			
GHz	$\Delta\phi = 5°$		$\Delta\phi = 10°$		$\Delta\phi = 5°$		$\Delta\phi = 10°$	
	Resolution, m	SCR	Resolution, m	SCR	Resolution, m	SCR	Resolution, m	SCR
1.5–2.5	0.013	2.22	0.014	2.06	0.011	1.78	0.012	1.74
1–3	0.018	1.49	0.015	1.34	0.013	1.51	0.013	1.36

A further confirmation of such a finding may be drawn through a visual inspection of Figures 6i,j and 7i,j and Figures 6k,l and 7k,l, which show the images employing the bandwidth of 1 and 2 GHz for both the bone marrow lesion and bone fracture, respectively.

The highest value of SCR, which was equal to 2.22, was obtained through the HP procedure using a bandwidth of 1 GHz (1.5 GHz to 2.5 GHz). It should be emphasized that our obtained linear value of SCR 2.22 corresponded to 6.9 dB, which was in excellent agreement with [28,29]. According to Table 3, the resolution that came from the experiments (1.1 cm) was in excellent agreement with the optical resolution limit of $\lambda_{1,f_{\mathrm{max}}}/4$, where $\lambda_{1,f_{\mathrm{max}}}$ represents the wavelength in the scenarios by considering a dielectric constant equal to the arithmetical average of the two layers calculated at the highest frequency of 3 GHz [28,29].

Concerning the impact of the transmitter $\Delta\phi$ in artefact removal, we observed that for a large lesion size (radius = 0.7 cm), similar SCR values were obtained when using both $\Delta\phi = 5°$ and $10°$. Instead, for a small lesion size (radius = 0.3 cm), SCR obtained when using $\Delta\phi = 5°$ was higher than that obtained for $10°$. It is worthwhile pointing out that $\Delta\phi = 5°$ corresponded to a spatial displacement of 0.87 cm, while $\Delta\phi = 10°$ corresponded to a spatial displacement of 1.74 cm. Thus, we may conclude that optimal artefact removal was obtained with a transmitting $\Delta\phi = 5°$, leading to a spatial displacement approximately of the same magnitude as the dimension of the lesion.

4. Conclusions

This paper presented the application of a new radar based microwave imaging procedure based on the Huygens principle approach, which achieved promising results for bone lesion detection. The procedure was successfully tested inside an anechoic chamber on a dedicated multilayer phantom. Subtraction between S21 obtained using two transmitting positions was employed in order to remove the artefact. The quantification of the microwave images was calculated using two parameters, which were the resolution and SCR, achieving an SCR of 2.22 and a resolution of 1.1 cm when using a frequency range from 1.5 GHz to 2.5 GHz. A further bandwidth increase may lead to an enhancement of the residual clutter.

It should be empathized that S21 is a measure of the total field; thus, detection can be achieved only after artefact removal, which can cancel the image of the transmitter and the reflection of the first layers. Together with the subtraction between S21 obtained using two slightly displaced transmitting positions, i.e., rotation subtraction, there are also other techniques that may be used for artefact removal, both in the frequency and in time domain [30]. Among the techniques in the frequency domain, the rotation subtraction could be effective also for imaging highly asymmetric scenarios (such as human bones), since it assumes only the similarity of the first layers' reflection when displacing slightly the transmitting position.

Research is in progress to show the performance of the procedure in case of multiple fractures, investigating the impact of zeroing all intensity values below 0.5 in the detection of real defects. Moreover, a comparison of the performances between rotation subtraction artefact removal and (local) average subtraction artefact removal is also in progress.

Ultimately, this paper verified that the microwave scanning procedure, which was based on HP, can be used to perform bone imaging to detect lesions and fractures in bone layers successfully, negating the use of X-rays. Our proposed scanning procedure was simple and required only two antennas in free space, thus no matching liquid was needed. This paper may pave the way for the construction of a dedicated bone imaging system that is inexpensive, compact, and portable, since it resorts to two rotating antennas coupled through a VNA.

Author Contributions: Conceptualization, B.K., M.G. and G.T.; methodology, B.K., B.S., N.G. and G.T.; supervision, M.G. and S.D.; validation, B.K., B.S., N.G., M.G., S.D. and G.T.; writing, original draft preparation, B.K. and N.G.; writing, review and editing, M.G., S.D. and G.T.

Funding: The project leading to this application received funding from the European Union's Horizon 2020 research and innovation programme under the Marie Sklodowska-Curie Grant Agreement No. 793449.

Conflicts of Interest: The authors declare no conflict of interest.

References

1. Oryan, A.; Monazzah, S.; Bigham- Sadegh, A. Bone Injury and Fracture Healing Biology. *Elsevier Biomed. Environ. Sci.* **2018**, *28*, 57–71.
2. Meaney, P.M.; Goodwin, D.; Golnabi, A.; Pallone, S.; Geimer, A.; Paulsen, K.D. 3D Microwave bone imaging. In Proceedings of the 6th European Conference on Antennas and Propagation (EUCAP), Prague, Czech Republic, 26–30 March 2012; pp. 1770–1771. [CrossRef]

3. Pham, D.L.; Xu, C.; Prince, J.L. Current methods in medical image segmentation. *Annu. Rev. Biomed. Eng.* **2000**, *2*, 315–337. [PubMed]

4. Al Nahid, A.; Khan, T.M.; Kong, Y. Hardware Implementation of Bone Fracture Detector Using Fuzzy Method Along with Local Normalization Technique. *Ann. Data Sci.* **2017**, *4*, 533–546. [CrossRef]

5. Mercuri, M.; Sheth, T.; Natarajan, M. Radiation exposure from medical imaging: A silent harm? *CMAJ* **2011**, *183*, 413–414. [CrossRef] [PubMed]

6. Kak, A.C.; Slanery, M. *Principle of Computerized Tomography*; IEEE Free Press: New York, NY, USA, 1987.

7. Staderini, E.M. UWB radars in medicine. *IEEE Aerosp. Electron. Syst. Mag.* **2002**, *17*, 13–18. [CrossRef]

8. Joines, W.T.; Jirtle, R.L.; Rafal, M.D.; Schaefer, D.J. Microwave power absorption differences between normal and malignant tissue. *Int. J. Rad. Oncol. Biol. Phys.* **1980**, *6*, 681–687. [CrossRef]

9. Meaney, P.M.; Paulsen, K.D. Nonactive antenna compensation for fixed-array microwave imaging: Part II imaging results. *IEEE Trans. Med. Imag.* **1980**, *18*, 508–518. [CrossRef]

10. Winters, D.W.; Van Veen, B.D.; Hagness, S.C. A sparsity regularization approach to the electromagnetic inverse scattering problem. *IEEE Trans. Antennas Propag.* **2010**, *158*, 145–154. [CrossRef]

11. Li, X.; Bond, E.J.; Van Veen, B.D.; Hagness, S.C. An overview of ultra-wideband microwave imaging via space-time beamforming for early-stage breast-cancer detection. *IEEE Antennas Propag. Mag.* **2005**, *47*, 19–34.

12. Porter, F.; Kirshin, E.; Santorelli, A.; Coates, M.; Popović, M. Time-domain multistatic radar system for microwave breast screening. *IEEE Antennas Wirel. Propag. Lett.* **2013**, *12*, 229–232. [CrossRef]

13. Santorelli, A.; Porter, E.; Kang, E.; Piske, T.; Popović, M.; Schwartz, J.D. A time-domain microwave system for breast cancer detection using a flexible circuit board. *IEEE Trans. Instrum. Meas.* **2015**, *64*, 2986–2994. [CrossRef]

14. Nikolova, A. Microwave imaging for breast cancer. *IEEE Microw. Mag.* **2011**, *12*, 78–94. [CrossRef]

15. Semenov, S.Y.; Corfield, D.R. Microwave tomography for brain imaging: Feasibility assessment for stroke detection. *Int. J. Antenna Propag.* **2008**, *2008*, 254830. [CrossRef]

16. Porter, E.; Coates, M.; Popović, M. An early clinical study of time-domain microwave radar for breast health monitoring. *IEEE Trans. Biomed. Eng.* **2016**, *63*, 530–539. [CrossRef]

17. Salvador, S.M.; Fear, E.C.; Okoniewski, M.; Matyas, J.R. Exploring joint tissues with microwave imaging. *IEEE Trans. Microw. Theory Tech.* **2010**, *58*, 2307–2313. [CrossRef]

18. Ruvio, G.; Cuccaro, A.; Solimene, R.; Brancaccio, A.; Basile, B.; Ammann, M.J. Microwave bone imaging: A preliminary scanning system for proof-of-concept. *Healthc. Technol. Lett.* **2016**, *3*, 218–221. [CrossRef]

19. Mirbeik-Sabzevari, A.; Tavassolian, N. Tumor Detection Using Millimeter-Wave Technology: Differentiating Between Benign Lesions and Cancer Tissues. *IEEE Microw. Mag.* **2019**, *20*, 30–43. [CrossRef]

20. Gabriel, C.; Gabriel, S.; Corthout, E. The dielectric properties of biological tissues: I. literature survey. *Phys. Med. Biol.* **1996**, *41*, 2231–2249. [CrossRef]

21. Zurich Med Tech. 2019. Available online: https://zmt.swiss/validation-hw/tsm/tle5c-24-2450/ (accessed on 26 October 2019).

22. Meaney, P.M.; Fox, C.J.; Geimer, S.D.; Paulsen, K.D. Electrical Characterization of Glycerin: Water Mixtures: Implications for Use as a Coupling Medium in Microwave Tomography. *IEEE Trans. Microw. Theory Tech.* **2017**, *65*, 1471–1478. [CrossRef]

23. Sani, L.; Paoli, M.; Raspa, G.; Ghavami, N.; Sacchetti, F.; Saracini, A.; Ercolani, S.; Vannini, E.; Duranti, M. Initial Clinical Validation of a Novel Microwave Apparatus for Testing Breast Integrity. In Proceedings of the IEEE International Conference on Imaging Systems and Techniques (IST), Chania, Greece, 4–6 October 2016; pp. 278–282. [CrossRef]

24. Tiberi, G.; Ghavami, N.; Edwards, D.J.; Monorchio, A. Ultrawideband microwave imaging of cylindrical objects with inclusions. *IET Microw. Antennas Propag.* **2011**, *5*, 1440–1446. [CrossRef]

25. Enders, P. Huygens' principle as universal model of propagation. *Latin Am. J. Phys. Educ.* **2009**, *3*, 19–32.

26. Sani, L.; Ghavami, N.; Vispa, A.; Paoli, M.; Raspa, G.; Ghavami, N.; Sacchetti, F.; Vannini, E.; Ercolani, S.; Saracini, A.; et al. Novel microwave apparatus for breast lesions detection: Preliminary clinical results. *Biomed. Signal Process. Control* **2019**, *52*, 257–263.

27. Ghavami, N.; Tiberi, G.; Edwards, D.J.; Safaai-Jazi, A.; Monorchio, A. Huygens principle based imaging of multilayered objects with inclusion. *Prog. Electromagn. Res.* **2014**, *58*, 139–149.

28. Ghavami, N.; Tiberi, G.; Edwards, D.J.; Monorchio, A. UWB Microwave Imaging of Objects with Canonical Shape. *IEEE Trans. Antennas Propag.* **2012**, *60*, 231–239. [CrossRef]

29. Fear, E.C.; Li, X.; Hagness, S.C.; Stuchly, M.A. Confocal microwave imaging for breast cancer detection: Localization of tumors in three dimensions. *IEEE Trans. Biomed. Eng.* **2002**, *4*, 812–822.
30. Elahi, M.A.; Glavin, M.; Jones, E.; O' Halloran, M. Artifact Removal Algorithms for Microwave Imaging of the Breast. *Prog. Electromagn. Res.* **2013**, *141*, 185–200. [CrossRef]

 *electronics*

Article

Non-Invasive Blood Glucose Monitoring Using a Curved Goubau Line

Louis WY Liu [1,*,†], **Abhishek Kandwal** [2,*,†], **Qingsha Cheng** [3], **Hongjian Shi** [3], **Igbe Tobore** [2] and **Zedong Nie** [2,*]

1   Faculty of Engineering, Vietnamese German University, Thu Dau Mot 820000, Vietnam
2   Shenzhen Institutes of Advanced Technology, Chinese Academy of Sciences, Shenzhen 518000, China; victor@siat.ac.cn
3   Department of Electrical and Electronic Engineering, Southern University of Science and Technology, I Park A7, Nanshen, Shenzhen 518055, China; planetx4u@gmail.com (Q.C.); lurkerjones@yahoo.com (H.S.)
*   Correspondence: liu.waiyip@vgu.edu.vn (L.W.Y.L.); abhishek@siat.ac.cn (A.K.); zd.nie@siat.ac.cn (Z.N.)
†   These authors are first co-authors who have contributed equally to this work.

Received: 19 May 2019; Accepted: 9 June 2019; Published: 12 June 2019

Abstract: Non-invasive blood glucose monitoring at microwave frequencies is generally thought to be unreliable in terms of reproducibility of measurements. The failure to reproduce a blood glucose measurement from one experiment to another is in major part due to the unwanted interaction of leaky waves between the ambient environment and the blood glucose measuring device. In this work, we have overcome this problem by simply eliminating the leaky modes through the use of surface electromagnetic waves from a curved Goubau line. In the proposed methodology, a fixed volume of blood-filled skin tissue was first formed by vacuum suction and partially wound with a curved Goubau line which was coated with a 3 mm thick layer of gelatin/glycerin composite. Blood glucose levels were non-invasively determined using a network analyzer. At 4.5 GHz, a near-linear correlation exists between the measured S12 parameters and the blood glucose levels. The measured correlation was highly reproducible and consistent with the measurements obtained using the conventional invasive lancing approach. The findings of this work suggest the feasibility of non-invasive detection of left and right imbalances in the body.

Keywords: Goubau line; non-invasive blood glucose measurement; Acu-check; lancet; leaky waves; surface waves

---

1. Introduction

Blood glucose levels are an important bio-marker in the medical or health care industry. Almost all health problems are associated with an abnormal blood glucose level. These health problems include diabetics, candidiasis, amyloidosis, cancers, scleroderma diabeticorum, vitiligo, acanthosis nigricans and hypoglycemia. The blood glucose level, as well as the ways that it varies over time, is important information for the diabetic population, smokers wishing to quit and cancer patients. Conventional approaches to blood glucose measurement that require extraction of capillary blood through the use of a lancet are not suitable for prolonged blood glucose monitoring for two main reasons: (1) the amount of capillary blood extracted is too much for repeated measurements within a prolonged period; and (2) the wound punctured by the lancet is highly susceptible to infection.

At the time of writing, there has been a lot of negativity surrounding the idea of non-invasive blood glucose measurements. This negativity does not exist without any reason. The original idea of non-invasive blood glucose measurement involving an electromagnetic wave was based on the assumption that a change in blood glucose levels will induce a measurable change in the permittivity of the blood at high frequencies. By measuring the blood permittivity change, in theory, one can determine the blood glucose levels. However, this assumption has not factored in the uncertainties due to the following facts:

1. The blood permittivity is not a strong function of the blood glucose concentration in the blood [1]. At microwave frequencies, the real part of the blood permittivity does not change much regardless of what the blood glucose level is. A change in the blood glucose level induces a change in the imaginary part of the blood permittivity. The blood contains a variety of different substances. The blood glucose is just one of those. There is no reason to expect the relative permittivity of the blood in the body to change substantially in response to a change in the blood glucose.
2. The density of the blood underneath the skin can be different from one body part to another and from one experiment to another experiment. It is, for this obvious reason, not logical to measure the blood glucose level from the elbow area. The density of the blood underneath the skin depends on the volume. It is virtually impossible to obtain reproducible measurements if the volume of the object under test is not kept constant from experiment to experiment.
3. The blood glucose measuring device can act like an antenna at high frequencies, particularly at microwave or millimeter-wave frequencies. This antenna effect is also known as leaky wave effect. This antenna effect enables the blood glucose measuring device to release and absorb leaky electromagnetic radiations to and from the ambient environment. It can substantially dis-stabilize the measured results at microwave or millimeter wave frequencies, particularly in the presence of an electromagnetic noise source in the background.

Of all the factors that contribute to the measurement uncertainties, the leaky wave effect is the most impactful. At the time of writing, our research group has already evaluated many published methods. These methods include the traditional split ring resonator approach [2], measurement at 60 GHz [3], as well as a substrate integrated waveguide (SIW) approach [4]. These approaches rely on the nonlinear effects at the resonant frequencies where the magnitude change in the measured scattering parameters are enormously amplified in response to a very large change in blood glucose levels. Measuring the S-parameters right at the resonant frequencies is definitely a help but the energy gained or lost by the glucose measuring device at the resonant frequencies are most likely leaky modes, contributed to by the resonances of higher order modes. Depending on the environment, these ambient leaky waves can change from experiment to experiment.

On the other hand, almost all the published approaches were based on either a body part with an unknown density or volume of blood or a body part with an unknown blood volume. Most importantly, almost none of these approaches have taken the obvious effects of leaky waves into account. In all honesty, we were unable to reliably reproduce the measurements from one experiment to another with these published approaches until the measurements were conducted with the participant under test rigidly fixed in position in a completely isolated environment.

In this work, we found that the changes in the measured S-parameters as a result of a change in the blood glucose concentration can be enormously magnified not only at the resonant frequencies, but also at the cut-off frequencies where the propagating energy turns from a surface mode to a leaky mode. To take advantage of this S-parameter magnification effect, and to minimize the effects of leaky waves, our proposed glucose measuring device was designed to absorb the leaky radiations and turn them into a dielectric loss due to the presence of blood glucose in the blood right at the cut-off frequency, where the device was about to become a leaky wave antenna. The proposed approach is based on a curved Goubau line winding an artificially fixed volume of blood-filled skin tissue [5–7]. With the help of the suction effect of a vacuum suction aspirator, the volume of the blood-filled skin tissue will remain unchanged from experiment to experiment. The Goubau line has already been applied in applications of wireless power transfer [8–13]. The use of surface electromagnetic waves associated with a curved Goubau line for blood glucose measurement is an unprecedented attempt. However, our in-vivo experimental results have proven beyond any doubt that the measured results from the proposed approach and the traditional lancing based approach have almost 100% agreement with each other.

2. Conceptual Background

Electromagnetic waves can be broadly classified into two categories: (a) surface electromagnetic waves and (b) leaky electromagnetic waves. Leaky waves propagate at a group velocity faster than the speed of light. The transverse electromagnetic nature of a leaky electromagnetic wave makes it possible to radiate in all directions, according to Poynting's theorem [14]. The process of radiations involves an exchange of electromagnetic energy between an antenna and the ambient environment. If a blood glucose measuring device acts like an antenna, then the accuracy and the stability of the measurements using this device will be negatively affected by the electromagnetic interference from the environment. On the other hand, surface electromagnetic waves propagate along the interface between two neighboring mediums at a group velocity below the speed of light, with virtually no radiation loss. Unlike leaky electromagnetic waves, surface electromagnetic waves tend to propagate with a relatively stable amplitude because they are relatively unaffected by the ambient electromagnetic interference. Therefore, there is no point of basing our blood glucose measurements on leaky electromagnetic waves. Whether an electromagnetic energy is leaky or not also depends on its frequencies. To ensure reproducibility of measurement, the blood glucose measurements should be conducted at frequencies where the group velocity of the propagating wave is less than the speed of light.

A Goubau line is a slow-wave guiding structure which, at frequencies below the cut-off frequency, primarily supports propagation of a surface electromagnetic wave at a group velocity below the speed of light (see Figure 1a). It is basically a groundless single-wire transmission line coated with a layer of dielectric material. This dielectric coating confines the evanescent electromagnetic energy on the conducting surface, which would otherwise turn the evanescent electromagnetic energy into a leaky wave. When both ends of the Goubau line are connected to a vector network analyzer for measurements of scattering parameters, we will either end up with a large magnitude of the S12 (or S21) parameter together with a small magnitude of the S22 (or S11) parameter, or the other way round. The former means that the characteristic impedance of the Goubau line is close to 50 Ω, whilst the latter means the opposite. In either scenario, the radiation efficiency of the leaky waves is low with virtually no antenna effect. The Goubau line in either of these two scenarios has vertically little or no interaction with the ambient electromagnetic interference. In the absence of any interaction with the ambient electromagnetic interference, the measured scattering parameters will be stable and reproducible.

The Goubau line, as shown in Figure 1a, is assumed to be a straight Goubau line. The radii of the dielectrically coated Goubau line and its center conductor are respectively a and b. The relative permittivity of the dielectric layer coating the center conductor is ε_r. Then the group velocity of a surface electromagnetic wave can be computed from the propagation constant using the modified version of the approximation formula proposed by Jaisson [15]:

$$k_z = k_0 \sqrt{1 + \frac{\varepsilon_r - 1}{1 + 6\varepsilon_r K_e/G}} \tag{1}$$

where

$$k_0 = \frac{2\pi f}{c} \tag{2}$$

$$K_e = 0.11593148 - 0.5 \ln(-q) - 0.5 \ln(-\ln(-q)) + \frac{0.5 \ln(-\ln(-q))}{\ln(-q)} \tag{3}$$

$$q = \left(\frac{1}{\varepsilon_r} - 1\right)\frac{G}{3}(k_0 b)^2 \tag{4}$$

$$G = \left(1 - \frac{a}{b}\right)\left[6 + \left(1 - \frac{a}{b}\right)\left(5 - \frac{2a}{b}\right)\right] \tag{5}$$

The group velocity is therefore given by:

$$V_g = \frac{2\pi f}{real\{k_z\}} \tag{6}$$

In order to eliminate leaky waves, which will render the measured scattering parameters non-reproducible from one experiment to another, the frequency at which the blood glucose measurement should be conducted at should satisfy the condition given by Equation (7):

$$f < \frac{c}{5.9809\,b\,\sqrt{G\left(1 - \frac{1}{\varepsilon_r}\right)}} \tag{7}$$

where the frequency f is also known as the cut-off frequency for the fundamental transverse magnetic mode of the Goubau line.

In Equations (1)–(7), it is assumed that the Goubau line is a straight insulated wire. It is difficult to determine the line characteristics by measuring the scattering parameters between both of the terminals of a straight Goubau line because there is a direct port-to-port coupling between these two terminals. The direct port-to-port coupling can be eliminated by measuring a curved Goubau line that has been bent by 180° as illustrated in Figure 1b. However, the curved portion of the Goubau line as shown in Figure 1b will radiate an additional amount of leaky waves in the directions as indicated by the pink arrows. This radiation loss does not depend on the cut-off frequency as given in Equation (7). Instead, it depends on the sharpness of the bend, or the radius of the bent portion of the Goubau line. In the bent portion of the Goubau line, the outer and inner radii of the curved portion of the Goubau line are respectively $R1 + (b - a)$ and $R2 - (b - a)$. This means the radiation loss due to the leaky waves at the curved Goubau line will be a function of the difference in distance between the outer edge and the inner edge of the Goubau line, which is $(R2 + (a - b) - R1 + (a - b))\pi$. According to the results of our repeated experimentation, the value of b in Equation (7) can be replaced with $b^{(R2+a-b)/(R1-a+b)}$ in order to factor in the additional radiation loss due to the bending effect at the curved Goubau line:

$$f < \frac{c}{5.9809\,b^{\frac{R2+a-b}{R1-a+b}}\,\sqrt{G\left(1 - \frac{1}{\varepsilon_r}\right)}} \tag{8}$$

The radiation loss due to the bent section of the curved Goubau line can be effectively minimized at certain frequencies by letting the curved portion of the Goubau line be partially wound over a cylindrical dielectric resonator of high permittivity, as is illustrated in Figure 1c. Because of its higher permittivity, this cylindrical dielectric resonator reduces the effective wavelength along the inner edge of the curved Goubau line, thereby reducing the overall distance difference between the outer edge and the inner edge of the curved Goubau line [7]. In the neighborhood of the cut-off frequency as given in Equation (8), whilst some of the surface electromagnetic waves in the curved portion of the Goubau line are radiating outwards as leaky waves, the cylindrical dielectric resonator is simultaneously absorbing the electromagnetic energy backward and inwards. The radiation loss in this scenario will be nullified or minimized as a result of the permittivity of the dielectric resonator. The electromagnetic energy being absorbed back to the cylindrical resonator will eventually manifest as a dielectric loss but this dielectric loss is totally independent of the ambient electromagnetic interference. On the other hand, the cylindrical dielectric resonator will also yield a series of stop bands at different resonant frequencies. These resonant frequencies can be approximately determined using the formula proposed by Siart [16]:

$$f_{m,TE} = \frac{0.5m}{b - a}\,\frac{c}{\sqrt{\mu_0(\varepsilon_r - \varepsilon_0)}} \tag{9a}$$

$$f_{m,TM} = \frac{0.5(m + 0.5)}{b - a} \frac{c}{\sqrt{\mu_0(\varepsilon_r - \varepsilon_0)}} \tag{9b}$$

where m is the order of mode. $f_{m,TE}$ and $f_{m,TM}$ are respectively the cut off frequencies of the m-th transverse electric mode and the m-th transverse magnetic mode. According to Equations (9a) and (9b), a change in the blood permittivity will induce a larger increase at the resonance of a higher order mode. If this resonance occurs at the frequencies close to the cut-off frequency as given in Equation (8), the leaky modes will be absorbed and converted into a dielectric loss that is highly dependent on permittivity of the cylindrical dielectric resonator.

Reducing the radiation loss of the Goubau line through the use of a cylindrical dielectric resonator implies that, at the cut-off frequency as given Equation (9a) or (9b), the proposed glucose measuring device will be less susceptible to the ambient electromagnetic interference. The main goal of this work is to take advantage of the effects due to the presence of the cylindrical dielectric resonator to determine the blood glucose levels. If the cylindrical dielectric resonator is a cylindrical volume of skin tissue filled with blood, and if the geometry of the cylindrical dielectric resonator is fixed, then we should be able to determine the exact permittivity due to the glucose concentration in the blood. In this work, this cylindrical volume of a blood-filled skin tissue is arbitrarily formed by vacuum suction using a twisted vacuum suction aspirator (see Figure 1d).

Figure 1d shows how the cylindrical volume of the blood-filled skin tissue is formed by vacuum suction. Due to the low pressure inside the vacuum suction aspirator, a large amount of blood will flow into the area of the skin where a blood glucose measurement is conducted. The proposed technique will not cause any injury to the skin or the body tissue even after several hours of use. Unlike other invasive techniques involving the use of a lancet, the proposed technique will allow the blood glucose levels to be continuously and painlessly monitored over a prolonged period. As the proposed glucose sensor releases or absorbs very little ambient electromagnetic energy, the measured scattering parameters will be highly stable. On the other hand, because the volume of the blood-filled skin tissue is very much fixed from one measurement to another, issue #2 as stated in the introduction section is very much a non-issue. In fact, the measured results have been proven to be accurate and reproducible.

The permittivity change of the blood is known to be a weak function of the glucose concentration in the blood at microwave frequencies. However, regardless of how small this blood permittivity change is, this blood permittivity change in the proposed methodology will induce an enormous change in the scattering parameters at frequencies around the cut-off frequency given by Equation (8).

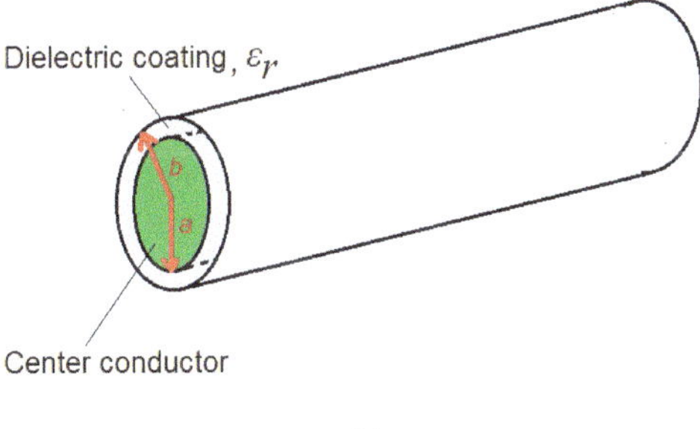

(a)

Figure 1. *Cont.*

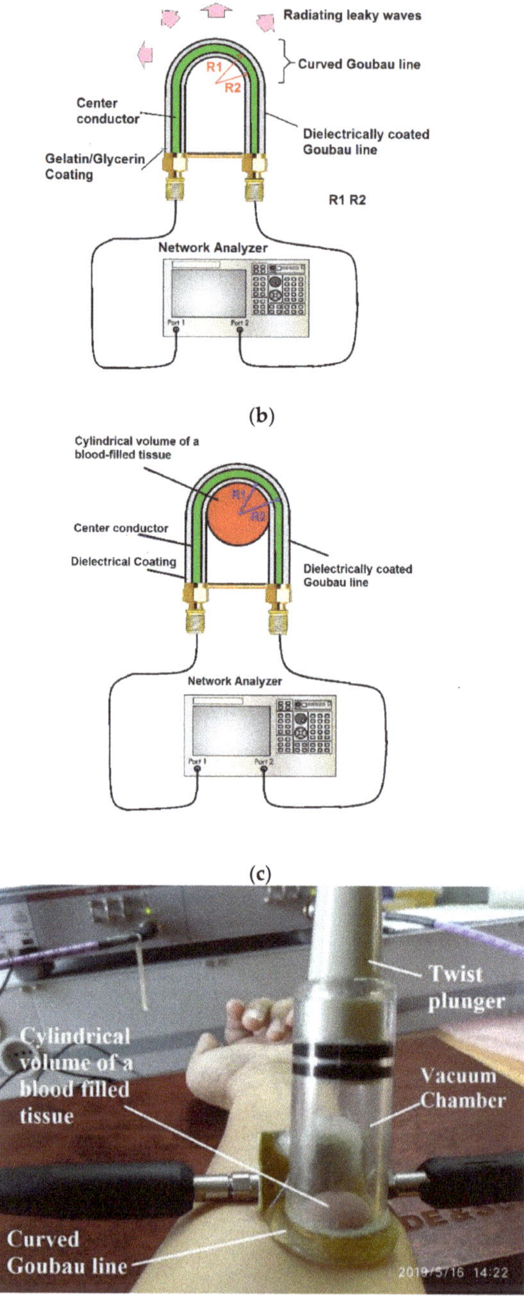

(b)

(c)

(d)

Figure 1. Concept of the proposed blood glucose measurement. (**a**) a straight dielectrically coated Goubau line; (**b**) a measurement setup for a curved Goubau line; (**c**) a measurement setup for the proposed blood glucose measurement; and (**d**) a photo showing how the cylindrical volume of a blood-filled skin tissue is formed using a vacuum suction aspirator.

3. Materials and Methods

To validate the proposed concept, we conducted a series of in-vivo blood glucose measurements in the Vietnamese German University of Vietnam as well as the Chinese Academy of Sciences in China. In each in-vivo blood glucose measurement, a participant was guided by a nurse to undergo an invasive blood glucose measurement involving the use of a lancing device known as Accu Chek Flaxclix^{TM} (Roche Diabetes Care, Inc., Indianapolis, IN, United States) at a particular time of the day. Within 15 m after this invasive blood glucose measurement, the participant was guided through a non-invasive blood glucose measurement on both left and right hands using the proposed method as illustrated in Figure 1c. The network analyzer used in this work was ZVA45B manufactured by Rohde & Schwarz R&S (in Mühldorfstraße 15, 81671 Munich Germany). During the measurements, the participants were allowed to move their body violently. Figure 2a shows the photograph of the actual setup for each non-invasive blood glucose measurement.

The data samples were collected over different blood glucose levels from 41 volunteers who were university staff members and school-aged students. Altogether, we managed to measure over 45 blood samples with blood glucose levels ranging from 60 mg/dl to 130 mg/dl. The measured results from the proposed non-invasive blood glucose measurements were then analyzed and compared with the results obtained using the invasive method.

All subjects gave their informed consent for inclusion before they participated in the study. The study was conducted in accordance with the Declaration of Helsinki, and the protocol was approved by the Ethics Committee of Vietnamese German University (VGU-EEIT-308).

The vacuum suction aspirator was an online shopping mall product known as a vacuum plunger, purchased from a cross-border seller in Lazada-Taobao. On the exterior wall of the vacuum suction aspirator as shown in Figure 1d, the area where the curved Goubau line was mounted was milled with a groove using a 5-axis precision CNC (Computerized Numerically Controlled) machine. The milling depth of the groove was maximized to minimize the separation between the Goubau line and the skin tissue to just 0.3 mm.

The center conductor of the curved Goubau line has a cross-sectional radius of 1mm. The dielectric material coating the Goubau line is a 3.8 mm thick layer of thermally cured gelatin/glycerin composite which has a dielectric constant of 60. Using the formula given in Equation (7), the cut-off frequency of the Goubau line was determined to be 4.36 GHz. Figure 2b shows the group velocity of the Goubau line as a function of frequency computed using the formula in Equation (6).

The diameter of the cylindrical blood-filled skin tissue is 10 mm. According to Equation (9b), there was supposed to be a transverse magnetic resonant in the neighborhood of 4.5 GHz when $m = 4$.

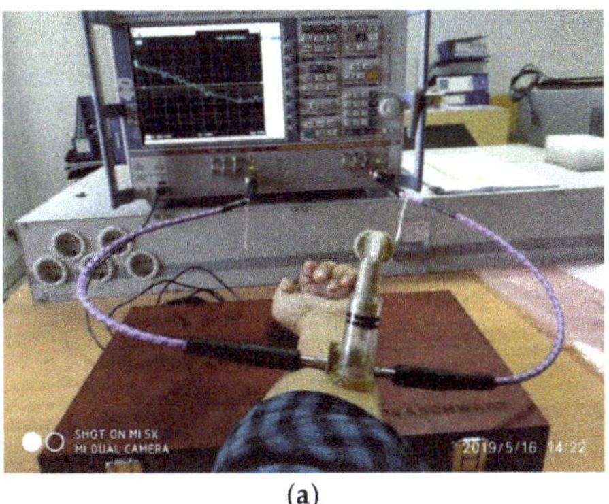

(a)

Figure 2. *Cont.*

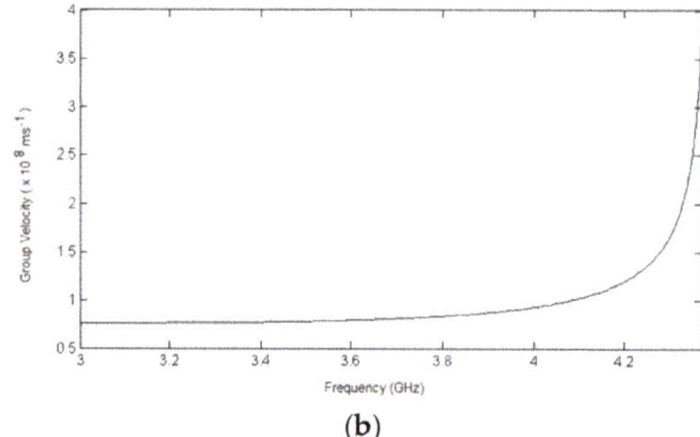

(b)

Figure 2. Details of the experiment. (**a**) The photo showing the actual measurement setup for measuring the blood glucose on the left hand; and (**b**) calculated group velocity of a surface wave propagating in the curved Goubau line (of which the diameter of the central conductor is 1 mm and the thickness of the gelatin coating is 3.8 mm).

4. Results

This section presents the results of our in-vivo measurement from the data samples from the participants with a normal blood pressure from the 41 volunteers. These participants include the university staff members and school-aged students. Most of the data samples have repeated blood glucose levels. The measurements were obtained from different times of the day.

Figure 3a shows the scattering parameters as a function of frequency in the dB scale when the blood glucose level was 77 mg/dl. A similar plot was obtained for each of the other blood glucose levels. It can be observed from Figure 3a that both S11 and S12 parameters started to trend downwards at frequencies higher than 4.35 GHz. The simultaneous downtrend of S11 and S12 parameters is an indication of an increase in leaky wave radiation efficiency.

During the measurements, the measured scattering parameters were very much unchanged at frequencies below 4.5 GHz even though the participants moved their bodies. The measured scattering parameters remained stable even when the nearby metal objects were moved. However, when the participants moved their bodies, the instability in terms of measured scattering parameters started to become obvious at frequencies higher than 4.6 GHz.

Figure 3c shows that there is a positive correlation between the blood glucose levels and the S12 parameters. This near-linear correlation was obtained by extracting the data of Figure 3b right at 4.57 GHz. The linearity of this correlation was almost completely lost at frequencies higher than 4.8 GHz.

During the non-invasive blood glucose measurement, we also observed an imbalance between the left and right hands on two participants. The measured differences in scattering parameters were small and, in terms of percentage, the differences were within 3%. We repeated the measurements several times but the observed phenomena remained.

We also used the same experimental setup to conduct a series of similar in-vivo measurements at frequencies very close to 60 GHz. The measured S21 or S12 parameters were between −65 dB and −80 dB. However, the measured scattering parameters at around 60 GHz were found to be unstable and non-reproducible from one experiment to another.

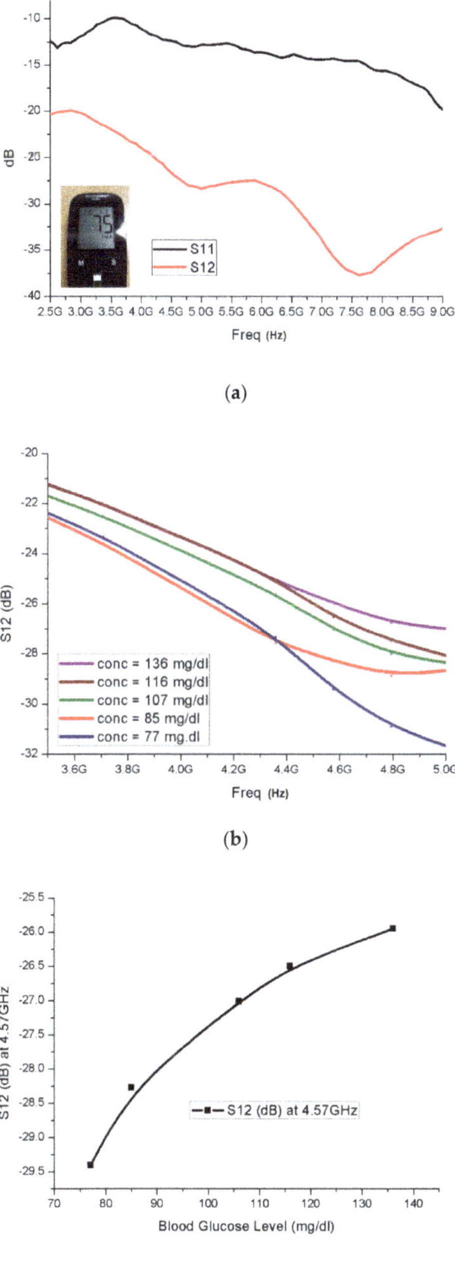

(a)

(b)

(c)

Figure 3. Measured results. (**a**) The measured S11 and S12 parameters against frequency when the blood glucose level was 75 mg/dl; (**b**) the measured S12 parameters as a function of frequency for different blood glucose levels (as denoted as "conc" in the graph). It is important to note that the blood glucose levels in this graph were obtained using the traditional lancing method within 15 m before the measurement of the S-parameters was conducted; (**c**) a graph showing the near-linear relationship between the blood glucose levels and the S12 parameters right at 4.57 GHz.

5. Discussion

The results of this investigation have proven beyond any doubt that the non-invasive blood glucose measurement using the proposed approach was reliable only when the interference from ambient leaky waves was significantly suppressed. The equations as given in Section 2 predicted the existence of the fundamental cut-off frequency at around 4.35 GHz, whilst the measured results reveal that the measurements started to be erratic and unstable at frequencies higher than 4.6 GHz, particularly at the time when the participant moved his/her body. It is possible that, at around 4.57 GHz, the cylindrical volume of the blood-filled skin tissue absorbed most of the leaky waves that would otherwise be lost to the surrounding. At 4.8 GHz, the effects of leaky waves of the blood glucose monitoring device completely took over those of the surface electromagnetic waves. At frequencies between 4.35 and 4.6 GHz, the measured scattering parameters were not only stable but also reproducible from experiment to experiment.

The effects of leaky waves were clearly counter-productive as far as non-invasive blood glucose measurements are concerned. The proposed blood glucose sensor as illustrated in Figure 1d was designed to minimize all forms of leaky waves. However, at each of the SMA terminals, the discontinuity between the coaxial line and the Goubau line did encourage excitation of higher order modes, which were mostly leaky in nature. In order for the measurement errors to be further minimized, there should be a smooth tapered transition between the coaxial cable and the Goubau line. We expect to implement this improvement in the next stage of the project.

As pointed out in Sections 1 and 2, the change in S-parameters due to a change in the blood glucose concentration can be magnified right at the cut-off frequency, which was 4.35 GHz in the proposed glucose measuring device. In Figure 3b, it can be seen that, at 3.5 GHz, a blood glucose level change from 75 mg/dl to 136 mg/dl produced an overall magnitude change of 1.42 dB in the S12 parameter. This magnitude change in the S12 parameters gradually increased as the frequency went higher. When the frequency was at or higher than 4.35 GHz, the magnitude of the S12 parameter for each concentration of blood glucose became noticeably observable. At frequencies in the neighborhood of 4.57 GHz, the same change in the blood glucose level resulted in an overall magnitude change of 4 dB in the S12 parameter. This amplified magnitude change in the S12 parameter was due in major part to the blood glucose levels when the propagating energy was changed from the transverse fundamental magnetic mode into the higher order transverse electromagnetic modes at the cut off frequency (i.e., 4.35 GHz). In this device, which is fundamentally different from other published counterparts [3,17–20], the majority of the energy from these higher order transverse electromagnetic modes was absorbed back as dielectric losses by the cylindrical volume of blood filled skin tissue, which, according to Equation (9a), also resonated at around the cut-off frequency.

One of the potential contributing factors to the uncertainties in the non-invasive blood glucose measurement was the accuracy of the calibration of a vector network analyzer. We were mindful of this issue. In this work, the TRL (Thru, Reflect, Line) calibration we used was based on the ZV-Z229 CALIBRATION KIT 2.92 mm from Rohde & Schwarz R&S (in Mühldorfstraße 15, 81671 Munich Germany). So far, we have not encountered any problem with the calibration. The measured S-parameters were consistently reproducible with little loss of accuracy from one experiment to another even though the calibration was re-done again and again.

During the non-invasive blood glucose measurement, the S12 parameters for 85 mg/dl and 77 mg/dl also overlapped at frequencies below 4.35 GHz. This was due to the fact that the participants' blood pressures were not 100% stable during the time when they were going through the non-invasive blood glucose measurements.

During our non-invasive blood glucose measurement, we observed an imbalance in terms of the measured scattering parameters in two individuals. This imbalance did not appear to be removable even after several attempts. The amount of blood flowing into the skin area where the proposed sensor was applied was dependent upon the combined effect of the individual's blood pressure and the air pressure inside the vacuum suction aspirator. If there is any difference in terms of the measured

S12 parameters between the left and right hands, then this imbalance was most likely due to the issue of blood circulation in their bodies. This imbalance cannot be detected using the conventional invasive Acuu-chek lancing approach. Although our observation is inconclusive at this stage, there is an implication that the proposed methodology can be further applied in detecting a left and right imbalance in the human body.

The results of this investigation have clearly proven beyond any doubt that the proposed methodology can be used to conduct in-vivo measurements with accurate and reproducible results. At the time of this writing, there has been a widespread misconception in the scientific community that non-invasive blood glucose monitoring is not possible. The findings of this work have demystified this misconception not only in terms of accuracy but also in terms of reproducibility and reliability.

6. Conclusions

This paper has presented an approach for non-invasive blood glucose monitoring at microwave frequencies based on surface electromagnetic waves. According to the results of our investigation, failure to obtain a reproducible blood glucose measurement was mainly due to the leaky wave interaction between the blood glucose measuring device and the ambient environment. This problem has been overcome through the use of surface electromagnetic waves propagating along a curved Goubau line. In the proposed methodology, a fixed volume of blood-filled skin tissue was created by vacuum suction and partially wounded with the said curved Goubau line. Non-invasive blood glucose measurements were conducted by measuring the S12 or S21 parameters between two terminals of the curved Goubau line using a network analyzer. At 4.5 GHz, there is a near-linear correlation between the measured S12 parameters and the blood glucose levels. The measured results at 4.5 GHz were both reproducible and consistent with the measurement obtained using the conventional invasive lancing approach. Implicated in the findings of this work is the feasibility of non-invasive detection of a left and right imbalance in the body.

Author Contributions: L.W.Y.L. has the original idea. L.W.Y.L. and A.K. participated in organizing the in-vivo and ex-vivo experiments using the proposed methodology. All other authors had a varying degree of intellectual contribution to this work. All authors helped in improving the article.

Funding: This research is internally funded by the authors' universities. This research received no external funding.

Acknowledgments: The authors would like to acknowledge the help from the nurse at Vietnamese German University, Nguyễn Thị Hoa.

Conflicts of Interest: The authors declare no conflict of interest.

References

1. Karacolak, T.; Moreland, E.C.; Topsakal, E. Cole-cole model for glucose-dependent dielectric properties of blood plasma for continuous glucose monitoring. *Microw. Opt. Technol. Lett.* **2013**, *55*, 1160–1164. [CrossRef]
2. Choi, H.; Nylon, J.; Luzio, S.; Beutler, J.; Porch, A. Design of continuous non-invasive blood glucose monitoring sensor based on a microwave split ring resonator. In Proceedings of the 2014 IEEE MTT-S International Microwave Workshop Series on RF and Wireless Technologies for Biomedical and Healthcare Applications (IMWS-Bio2014), London, UK, 8–10 December 2014.
3. Cano-Garcia, H.; Gouzouasis, I.; Sotiriou, I.; Saha, S.; Palikaras, G.; Kosmas, P.; Kallos, E. Reflection and transmission measurements using 60 GHz patch antennas in the presence of animal tissue for non-invasive glucose sensing. In Proceedings of the 2016 10th European Conference on Antennas and Propagation (EuCAP), Davos, Switzerland, 10–15 April 2016.
4. Xie, K.; Liu, L.W.; Zhang, Q.; Cheng, Q. Monitoring blood glucose fluctuation using SIW cavity with a coupling slot. In Proceedings of the 2017 Sixth Asia-Pacific Conference on Antennas and Propagation (APCAP), Xi'an, China, 16–19 October 2017.
5. Goubau, G. Surface Waves and Their Application to Transmission Lines. *J. Appl. Phys.* **1950**, *21*, 1119. [CrossRef]
6. Goubau, G.J.E. Surface Wave Transmission Line. U.S. Patent 2,685,068, 27 July 1954.

7. Chiron, B.P. Surface Wave Transmission Line Element. U.S. Patent 2,807,786, 21 May 1953.
8. Liu, L.W.Y.; Zhang, Q.; Chen, Y. Harvesting Atmospheric Ions Using Surface Electromagnetic Technology. *Adv. Technol. Innov.* **2017**, *2*, 99–104.
9. Liu, L.W.Y.; Ge, S.; Zhang, Q.; Chen, Y. Capturing surface electromagnetic energy into a dc through single-conductor transmission line at microwave frequencies. *Prog. Electromagn. Res. M* **2017**, *54*, 29–36. [CrossRef]
10. Liu, L.W.; Zhang, Q.; Chen, Y.; Teeti, M.A.; Das, R. Wireless Energy Harvesting by Direct Voltage Multiplication on Lateral Waves from a Suspended Dielectric Layer. *IEEE Access* **2017**, *5*, 21873–21884. [CrossRef]
11. Liu, L.W.; Kandwal, A.; Eremenko, Z.E.; Zhang, Q. Open-ended voltage multipliers for wireless transmission of electric power. *J. Microw. Power Electromagn. Energy* **2017**, *51*, 187–204. [CrossRef]
12. Liu, L.W.; Zhang, Q.; Chen, Y. Avramenko Diode Circuit Topology for Microwave Energy Harvesting in Goubau Line and Wireless. *IEEE Access* **2018**, *6*, 18883–18893. [CrossRef]
13. Louis, W.L.; Abhishek, K.; Hongjian, S.; Qingsha, S.C. Wireless Power Transfer using An RF Plasma. *IEEE Access* **2018**, *6*, 73905–73915.
14. Poynting, J.H. On the Transfer of Energy in the Electromagnetic Field. *Proc. R. Soc. Lond.* **1883**, *36*, 186–187. [CrossRef]
15. Jaisson, D. Simple formula for the wave number of the Goubau line. *J. Electromagn.* **2014**, *34*, 85–91. [CrossRef]
16. Siart, U.; Adrian, S.; Eibert, T. Properties of axial surface waves along dielectrically coated conducting cylinders. *Adv. Radio Sci.* **2012**, *10*, 79–84. [CrossRef]
17. Saha, S.; Cano-Garcia, H.; Sotiriou, I.; Lipscombe, O.; Gouzouasis, I.; Koutsoupidou, M.; Palikaras, G.; MacKenzie, R.; Reeve, T.; Kosmas, P.; et al. A Glucose Sensing System Based on Transmission Measurements at Millimetre Waves using Micro strip Patch Antennas. *Sci. Rep.* **2017**, *7*, 6855. [CrossRef] [PubMed]
18. Shaker, G.; Smith, K.; Omer, A.E.; Liu, S.; Csech, C.; Wadhwa, U.; Safavi-Naeini, S.; Hughson, R. Non-Invasive Monitoring of Glucose Level Changes Utilizing a mm-Wave Radar System. *Int. J. Mob. Hum. Comput. Interact.* **2018**, *10*, 10–29. [CrossRef]
19. Shaker, G.; Liu, S.; Wadhwa, U. Non-invasive glucose monitoring utilizing electromagnetic waves. In Proceedings of the ACM 19th International Conference on Human-Computer Interaction with Mobile Devices and Services, Vienna, Austria, 4–7 September 2017.
20. Torii, T.; Chiba, H.; Tanabe, T.; Oyama, Y. Measurements of glucose concentration in aqueous solutions using reflected THz radiation for applications to a novel sub-THz radiation non-invasive blood sugar measurement method. *Digit. Health* **2017**, *3*, 2055207617729534. [CrossRef] [PubMed]

Article

Imaging of the Internal Structure of Permafrost in the Tibetan Plateau Using Ground Penetrating Radar

Yao Wang [1,2], Zhihong Fu [1,2,*], Xinglin Lu [1,2], Shanqiang Qin [1,2], Haowen Wang [1,2] and Xiujuan Wang [1,2]

1   State Key Laboratory of Power Transmission Equipment and System Security and New Technology, Chongqing University, Chongqing 400044, China; 20161101003@cqu.edu.cn (Y.W.); 20191101379@cqu.edu.cn (X.L.); shqiangqin@cqu.edu.cn (S.Q.); 20097482@cqu.edu.cn (H.W.); 20111101002@cqu.edu.cn (X.W.)
2   School of Electrical Engineering, Chongqing University, Chongqing 400044, China
*   Correspondence: fuzhihong@cqu.edu.cn; Tel.: +86-130-6235-2738

Received: 23 November 2019; Accepted: 26 December 2019; Published: 30 December 2019

Abstract: The distribution of the permafrost in the Tibetan Plateau has dramatically changed due to climate change, expressed as increasing permafrost degradation, thawing depth deepening and disappearance of island permafrost. These changes have serious impacts on the local ecological environment and the stability of engineering infrastructures. Ground penetrating radar (GPR) is used to detect permafrost active layer depth, the upper limit of permafrost and the thawing of permafrost with the season's changes. Due to the influence of complex structure in the permafrost layer, it is difficult to effectively characterize the accurate structure within the permafrost on the radar profile. In order to get the high resolution GPR profile in the Tibetan Plateau, the reverse time migration (RTM) imaging method was applied to GPR real data. In this paper, RTM algorithm is proven to be correct through the groove's model of forward modeling data. In the Beiluhe region, the imaging result of GPR RTM profiles show that the RTM of GPR makes use of diffracted energy to properly position the reflections caused by the gravels, pebbles, cobbles and small discontinuities. It can accurately determine the depth of the active layer bottom interface in the migration section. In order to prove the accuracy of interpretation results of real data RTM section, we set up the three dielectric constant models based on the real data RTM profiles and geological information, and obtained the model data RTM profiles, which can prove the accuracy of interpretation results of three-line RTM profiles. The results of three-line RTM bears great significance for the study of complex structure and freezing and thawing process of permafrost at the Beiluhe region on the Tibetan Plateau.

Keywords: ground penetrating radar; reverse time migration; Tibetan Plateau; permafrost active layer; internal structure

---

1. Introduction

The Tibetan Plateau is known as the "third pole of the world." Its average altitude is higher than 4500 m, which gives it the highest and most complex terrain of a plateau with permafrost regions in the world. The character of Tibetan Plateau permafrost has obvious vertical zoning. With the increase of altitude, the freezing depth of permafrost obviously ascends. Compared with North America and Russia's arctic permafrost [1], it is of high temperature, high ice content, thin thickness and poor stability [2,3]. The permafrost is very sensitive to the change of ecological environment and human activities due to global warming [4–9]. Acceleration of permafrost degradation, deepening of permafrost active depth and disappearance of island permafrost are reported [10–12]. Meanwhile, construction of engineering infrastructures, such as the Qinghai–Tibetan highway, has dramatically altered the original regime of groundwater and surface water resources. Surface runoff and roadside water has directly or indirectly affected the stability of permafrost. It bears great significance to the engineering/construction and the protection of the ecological environment that the research of the permafrost's distribution state, active depth and the fine structure is conducted.

Ground penetrating radar (GPR) is the most powerful and widely used geophysical tool in permafrost studies [13–18]. Combined with other invasive geological explorations, such as trenching, coring and boring, it is an efficient and meticulous way to study the permafrost distribution characteristics, burial location and the evolution process [10–22]. There are many reasons which affect the process of the freezing and thawing of permafrost in Tibetan Plateau [23,24], such as slope direction, slope, vegetation, the thickness of the snow cover and permafrost duration, organic layer and the soil properties, human engineering activities, etc. The thickness of permafrost layer, water content, ice content and the activity layer depth have lateral changes, obviously within the local scale, which changes the layers' electrical structure obviously. The ice, the position of the small fault and the local fine structure in the internal of permafrost, which can indicate the change of the state of internal freezing and thawing process, affect the freezing and thawing of permafrost in the context of climate change. However, it is difficult to identify the internal structure of the permafrost in complex area from the raw radar profile. In addition, the shallow permafrost layer has repeated freezing and thawing. It is difficult to effectively reflect the state of active permafrost with seasons from the radar section.

The process of migration returns the underground reflection point information back to properly positioned reflections, and the reflection wave, simultaneously with the diffraction wave, automatically converges and interferes with automatic decomposition; better migration methods can provide high resolution interpretation of imaging. Compared with Kirchhoff migration method [25,26], reverse time migration (RTM) can effectively use the full wave field information [27–29]. While compared with one-way wave equation migration [30], RTM has no limitations in propagation direction and dip angle, since RTM does not need to separate the wave field, and can better use the reversed branch and the multiple waves; it is quite adaptive to lateral velocity variations, too. RTM is the state of the art in high precision migration methods.

When the electromagnetic (EM) impulse has a central frequency well above the transition frequency (the ratio of the dielectric constant to the electric conductivity of the media) the EM filed propagation is essentially a wave field; the kinematic characteristics of the propagation of the EM wave and the elastic wave are quite similar, such that all the seismic data processing techniques can be used to process the radar data [31]. The conventional migrations have been widely used in radar data imaging [32,33]. Fisher et al. was the first to apply the finite difference (FD) RTM to GPR [31]. They simply used the acoustic wave equation to the post-stacking GPR data. Sanada and Ashida continued the work to develop the RTM algorithm directly from the Maxwell's equation with the consideration of finite conductivity in the algorithm [34]. Leuschen and Plumb realized the FD-RTM on both multi-offset and zero-offset data [35]. Zhou conducted the FD-RTM based on Maxwell's equations and achieved the common shot point prestack RTM of GPR data [36]. Liu combined the FD-RTM algorithm and the full waveform inversion to estimate migration velocity [37].

In this study, we introduced the principle of GPR and applied the FD-RTM algorithm to the GPR data acquired at the site of Beiluhe in the Tibetan Plateau. Combining the RTM imaging sections from real data and forward modeling, we demonstrated that the FD-RTM can be applied to determine the active layer depth and characterize the fine structures in the active layer in the Tibetan Plateau's permafrost regions. Through the design a series of active layer models, we generated the forward modeling synthetic data and demonstrated that RTM is a powerful tool to study the structure of the active layer of the permafrost. In the Beiluhe region the imaging result of GPR RTM profile shows that the RTM technique can clearly characterize the positions of the fine structures in the migrated section. It has great significance of using RTM section to research complex structure and freezing and thawing process in the active layer at the Beiluhe region.

2. Materials and Methods

2.1. Principles of GPR

The GPR system consists of transmitter antenna, receiver antenna, controller, data logger and display, as shown in Figure 1. Electromagnetic waves from the transmitting antenna propagate into the surrounding medium at a certain velocity which is dependent mainly on the dielectric constant of the medium [38,39]. When the electromagnetic wave encounters the interface of the medium, part of the energy is transmitted to the second layer of medium, and the remaining energy is reflected according to the reflection coefficient R which is given by:

$$R = \frac{\sqrt{\varepsilon_1} - \sqrt{\varepsilon_2}}{\sqrt{\varepsilon_1} + \sqrt{\varepsilon_2}} \tag{1}$$

where the ε_1 and ε_2 are the dielectric constants of the first layer and the second layer of the medium, respectively [40]. The reflected electromagnetic wave is received by the receiving antenna. The travel time and amplitude information of the signal can be used to image the medium.

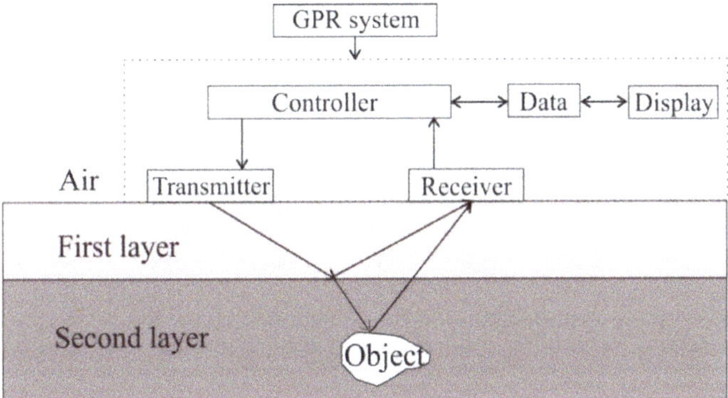

Figure 1. Ground penetrating radar (GPR) system and detection diagram.

2.2. Reverse-Time Migration (RTM)

The RTM procedure consists of three steps: forward continuation, reverse continuation and imaging. First, the forward continuation propagates the wave energy to the maximum moment along time axis; the direction of wave field and the result are preserved. Second, reverse continuation is propagated the wave field to zero moment along the time axis in reverse direction, and then, the positive wave field is read, which has same time as the forward continuation. In the third step, appropriate imaging condition is applied to get the information of underground structure. The detailed realization process is shown in Figure 2.

In the process of RTM, we choose the zero-lag cross-correlation imaging condition and the Laplace filter to remove the low frequency noise at the subsurface [41,42].

$$image(x, y) = \sum_{time} S(x, y, t)R(x, y, t) \tag{2}$$

where $image(x, y)$ is the result of imaging; $S(x, y, t)$ is the wave-filed of forward continuation in time domain, which is calculated from forward modeling data. $R(x, y, t)$ is the wave-field of reverse continuation in time domain, which is obtained from the acquired GPR data.

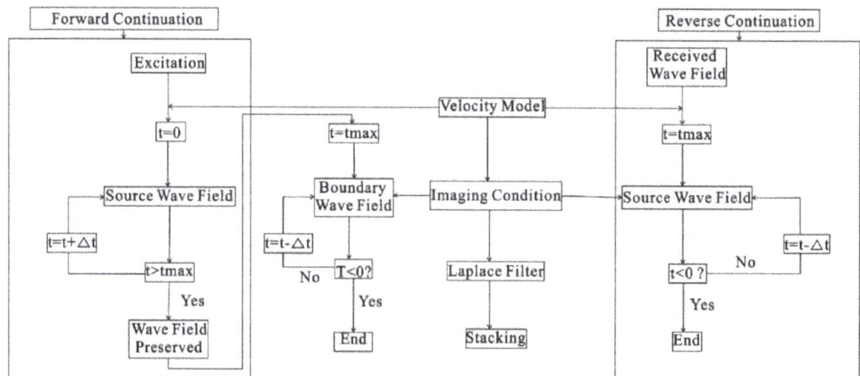

Figure 2. The realization process of reverse time migration (RTM) algorithm.

2.3. Two-Dimensional Radar Wave Forward Modeling

Efficiently solving the forward problem is the key to RTM imaging. The major task of the forward modeling algorithm is solving 2D wave equation from the original Maxwell's equations in rectangular coordinate system. The original Maxwell curl equations can be written as

$$\nabla \times H = \frac{\partial D}{\partial t} + J \tag{3}$$

$$\nabla \times E = -\frac{\partial B}{\partial t} - Jm \tag{4}$$

where E is the electric field; D is electric displacement; H is the magnetic field; B is the magnetic flux; J is electric current density; J_m is magnetic electric current density. For the 2D TE (Transverse electric) mode $\partial/\partial_z = 0$ and $H_z = E_x = E_y = 0$. Therefore, in the Cartesian coordinates we have:

$$\frac{\partial E_z}{\partial y} = -\mu\frac{\partial H_x}{\partial t} - \sigma_m H_x \tag{5}$$

$$\frac{\partial E_z}{\partial x} = \mu\frac{\partial H_y}{\partial t} + \sigma_m H_y \tag{6}$$

$$\frac{\partial H_y}{\partial x} - \frac{\partial H_x}{\partial y} = \varepsilon\frac{\partial E_z}{\partial t} + \sigma E_z \tag{7}$$

where ε is the dielectric constant; μ is unit permeance; σ_m is magneto conductivity; σ is the electric conductivity; Hx and Hy are the x and y components of the magnetic field, respectively; and Ez is the electric field oriented in z direction.

2.4. The Forward Modeling Data RTM

Since groove is a common geological structure in the field, the propagation of electromagnetic waves in groove is complicated, so a groove model is often used to verify the accuracy of radar or seismic imaging algorithms. In order to examine the correctness of the RTM algorithm, we designed a groove model (Figure 3a) for forward modeling. The dielectric constants of the upper and lower media were 2.0 and 4.0, respectively. The model size was 4×4 m. The grid spacing was 0.02 m. Ricker wavelet was regarded as source wavelet, for which the peak frequency was 400 MHz. The time step was 2.0×10^{-11} s and the recording length was 50 ns. We used finite-difference forward algorithm calculate to radar profile (Figure 3b) for groove model and used RTM algorithm calculate to migration profile (Figure 3c). From the result of the migration, we can see the reflection and diffraction (the yellow circle in Figure 3b) caused by the uneven interface in radar profile (Figure 3b). The groove interfaces (the red circle in Figure 3c) of the shape and depth of the migration profile (Figure 3c) are similar to the original velocity model (Figure 3a). The interface reflection of two intersecting axes (the yellow arrows in Figure 3c) is caused by diffraction of the jump point on the boundary at the imaging process in migration profile (Figure 3c).

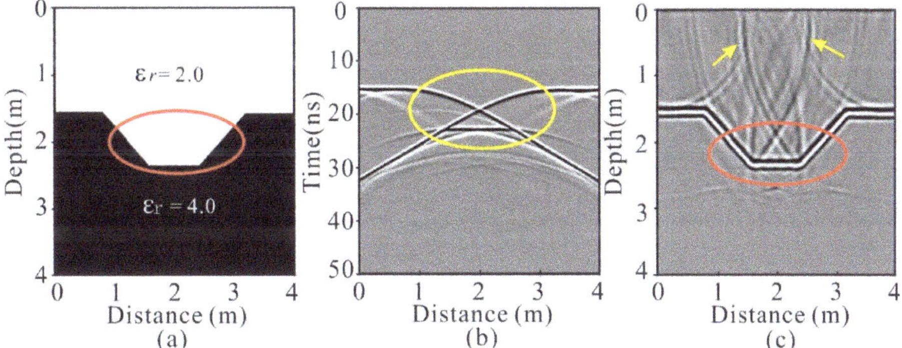

Figure 3. The imaging of RTM of groove model: (**a**) dielectric constant mode, (**b**) radar profile after removing the direct wave and (**c**) RTM profile.

2.5. RTM of GPR Profiles in the Beiluhe Region

2.5.1. Brief Description of the Study Area

The Beiluhe region (E92.9318°, N34.8214°, and see Figure 4) is located in high alluvial plain in the north part of the Tibetan Plateau. In this area, terrain is gentle and the change of elevation is from 4600 to 4700 m. The layers are mainly composed of Quaternary salt, pluvial fine sand, silt and a silty clay layer. It is covered with a peat layer in local region and made of tertiary mudstone and sandstone below 2 m. The average of annual rainfall is 300 mm, the average of annual temperature is −5.0–3.8 °C. The average of annual ground temperature is 2.0–0.5 °C in this region, respectively. Frozen soil types mainly comprise rich ice and frozen soil in the local region; the thickness of permafrost is 50–80 m; the upper limit of permafrost was −1.8 to −2.2 m over the last 30 years. The average temperatures have risen by about 0.3 to 0.4 °C in the Tibetan Plateau with the global warming. That caused degradation of permafrost region, increased the activity layer's thickness and caused the disappearance of local island permafrost [7,8,43]. The influence of human engineering activities such as the Tibetan highway, make the lithology of subsurface and coverage of vegetation is not uniform; the thickness of the active layer in the local scope has obviously changed, and the change of permafrost is transverse discontinuous.

2.5.2. GPR Data Acquisition and Processing

In May 2018, we used the GSSI SIR30 GPR apparatus for the characterization of the permafrost on both sides of the Qinghai–Tibetan highway. The GPR antennas are all bow-tie antennas, and receiver antenna and transmitter antennas are perpendicular to the survey line. We designed three survey lines with length of 200 m (the three red lines in Figure 4) and applied the common midpoint (CMP) method to obtain the formation velocity. The antennas' frequencies are 100 MHz and 400 MHz, respectively; the polarization directions of the transmitter antenna and the receiver are parallel. The sampling interval is 0.047 ns, with the total number of 1024 samples in one trace. The method of CMP is mainly used for detecting propagation velocity of radar wave. This method of profile is mainly used for detecting small structure changes at the shallow surface layer. We used 100 MHz antennas with a step size of 10 cm to move in the opposite direction of other antennas. Based on CMP velocity analysis [44,45], we measured the depth of active layer to be about 50 cm; the average found that the bottom interface depth propagation speeds are 0.10 m/ns and 0.12 m/ns on the upper and lower parts of the bottom interface of active layer, respectively. We used 400 MHz antennas to collect high frequency GPR data along survey lines (Figure 5).

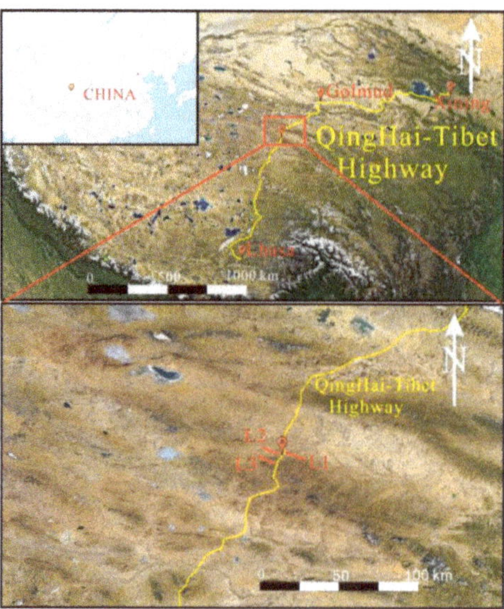

Figure 4. The location and lines layout diagram in Beiluhe permafrost region. Lines distribution on both sides of the Tibetan highway. We chose part of the section of on the east side of L1 line (the L1 red line) and on the west side of L3 line (the L3 red line) and L2 line (the L2 red line) to image RTM.

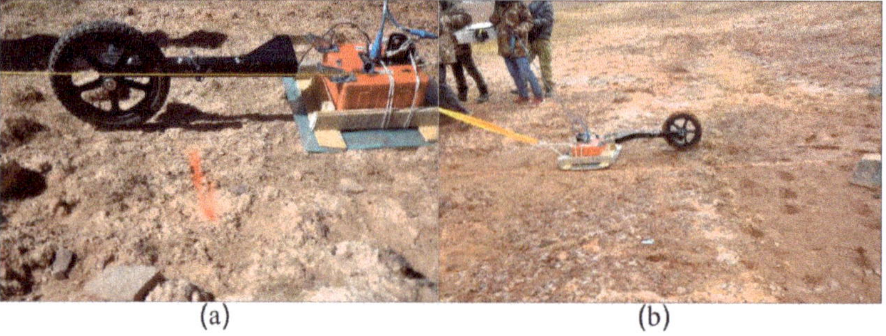

Figure 5. The picture of landscape: (**a**) near picture, (**b**) far picture.

When we interpreted radar data, we found they to have complex reflection and diffraction, and a small fault on the subsurface in the radar profile. In order to obtain fine structure of internal of active layer, we chose 250 trace radar datums in L1 line (the L1 line in Figure 6), 200 traces radar datums in L3 line (the L3 line in Figure 8) and L2 line (the L2 line in Figure 10) to image of RTM. In order to prove the accuracy of interpretation result of three-line RTM profiles, we designed three dielectric constant models which were based on the three-line RTM profiles and geological information of active and permafrost layers, respectively, and obtained the RTM profiles of three models.

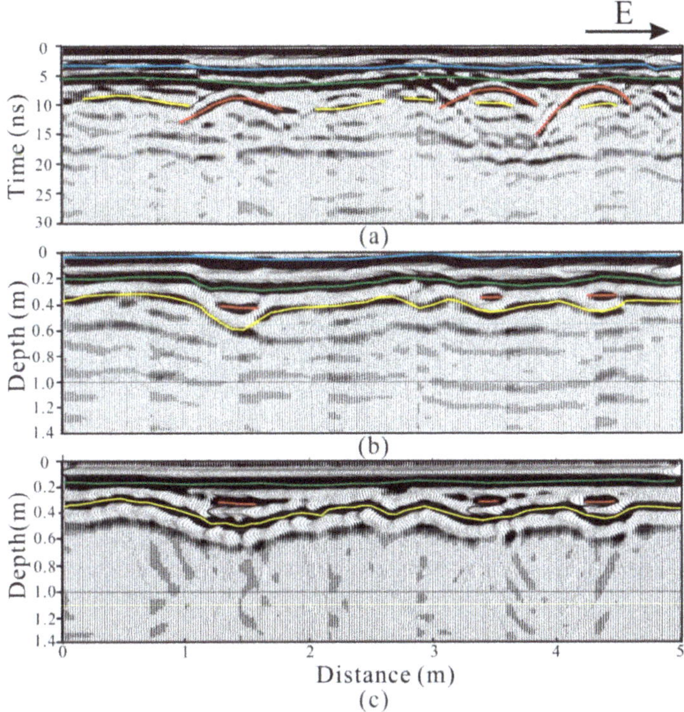

Figure 6. The imaging of migration of L1 line part profile. (**a**) Radar profile; (**b**) real data RTM profile; (**c**) model data RTM profile.

3. Results

3.1. Diffraction in Permafrost Internal

Figures 6 and 7 are the GPR data and dielectric constant models, respectively. The real data RTM profile in Figure 6b was calculated from the real GPR data in Figure 6a using the RTM algorithm. The dielectric constant model in Figure 7 was designed based on real data RTM profile (Figure 6b) and geological information. Then, we used the dielectric constant model to obtain forward modeling data, and calculated model data RTM profile in Figure 6c using RTM algorithm. For the sake of contrast, we put the real GPR profile, real data RTM profile and model data RTM profile into one (Figure 6). It has three obvious diffraction in the part of the permafrost profile of L1 line at the Beiluhe site (Figure 6a), which were probably caused by buried cobbles. It has a discontinuous reflection axis at 5 ns (the green line in Figure 6a), and a continuous reflection axis at 3 ns (the blue line in Figure 6a) on above of diffraction. It has intermittent reflection (the yellow line in Figure 6a), whose location at the bottom interface of the active layer is below the diffraction. Due to much attenuation, energy loss is too serious; the reflection on bottom of the energy is very weak (the yellow line in Figure 6a). It is difficult to effectively identify the position of the diffraction and determine the depth of the bottom interface of peat layer and active layer from the radar profile. The three obvious diffractions (the red lines in Figure 6a) move back to the original place after the migration (Figure 6b). The depth of the three cobbles is about 0.3 m. In combination with geological information of a shallow surface, we consider these diffractions are caused by cobbles with a size of about 10 cm. The 3 ns reflection axis (the blue line in Figure 6a) at the top of diffractions and the ground direct wave merged into a strong axis; we made sure the reflection axis was radar direct wave. The reflection axis at 5 ns on the top of diffraction (the green line in Figure 6a) heads back to a continuous reflection axis (the green line in Figure 6b), and we consider the reflection axis as being at the bottom interface of peat layer; the depth is 0.2 m. The intermittent reflection (the yellow line in Figure 6a) heads back to continuity reflection (the yellow line in Figure 6b), and we consider the reflection the bottom interface of active layer; the depth is about 0.45 m in the radar profile. Combinations of radar migration profiles from different seasons can effectively reflect that with the climate changes, so does the process of ice changing into permafrost.

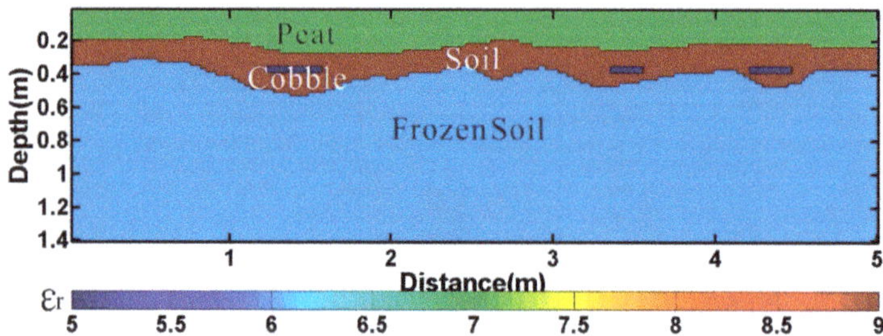

Figure 7. The relative dielectric constant model was designed using the L1 line RTM profile.

Based on the L1 line RTM profile, we designed a dielectric constant model of diffractions in the internal permafrost layer (Figure 7), and obtained the model data RTM profile (Figure 6c). We found exact correspondences between real data RTM profile and model data RTM profile. They have same feature and depth for the active layer (the green line in Figure 6b,c), permafrost table (the yellow line in Figure 6b,c) and the cobbles (the red line in Figure 6b,c) in the real data RTM profile and model data RTM profile.

3.2. Fine Structure in Internal Permafrost Layer

There are a lot of reflection axes crossing and messing in the part of the L3 line section of Beiluhe permafrost region (Figure 8a). About the time of 10 ns approximately, a very continuous reflection axis (the green lines in Figure 8a) appears, but with a lot of intermittent reflection axes (the yellow line in Figure 8a) at the top. The small reflections (the yellow line in Figure 8a) and the continuous reflection of around 10 ns (the green lines in Figure 8a) have difficulty reflecting the actual internal structure of the permafrost layer. It is difficult to reflect the fine structure of internal permafrost layer from a radar profile. The cross reflection and small reflection at subsurface (the yellow line in Figure 8a) cover the properly position, and show three distinct reflection layers (the red, yellow and green line in Figure 8b) which are significantly deeper than the depth of pit data. There are two continuous reflections on the top of active layer (the red and yellow lines in Figure 8b), which may be the interface of soil, and siltstone and fine sandstone, respectively. In the process of thawing in the RTM profile, we combine other factors with pit data and consider the depth of the bottom interface of active layer to be about 0.75 m (the green line in Figure 8b): the permafrost of subsurface, and the interface of between soil and siltstone (the red line in Figure 8b) and between siltstone and fine sandstone (the yellow line in Figure 8b) where the moisture content increased obviously, which was caused by a strong reflection in the radar profile. The depth is about 0.9 m in the RTM profile. It has weak reflection (the blue line in Figure 8b) which is caused by internal of freezing permafrost. It has an obviously lateral change the depth of interface of among the soil, siltstone, fine sandstone and active layers. Due to it having some vegetation, caused by the thawing of different levels in permafrost soil, siltstone and sandstone, it can be truly reflected internal state of freezing and thawing permafrost on the detection period.

Based on the L3 line RTM profile, we designed a dielectric constant model of fine internal structure in permafrost layer (Figure 9). Combining the geological information and characteristics of the permafrost layer, we inserted six permafrost layers in the dielectric constant model (Figure 9). After the calculation, we obtained the model RTM's profile data (Figure 8c). The layers had same characteristics and depth of active layers and a permafrost layer (the red, yellow, green and blue line in the Figure 8b,c) in the real data RTM profile (Figure 8b).

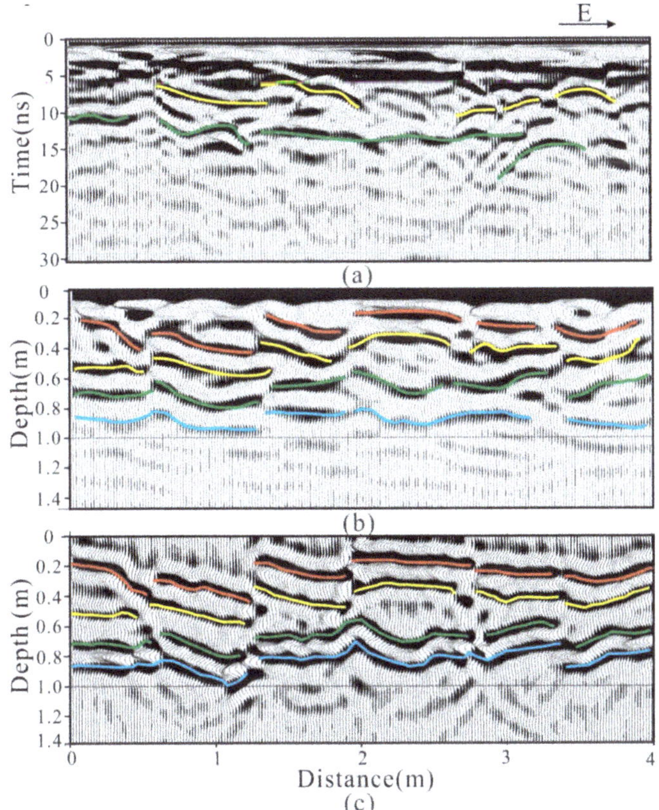

Figure 8. The imaging of migration of L3 line part profile. (**a**) Radar profile; (**b**) real data RTM profile; (**c**) model data RTM profile.

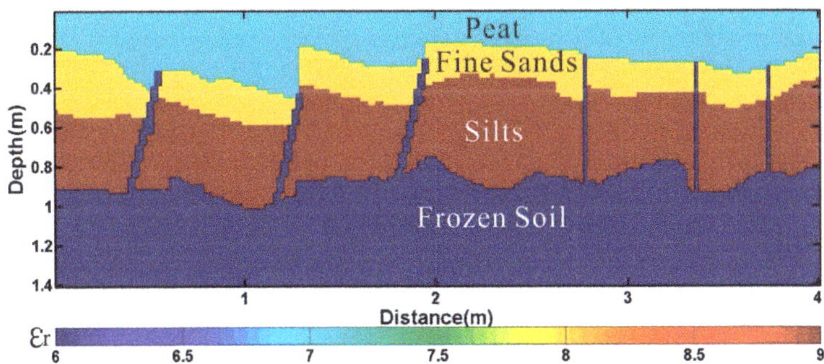

Figure 9. The relative dielectric constant model was designed using the L3 line RTM profile.

3.3. Small Lateral Fault Broken

There are two obvious small blocks at the shallow surface in part of L2 line radar profile and RTM profile (Figure 10a,b). Their depths were both about 0.2 m; both were about 5 × 10 cm. Due to the depth being small, they may have been stone. They had a small reflection at the 5 ns in the radar profile (the green line in Figure 10a). But the energy of reflection was at the bottom of small reflection. They obviously had three reflections (the green line C in Figure 10b) which combine into one reflection in the RTM profile. That may have been interface in the frozen state. When the seasonal frozen soil layer starts thawing, the interface separates three interfaces. It has obvious convergence diffraction at the bottom of small fault (the blue line in Figure 10a), and enhances continuity reflection where the depth of reflection is 80 cm (the yellow line in Figure 10b) or 100 cm (the blue line in Figure 10b). In combination with pit data, the two reflections (the yellow and blue lines in Figure 10b) are at the bottom interface of active layer. The depth of interface is about 90 cm and deeper than the depth of the interface of the other location significantly. The layers are broken by fault F2 in the regional of E and F. Combined with geological information, the two interfaces are interfaces between soil and siltstone (the green line at regional of E and F in Figure 10b), and between siltstone and sandstone (the red line at regional of E and F in Figure 10b). They have obvious uplift at the side of up fault (D and F in Figure 10b), which indicate fault F1 and F2 are reverse faults. The formation of reverse faults is caused by unequal stress when a permafrost thaws. The region of E is risen, and on either side of the extrusion regions D and F. Due to influences of temperature, soil and shallow permafrost soil, the thawing is unequal; there are complex structures in the shallow permafrost. That can clearly reflect the position of the small fault from the RTM profile (Figure 10b). In combination with RTM profiles of different seasons, we can provide scientific advice for the Tibetan highway engineering construction.

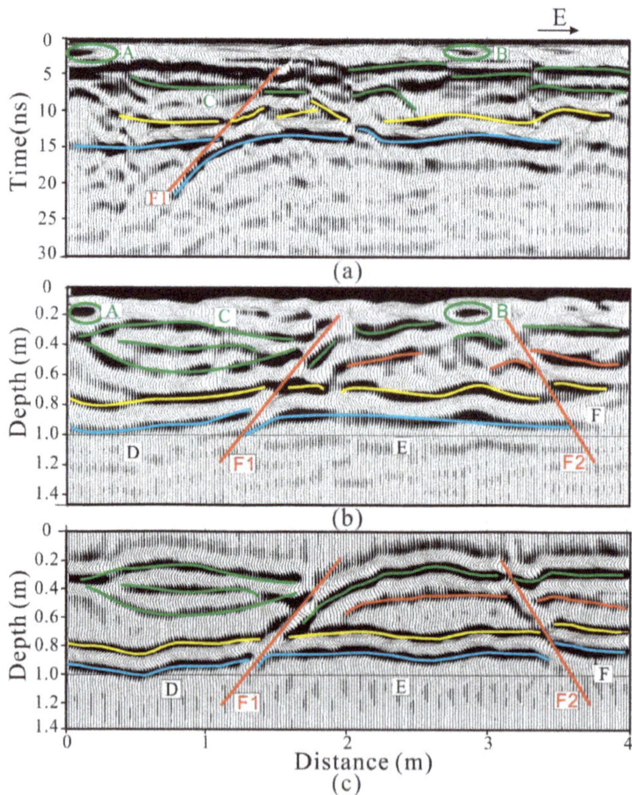

Figure 10. The imaging of migration of L2 line part profile. (**a**) Radar profile; (**b**) real data RTM profile; (**c**) model data RTM profile.

Based on the L2 line RTM profile and geological information of active and permafrost layers, we designed the dielectric constant model (Figure 11) of the small lateral fault, and obtained the model data RTM profile (Figure 10c). We found exact correspondence between the real data RTM profile and the model data RTM profile. They have the same features and depths for the small broken fault (F1 and F2 in Figure 10b,c) and the permafrost layer (the yellow and blue lines in Figure 10b,c).

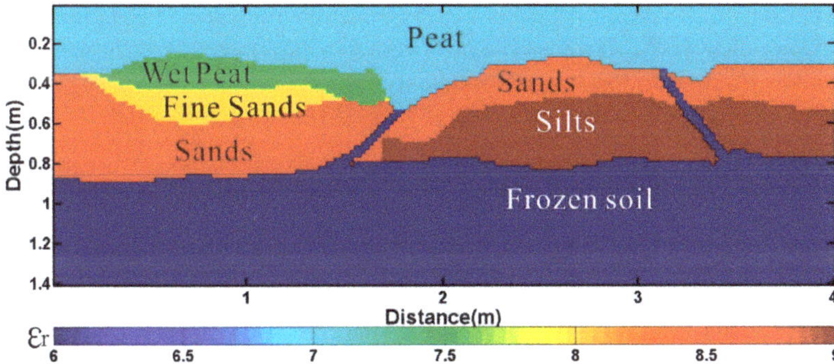

Figure 11. The relative dielectric constant model was designed using the L2 line RTM profile.

4. Discussion and Conclusions

With global warming, the Tibetan Plateau has experienced severe degradation of permafrost and disappearance of island permafrost, which has seriously affected the local ecology and stability of engineering infrastructures in the Tibetan Plateau. The place is mainly composed of a fine sand layer at the subsurface of the Beiluhe permafrost region in the Tibetan Plateau. The types of permafrost soil are mainly rich ice and frozen soil and ice containing soil at the location. Due to effects of slope, vegetation coverage and engineering activities, the active layer has obvious lateral changes in the process of freezing and thawing. There are obvious cross-overs of reflection, diffraction and energy inequality phenomena in the radar profile.

GPR provides a wealth of interpretive information about active and permafrost layer. The conductivity of the subsurface is lower (<10 mS/m) in the Beiluhe region. It has flat terrain, rarely affected by external distractions, and there is a high signal noise ratio (SNR) for radar data in this region. At the same time the CMP velocity analysis and formation depth data from pit excavation provide a precise velocity-depth model. This is the key to RTM in the Tibetan region.

Research of changes at the site of permafrost soil of Tibetan highway can provide scientific advice for the Tibetan highway's engineering construction. Previous studies have merely used the ground temperature and active layer thickness depth to research permafrost actives layer, and cannot show internal fine structure [46–48]. In this article, RTM is applied to process of permafrost GPR data of the Tibetan highway. The three RTM profiles clearly reflect the internal fine structure of permafrost and thawing state. The RTM profiles show large pieces of cobbles (Figure 6b). Due to the influence of soil at the shallow surface, the seasonal permafrost has obvious lateral changes and forms small discontinuities (Figure 8b). The depth of bottom interface of peat layer is about 0.2 m. The depth of bottom interface of the active layer has obviously changed; the deepest and shallowest parts are about 0.9 m and 0.45 m, respectively. In this region, the permafrost active layer thickness is about 1.8–2.2 m. The permafrost active layer will melt with the rise of temperature, and the maximum melting depth is in August. The survey was in May, and the maximum melting depth was about 0.9 m. Therefore, the data up to 1.4 m were sufficient.

In order to prove the accuracy of interpretation results of three-line RTM profiles (Figures 6b, 8b and 10b), we designed a dielectric constant models (Figures 7, 9 and 11) based on the three-line RTM profile and geological information, and obtained the model data RTM profiles (Figures 6c, 8c and 10c).

We found exact correspondences between real data RTM profiles and model data RTM profiles. From the three model data RTM profiles, we can prove the accuracy of interpretation results of three-line RTM profiles. In other words, the dielectric constant models of three-line (Figures 7, 9 and 11) can show the characteristics of diffraction, fine internal structure, and the small, broken lateral fault in the permafrost layer exactly.

The thawing of permafrost layer has influenced the Tibetan highway stability and indicates the change of climate. It can provide high resolution geological interpretation from the RTM profile. In combination with radar RTM profile of different seasons which can reflect the process of changing of seasonal freezing and thawing, it has great significance to researching the process of freezing and thawing of permafrost in the changing context of the Tibetan Plateau.

Author Contributions: Y.W., X.L., Z.F. and H.W. performed the field experiments; Y.W. and S.Q. performed the forward modeling; Y.W., X.L. and X.W. analyzed the radar data; Y.W., Z.F. and X.L. jointly wrote the paper. All authors have read and agreed to the published version of the manuscript.

Funding: This research was funded by the National Key R&D Program of China, grant number 2017YFC0601804; the National Natural Science Foundation of China, grant number 51777017; and the Chongqing Key Industries Common Key Technology Innovation Projects, grant number CSTC2017ZDCYZDYFX0045.

Conflicts of Interest: The authors declare no conflicts of interest.

References

1. Gregory, P.; Wayne, W.H.P.; Kevin, K.W. Geophysical mapping of ground ice using a combination of capacitive coupled resistivity and ground-penetrating radar, Northwest Territories. *J. Geophys. Res.* **2008**, *113*. [CrossRef]

2. Chen, S.; Liu, W.; Qin, X.; Liu, Y.; Zhang, T.; Chen, K.; Hu, F.; Ren, J.; Qin, D. Response characteristics of vegetation and soil environment to permafrost degradation in the upstream regions of the Shule River Basin. *Environ. Res. Lett.* **2012**, *7*, 45406–45416. [CrossRef]

3. Luo, J.; Niu, F.J.; Lin, Z.J.; Lu, J.H. Permafrost Features around a Representative Thermokarst Lake in BeiLuhe on the Tibetan Plateau. *J. Glaciol. Geocryol.* **2012**, *34*, 1110–1117.

4. Brosten, T.R.; Bradford, J.H.; McNamara, J.P.; Gooseff, M.N.; Zarnetske, J.P.; Bowden, W.B.; Johnston, M.E. Estimating 3D variation in active-layer thickness beneath arctic streams using ground-penetrating radar. *J. Hydrol.* **2009**, *373*, 479–486. [CrossRef]

5. Lupascu, M.; Welker, J.M.; Seibt, U.; Maseyk, K.; Xu, X.; Czimczik, C.I. High Arctic wetting reduces permafrost carbon feedbacks to climate warming. *Nat. Clim. Chang.* **2014**, *4*, 51–55. [CrossRef]

6. Schuur, E.A.G.; Mcguire, A.D.; Schadel, C. Climate change and the permafrost carbon feedback. *Nature* **2015**, *520*, 171–179. [CrossRef]

7. Lin, Z.J.; Niu, F.J.; Fang, J.H.; Luo, J.; Yin, G.A. Interannual variations in the hydrothermal regime around a thermokarst lake in Beiluhe, Qinghai-Tibet Plateau. *Geomorphology* **2017**, *276*, 16–26. [CrossRef]

8. Yin, G.A.; Niu, F.J.; Lin, Z.J.; Luo, J. Effects of local factors and climate on permafrost conditions and distribution in Beiluhe basin, Qinghai-Tibet Plateau, China. *Sci. Total Environ.* **2017**, *581*, 472–485. [CrossRef]

9. Yang, C.; Wu, T.H.; Wang, J.M.; Yao, J.M.; Li, R.; Zhao, L.; Xie, C.W.; Zhu, X.F.; Ni, J.; Hao, J.M. Estimating Surface Soil Heat Flux in Permafrost Regions Using Remote Sensing-Based Models on the Northern Qinghai-Tibetan Plateau under Clear-Sky Conditions. *Remote Sens.* **2019**, *11*, 416. [CrossRef]

10. Li, R.; Ding, Y.Z.; Wu, T.H.; Yao, X.; Du, E.; Liu, G.Y.; Qiao, Y.P. Temporal and spatial variations of the active layer along the Qinghai Tibet Highway in a permafrost region. *Chin. Sci. Bull.* **2012**, *57*, 4609–4616. [CrossRef]

11. Peng, H.; Ma, W.; Mu, Y.H.; Jin, L.; Yuan, K. Degradation characteristics of permafrost under the effect of climate warming and engineering disturbance along the Qinghai-Tibet Highway. *Nat. Hazards* **2015**, *75*, 2589–2605. [CrossRef]

12. Xiao, J.; Liu, L. Permafrost subgrade condition assessment using extrapolation by deterministic deconvolution on multi-frequency GPR data acquired along the Qinghai-Tibet railway. *IEEE J. Sel. Top. Appl. Earth Obs. Remote Sens.* **2015**, *9*, 83–90. [CrossRef]

13. Qing, W.; Yupeng, S. Calculation and Interpretation of Ground Penetrating Radar for Temperature and Relative Water Content of Seasonal Permafrost in Qinghai-Tibet Platea. *Electronics* **2019**, *8*, 731.

14. Schwamborn, G.J.; Dix, J.K.; Bull, J.M.; Rachold, V. High resolution Seismic and Ground Penetrating radar Geophysical Profiling of a Thermokarst Lake in the Western Lena Delta, Northern Siberia. *Permafr. Periglac. Process.* **2002**, *13*, 259–269. [CrossRef]

15. Jorgensen, A.S.; Andreasen, F. Mapping of permafrost surface using ground-penetrating radar at Kangerlussuaq Airport, Western Greenland. *Gold Reg. Sci. Technol.* **2007**, *48*, 64–72. [CrossRef]

16. Wainstein, P.; Moorman, B.; Whitehead, K. Glacial conditions that contribute to the regeneration of Fountain Glacier proglacialicing, Bylot Island, Canada. *Hydrol. Process.* **2014**, *28*, 2749–2760. [CrossRef]

17. Merz, K.; Maurer, H.; Buchli, T.; Horstmeyer, H.; Green, A.G.; Springman, S.M. Evaluation of Ground Based and Helicopter Ground Penetrating radar Data Acquired Across an Alpine Rock Glacier. *Permafr. Periglac. Process.* **2015**, *26*, 13–27. [CrossRef]

18. Liu, L.; Qian, R.Y. Ground Penetrating Radar: A critical tool in near-surface geophysics. *Chin. J. Geophys.* **2015**, *58*, 2606–2617.

19. Thomas, M.U.; Jeffrey, T.R.; Claire, A.; Douglas, D.A.; Sturt, W.M.; Owen, K.M.; Andrew, H.T.; Christopher, B.W. Frozen: The Potential and Pitfalls of Ground-Penetrating Radar for Archaeology in the Alaskan Arctic. *Remote Sens.* **2016**, *8*, 1007.

20. Wang, Y.P.; Jin, H.J.; Li, G.Y. Investigation of the freeze–thaw states of foundation soils in permafrost areas along the China–Russia Crude Oil Pipeline (CRCOP) route using ground-penetrating radar (GPR). *Cold Reg. Sci. Technol.* **2016**, *126*, 10–21. [CrossRef]

21. Campbell, S.; Affleck, R.T.; Sinclair, S. Ground-penetrating radar studies of permafrost, periglacial, and nearsurface geology at McMurdo Station, Antarctica. *Cold Reg. Sci. Technol.* **2018**, *148*, 38–49. [CrossRef]

22. Shen, Y.P.; Zuo, R.F.; Liu, J.K.; Tian, Y.H.; Wang, Q. Characterization and evaluation of permafrost thawing using GPR attributes in the Qinghai-Tibet Plateau. *Cold Reg. Sci. Technol.* **2018**, *151*, 302–313. [CrossRef]

23. Wu, T.; Li, C.G.; Nan, Z. Using ground penetrating radar to detect permafrost degradation in the northern limit of permafrost on the Tibetan Plateau. *Cold Reg. Sci. Technol.* **2005**, *41*, 211–219. [CrossRef]

24. Pang, Q.Q.; Zhao, L.; Li, S.X.; Ding, Y. Active layer thickness variations on the Qinghai-Tibet Plateau under the scenarios of climate change. *Environ. Earth Sci.* **2012**, *66*, 849–857. [CrossRef]

25. Sun, R.; Mcmechan, G.A. Prestack 2D parsimonious Kirchhoff depth migration of elastic seismic data. *Geophysics* **2011**, *76*, 157–164. [CrossRef]

26. Li, S.; Fomel, S. Kirchhoff migration using Eikonal based computation of traveltime source derivatives. *Gephysics* **2013**, *78*, 211–219. [CrossRef]

27. Sun, R.; Mcmechan, G.A.; Lee, C.S.; Chow, J.; Chen, C.H. Prestack scalar reverse time depth migration of 3D elastic seismic data. *Geophysics* **2006**, *71*, 190–207. [CrossRef]

28. Guitton, A.; Kaelin, B.; Biondi, B. Least-squares attenuation of reverse time migration artifacts. *Geophysics* **2007**, *72*, 19–23. [CrossRef]

29. Shragge, J. Reverse time migration from topography. *Geophysics* **2014**, *79*, 1–12. [CrossRef]

30. Zhang, J.H.; Wang, W.M.; Fu, L.Y.; Yao, Z.X. 3D Fourier finite-difference migration by alternating direction implicit plus interpolation. *Geophys. Prospect.* **2008**, *56*, 95–103. [CrossRef]

31. Fisher, E.; Mcmechan, G.A.; Annan, A.P.; Cosway, S.W. Examples of reverse-time migration of single-channel, ground-penetrating radar profiles. *Geophysics* **1992**, *57*, 577–586. [CrossRef]

32. Almeida, E.R.; Porsani, J.L.; Catapano, I.; Gennarelli, G.; Soldovieri, F. Microwave Tomography Enhanced GPR in Forensic Surveys: The Case study of a Tropical Environment. *IEEE J. Sel. Top. Appl. Earth Obs. Remote Sens.* **2016**, *1*, 115–124. [CrossRef]

33. Sakamoto, T.; Sato, T.; Aubry, P.; Yarovoy, A. Ultra-Wideband radar Imaging Using a Hybrid of Kirchhoff Migration and Stolt F-K Migration with an Inverse Boundary Scattering Transform. *IEEE Trans. Antennas Propag.* **2015**, *63*, 3502–3512. [CrossRef]

34. Sanada, Y.; Ashid, Y. An imaging algorithm for GPR data. In *Symposium on the Application of Engineering and Environmental Problems*; Environmental & Engineering Geophysical: Denver, CO, USA, 1999; pp. 564–573.

35. Leuschen, C.J.; Plumb, R.G. A matched-filter-based reverse time migration algorithm for ground-penetrating radar data. *IEEE Trans. Geosci. Remote Sens.* **2001**, *39*, 929–936. [CrossRef]

36. Zhou, H.; Stao, M.; Liu, H. Migration velocity analysis and prestack migration of common-transmitter GPR data. *IEEE Trans. Geosci. Remote Sens.* **2005**, *43*, 86–91. [CrossRef]

37. Liu, S.; Lei, L.L.; Fu, L.; Wu, J. Application of prestack reverse time migration based on FWI velocity Estimation to ground penetrating radar data. *J. Appl. Geophys.* **2014**, *107*, 1–7. [CrossRef]

38. Daniel, D.J. *Ground Penetrating Radar*, 2nd ed.; The Institution of Electrical Engineers: London, UK, 2004.

39. Luca, B.C.; Fabio, T.; Nikos, E.; Francesco, B. Signal Processing of GPR Data for Road Surveys. *Geosciences* **2019**, *9*, 96.

40. Ukaegbu, I.K.; Gamage, K.A.; Aspinall, M.D. Nonintrusive Depth Estimation of Buried Radioactive Wastes Using Ground Penetrating Radar and a Gamma Ray Detector. *Remote Sens.* **2019**, *11*, 141. [CrossRef]

41. Yan, H.Y.; Liu, Y. Acoustic prestack reverse time migration using the adaptive high-order finite-difference method in time-space domain. *Chin. J. Geophys.* **2013**, *56*, 181–195.

42. Zhang, Y.; Sun, J. Practical issues in reverse time migration: True amplitude gathers noise removal and harmonic source encoding. *EAGE Ext. Abstr.* **2009**, *1*, 1–5.

43. Luo, J.; Yin, G.A.; Niu, F.J.; Lin, Z.J.; Liu, M.H. High Spatial Resolution Modeling of Climate Change Impacts on Permafrost Thermal Conditions for the Beiluhe Basin, Qinghai-Tibet Plateau. *Remote Sens.* **2019**, *11*, 1294. [CrossRef]

44. Sham, J.F.; Lai, W.W. Development of a new algorithm for accurate estimation of GPR's wave propagation velocity by common-offset survey method. *NDT E Int.* **2016**, *83*, 104–113. [CrossRef]

45. Cui, F.; Li, S.Y.; Wang, L.B. The accurate estimation of GPR migration velocity and comparison of imaging methods. *J. Appl. Geophys.* **2018**, *159*, 573–585. [CrossRef]

46. Sun, Z.; Wang, Y.B.; Sun, Y.; Niu, F.J.; Li, G.Y.; Gao, Z.Y. Creep characteristics and process analyses of a thaw slump in the permafrost region of the Qinghai-Tibet Plateau, China. *Geomorphology* **2017**, *293*, 1–10. [CrossRef]

47. Wollschlager, U.; Gerhards, H.; Yu, Q.; Roth, K. Multi-channel ground-penetrating radar to explore spatial variations in thaw depth and moisture content in the active layer of a permafrost site. *Cryosphere* **2010**, *4*, 269–283. [CrossRef]

48. Wu, Q.B.; Hou, Y.D.; Yun, H.B.; Liu, Y.Z. Changes in active-layer thickness and near-surface permafrost between 2002 and 2012 in alpine ecosystems, Qinghai-Xizang (Tibet) Plateau, China. *Glob. Planet. Chang.* **2015**, *124*, 149–155. [CrossRef]

Article

Design of X-Bandpass Waveguide Chebyshev Filter Based on CSRR Metamaterial for Telecommunication Systems

Mahmoud AbuHussain * and Ugur C. Hasar

Department of Electrical and Electronics Engineering, Gaziantep University, Gaziantep 27310, Turkey; uchasar@gantep.edu.tr
* Correspondence: ma41147@mail2.gantep.edu.tr

Received: 16 November 2019; Accepted: 25 December 2019; Published: 6 January 2020

Abstract: This paper presents a new design of a fifth order bandpass waveguide filter with Chebyshev response which operates in the X-band at 10 GHz center frequency. By using a complementary split ring resonator (CSRR) upper and lower sections that are placed on the same transverse plane and are not on the same parallel line, CSRR sections are shifted from each other. A simple model of lumped elements RLC is introduced and calculated as well. The model of the proposed bandpass waveguide filter is synthesized and designed by using computer simulation technology (CST). Hereafter, by selecting proper physical parameters and optimizing the overall CSRR geometrical dimensions by taking into consideration the coupling effect between resonators, a shortened length of the overall filter, and a wider bandwidth over the conventional one are obtained. As a result, the proposed filter is compared with the conventional bandpass waveguide filter that is coupled by inductive irises with Chebyshev response, in addition to other studies that have used the metamaterial technique. The proposed filter reduces the overall physical length by 31% and enhances the bandwidth up to 37.5%.

Keywords: Chebyshev filter; cavity; metamaterial; waveguide; X-band; meta-resonator

---

1. Introduction

Waveguides are in general hollow metallic wave guiding structures used as filters, couplers, combiners, and amplifiers in various microwave and millimeter wave applications due to their less power consumption and high power handling capacity [1]. Recent advents in wireless communication applications have enormously increased the demand on antennas and filters with improved frequency selectivity as well as decreased overall physical size. For instance, such a demand results in various forms of designs and implementations of waveguide filters with improved characteristics of sharpness, bandwidth, and physical size [2–4].

Filters are components designed with the purpose of selecting or transmitting signals over specified band and rejecting signals over the other bands. Although various forms of microwave filters are available in the literature, they all have the same common point by which their performance is evaluated using the distributed network concept, assuming that they usually consist of periodic structures exhibiting passband and stopband characteristics in various frequency bands [1]. A variety of waveguide filters with different characteristics of low insertion loss, high Q factor, and sharp frequency selectivity have been proposed for various applications [5–8]. For example, inductive irise-coupled resonators are implemented into waveguide structures to have a Chebyshev response with N-poles [9]. In addition, stepped-impedance resonators are employed in the design of an X-band waveguide filter at 10 GHz to have superior filter characteristics [10]. Furthermore, a co-planar waveguide filter by using the bended stub and folded structure at 2.4 GHz [11], E-plane waveguide filter with periodically loaded resonators [12–14], and T-shaped resonators with reduced size [15] are also implemented. On the other hand, the introduction of the substrate-integrated waveguide technique [16] has lead to the emergence of many microwave filters with quarter wavelength resonator based on impedance and admittance inverters' method [17]. An aperture engraved in the middle common ground of the vertically stacked cavities [18] and U-shape slots on substrate integrated waveguide (SIW) with low temperature co-fired ceramic (LTCC) [19] at 40 GHz and 3D printed inserted into waveguide filter at 10 GHz [20] are implemented. In a recent study, two quarter-wavelength stepped-impedance resonators are realized to suppress the effect of a third harmonic in a wideband bandpass filter design being presented [21].

After the implementation of engineered materials [22] coined as metamaterials (MMs) based on the study of Veselago [23], a new era has been opened in the design and implementation of microwave filters due to their unique, unnatural, and exotic electromagnetic properties. Split ring resonators (SRRs) played a vital rule in the design of microwave applications like filters. Because of resonance structures of (SRRs) and their resonance behavior, they can be used for miniaturizing overall microwave devices. Waveguides embedded with metamaterial in various forms with diverse electromagnetic properties are designed to reduce the size of the filter [24]; to have a narrow-band filter in a rectangular waveguide using complementary split ring resonator (CSRR) has been proved, but the authors didn't mention the type of response or the final size of the proposed filter [25]. Using different models of CSRRs inside rectangular waveguide to design bandpass filter for single and dual mode has been demonstrated [26], and a compact dual-band waveguide bandstop filter using double SRR structure is designed [27].

In addition to these studies, gradually, metamaterials have spread out in many syntheses and designs to improve performance whether cutoff frequency or bandwidth of filter by using coupled split ring resonators (SRRs) and a negative image of split ring resonator CSRRs [28–33] are proposed and designed. However, previous studies have discussed and focused on dimensions of meta-resonators without mentioning or discussing shifts between upper and lower rings. Due to the shift property between CSRR rings, reaching the center frequency in a faster way is expected.

In [34], the authors have designed and simulated a three poles Chebyshev bandpass waveguide filter by using single CSRR in the X-band at center frequency 11.95 GHz. Although the methodology used in this article is straightforward due to the usage of a single resonator, it gives a narrow bandwidth 500 MHz and is not able to control the bandwidth or the selectivity of filter well; particularly, they have implemented two filters with a different number of poles and given the same narrow bandwidth; furthermore, they didn't mention the overall physical size of the proposed filter. Whilst the authors in [33] have proposed a unit cell of CSRR and tuned it at 12.1 GHz center frequency without mentioning the type of filter or the number of filter's poles, they have just implemented their resonator based on optimization of its physical dimensions to get a desired center frequency with 2 GHz passband. In addition, their method has just depended on the changing physical dimensions of CSRR itself.

Here, in our work, in order to improve the miniaturization of Chebyshev bandpass waveguide filter based metamaterials over a conventional filter at 10 GHz center frequency with five resonators, and to improve its selectivity and insertion loss alongside bandwidth as well. Simple double symmetric unit cells with electric coupling between CSRRs which are engraved from the center to facilitate and tune the resonating frequency at 10 GHz are used. The CSRRs have been designed with shifting between the upper and lower sections. The shifting between two CSRRs—right or left from each other—means the changing of electric coupling between them, and, due to this change in coupling, the center frequency will be affected directly with respect to this shifting. In this work, with the change in CSRR's physical dimensions besides the shift between two loops, the optimal waveguide filter is obtained in a systematic manner as it will be shown in Section 3.1. Cutoff frequency, Q factor, and S-parameters are studied and calculated by using computer simulation technology (CST), and a circuit simulator is used to implement and simulate a lumped element bandpass filter as well.

Unlike the conventional filters, the proposed filter is designed based on the metamaterial resonators' technology that are coupled directly without H-plane waveguide junctions between resonators that give the design important factors, especially for telecommunication applications such as compact size, simple structure, and wide bandwidth. Thus, according to these features, the proposed filter could be employed inline with low noise amplifier (LNA) for X-band receiver circuits which works at 10 GHz center frequency (9.15 GHz–10.44 GHz) to suppress the unwanted signal band, and waveguide size is critical in such circuits as well [35]. In addition, the proposed filter could be employed inline with X-band power amplifier at 10 GHz center frequency to achieve the maximum power and extend the desired bandwidth [36]. An X-band RADAR transceiver module that works at the same frequency band (9.15 GHz–10.44 GHz) could be inserted with high selectivity and a high quality factor [28].

2. Fundamentals of Waveguide Resonators

The working principles of microwave filters operating at high frequencies are very similar to those of lumped elements series or parallel *RLC* resonators in circuit theory [1]. Microwave resonators are constructed by enclosing (short circuiting) both ends of waveguide structures as shown in Figure 1a by metallic walls. For instance, Figure 1b shows the geometry of a rectangular waveguide resonator ($a > b$) with length d shorted at both ends $z = 0$ and $z = d$. Here, it is assumed that waveguide is extending in the z-axis. Electromagnetic energy (in the electric and magnetic fields) is stored within the box enclosure as shown in Figure 2, and it reaches its maximum at good coupling, and it dissipates power in the metallic walls and/or within the filling dielectric material $\epsilon = \epsilon_0 \epsilon_r$, where ϵ_r is a medium's relative permittivity.

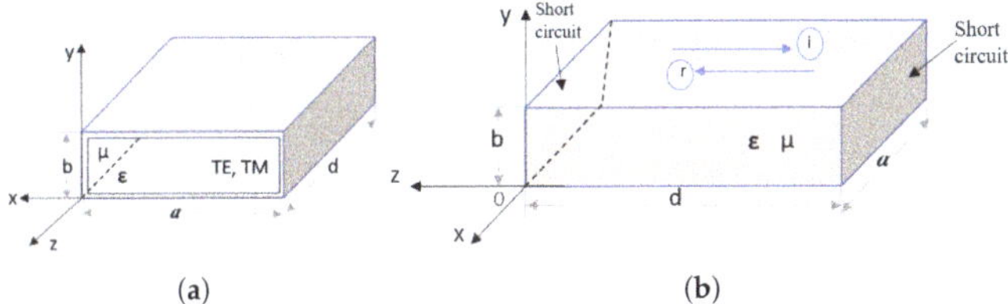

Figure 1. (**a**) rectangular waveguide $a > b$; (**b**) rectangular cavity, shorted and perfect electric conductor (PEC) for all sides $d > a > b$.

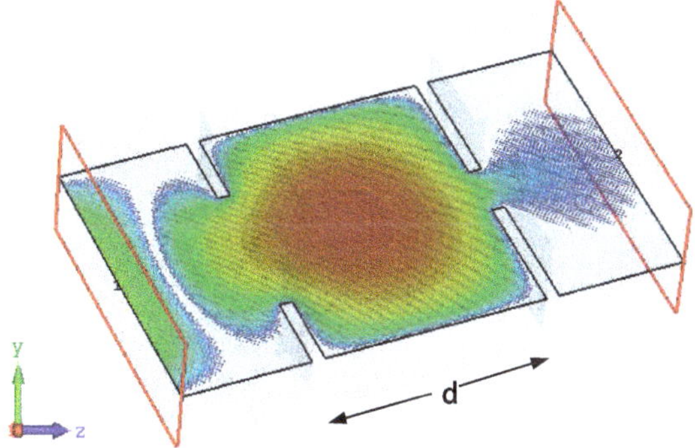

Figure 2. Electromagnetic energy stored in the rectangular waveguide resonator for X-band WR-90 with dimension a = 22.86 mm and b = 10.16 mm at center frequency 10 GHz with length d.

By applying the Poynting vector $\mathbf{P} = \mathbf{E} \times \mathbf{H}$ in a cavity in which applying conditions of modes, power will be purely imaginary, and time averaged power density (real part) will be zero. This means that there is no real power in/out the cavity will be transferred. However, the imaginary part means that there is a stored energy in electromagnetic fields inside the cavity [1]. Meanwhile, the quality factor (Q) is an important parameter and should be taken into consideration once design of the microwave filter is needed. The quality factor measures the strong coupling either in electric (aperture) or magnetic origin (loop) between resonators; in other words, it measures the losses in the resonator. Q is alternatively defined as the ratio of a resonator's center frequency to its 3-dB bandwidth (means bandwidth of the range of frequencies for which the energy is at least half its peak value) and given by the formula $Q = \dfrac{f_0}{BW_{3dB}}$ [1]. However, the external coupling factor (k factor) between resonators is defined by

$$k_{ij} = \mp \frac{f_2^2 - f_1^2}{f_2^2 + f_1^2} = \mp \frac{\omega_2^2 - \omega_1^2}{\omega_2^2 + \omega_1^2},\tag{1}$$

where f_1 and f_2 are the lower and higher frequencies of the filter, and i, j the adjacent cavities.

Under the condition that a cavity is lossless, transverse electric field intensity can be derived as [1]:

$$\bar{E}_t(x,y,z) = \bar{e}(x,y) \left(A^+ e^{-j\beta_{mn}z} + A^- e^{j\beta_{mn}z} \right),\tag{2}$$

where $\bar{e}(x,y)$ is the transverse variation of mode, A^+ and A^- are, respectively, the arbitrary amplitudes of the forward and backward traveling waves, and β_{mn} is the phase constant of wave. Application of boundary conditions on the side walls $(x = 0, a)$ and $(y = 0, b)$ of the cavity results in

$$\beta_{mn} = \sqrt{k^2 - \left(\frac{m\pi}{a}\right)^2 - \left(\frac{n\pi}{b}\right)^2},\tag{3}$$

where $k = \sqrt{\mu\epsilon}$ is the wave number of the material filling the waveguide, μ and ϵ are the permeability and permittivity of the filled material, and m and n denote the number of half-sine (or cosine) variations in x and y directions. Application of boundary conditions at the front and rear faces of the enclosed waveguide cavity at $(z = 0, d)$ yields the well-known following resonant frequency f_r

$$f_r(mnl) = \frac{1}{2\pi\sqrt{\mu\epsilon}} \sqrt{\left(\frac{m\pi}{a}\right)^2 + \left(\frac{n\pi}{b}\right)^2 + \left(\frac{l\pi}{d}\right)^2},\tag{4}$$

where l denotes the number of half-sine (or cosine) variations on the z-axis. Figure 3a shows the S_{21} parameter of the resonator at 10 GHz resonant frequency when the port 2 in Figure 2 is weak, and Figure 3b shows the real reference impedance to the same resonator $Z_{ref} = 499\ \Omega$ at 10 GHz.

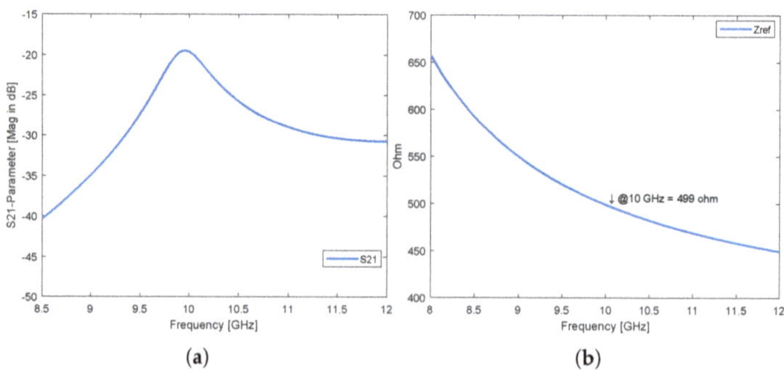

Figure 3. (**a**) S21 parameter of waveguide resonator, and (**b**) reference Impedance real part of waveguide resonator.

3. Design and Simulation

In the late 1960s, Veselago has paved the way for using artificial materials (unnatural material) [23] with negative μ and ϵ. Considerably later, the evidence of medium with simultaneous negative μ and ϵ was demonstrated experimentally [22]. As demonstrated in [25], one can use CSRR to design a compact waveguide filter by an engraving metallic sheet. Here, in this paper, a double ring meta-resonator engraved from the center of CSRR as shown in Figure 4a to reduce the physical size of the waveguide filter that is coupled by inductive or capacitive irises has been used. The engraved rings on the copper plane are not mutually aligned to each other, thus the upper ring and the lower ring are not on the same vertical line (shifted). Nevertheless, the shifting of rings plays another role besides selecting proper CSRR physical dimensions to move the resonant frequency as desired. Based on the coupling phenomena, shifting changes the coupling between two rings, so that selecting a good shift with a proper physical dimension of CSRR gives a quick response at desired resonant frequency as shown in the response in Figure 4b.

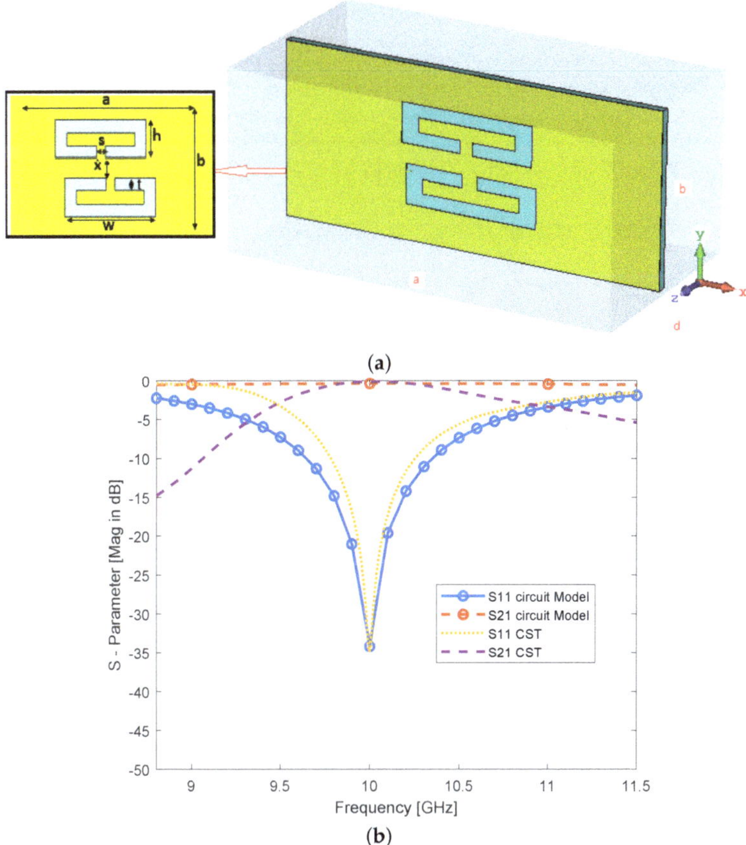

Figure 4. (**a**) rectangular meta-resonator design; (**b**) S-parameter response of waveguide meta-resonator S_{11} and S_{21} which is compared by using (CST) and a circuit simulator.

As shown in Figure 4, computer simulation technology (CST) is used to simulate rectangular waveguide resonator which initial length is $d = \lambda/2$ at 10 GHz for X-band applications with lower cutoff frequency $f_l = 9.15$ GHz, higher cutoff frequency $f_h = 10.44$ GHz, and 0.043 dB passband ripple. In addition, using substrate material (RT/Duroid) with a thickness of 0.5 mm, permittivity constant $\epsilon_r = 2.2$, and appropriate mesh density selection 10 steps per wavelength with 20 min number of steps are selected. Annealed copper (lossy material) with conductivity $\sigma = 5.8 \times 10^7$ [S/m], thickness 30 μm is chosen for metallic layer, and impedance at resonant frequency $Z_0 = 499\,\Omega$ as shown in Figure 3b.

The microwave resonant circuits in high frequency behave as *RLC* elements in low frequency circuits (including ohmic losses) that can be excited by external magnetic source. In other words, to obtain lumped elements model from distributed model, values of components that are calculated using Equations (5)–(7) as proposed in [26] will be used. The equivalent lumped elements model for the proposed meta-resonator (microstrip) is shown in Figure 5:

$$R_i = Z_0 \frac{|S_{21}(j\omega_0)|}{2(1 - |S_{21}(j\omega_0)|)}, \tag{5}$$

$$L_i = B_{3dB} Z_0 \frac{|S_{21}(j\omega_0)|}{2\omega_0^2}, \tag{6}$$

$$C_i = \frac{2}{B_{3dB} Z_0 |S_{21}(j\omega_0)|}, \tag{7}$$

where ω_0 is angular frequency (rad/s), B_{3dB} bandwidth at specific frequency, Z_0 port impedance, and S_{21} is S-parameter at considered resonant frequency. Based on what is proposed in [26] and the definition of circuit R, L, and C parameters, shown in Figure 5, and, because of symmetry between two CSRRs, the values are obtained at 10 GHz resonant frequency by using Equations (5)–(7): $Z_0 = 499\,\Omega$, $B_{3dB} = 0.07$ GHz, $R_{1,2} = 5989.65\,\Omega$, $C_{1,2} = 0.0508$ pF, $L_{1,2} = 4.9807$ nH.

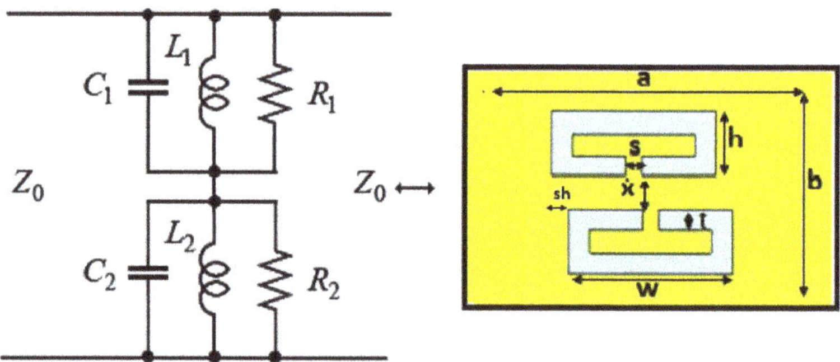

Figure 5. Rectangular meta-resonator and its simple lumped model RLC circuit.

3.1. Bandpass Waveguide Filter Using Symmetric CSRR

Up to now, a single waveguide meta-resonator by using CSRR and according to simulation results of meta-resonator as shown in Figure 4b has been analyzed, it can be noticed that the variation of parameters *sh*, *w*, and *t* introduce a shift of the resonant frequency and effect on bandwidth according to the main values of the same Figure 4a ($f_0 = 10$ GHz and $B_{3dB} = 0.07$ GHz), as listed in Table 1 and already has been shown in Figures 6–11. Changing the width (s) of the engraved center of two loops leads to changing resonant frequency according to the variation coupling capacitance as shown in Figure 7 and, once engraving width increases, the resonant frequency increases as well. Figure 8 shows the variations of vertical separation (x) between the two loops and its effect on the change electrical coupling between loops, as a result of increasing the separation, the resonant frequency will be decreased and will then move to the left. Figure 9 shows an increase in the height (h) of the loop, decreasing the resonant frequency and moving to the left. In Figure 10, alternation in the line width of loops has a direct effect on the resonant frequency, decreases (t) parameter, and increases resonant frequency according to a variation of currents that moves along a path of the loops. As well as changing the width of the loops (upper and lower), the bandwidth and the resonant frequency change accordingly, and an increase in the (w) parameter decreases the resonant frequency as shown in Figure 11. For the waveguide filter design, WR- 90 standard, $a > b$, $a = 22.86$ mm, $b = 10.16$ mm, and Chebyshev response are used here. The Chebyshev response that exhibits the equal-ripple passband and maximally flat stopband. The alternation of the Chebyshev response as well as changing the bandwidth and sharpness of a filter could be affected based on the number of its resonators. Meaning that an increase in filter order N will give more sharpness and enhance bandwidth, but, meanwhile, filter design will be more complicated. Figure 12 shows the sharpness of a filter with a Chebyshev response. The sharpness of a filter with a Chebyshev response gets better with an increase in filter order, as expected. The amplitude-squared transfer function that describes this type of response is given by [37]

$$|S_{21}(j\Omega)|^2 = \frac{1}{1 + \epsilon^2 T_n^2(\Omega)},$$

(8)

where $T_n(\Omega)$ is a Chebyshev function of the first kind of order n and the ripple constant ϵ is related to a given passband ripple L_{Ar} in dB by

$$\epsilon = \sqrt{10^{\frac{L_{Ar}}{10}} - 1}.$$

(9)

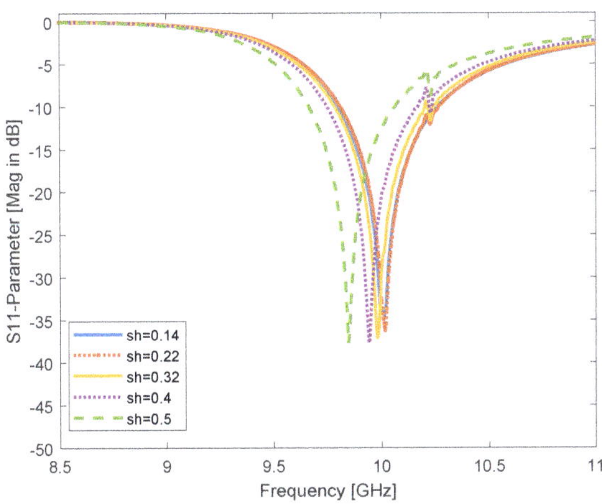

Figure 6. Alteration of horizontal shift between two CSRR rings (upper and lower loops) sh-parameter for sh = 0.14, 0.22, 0.32, 0.4, and 0.5 mm; resonant frequency is shifted to the left once the sh parameter is increased.

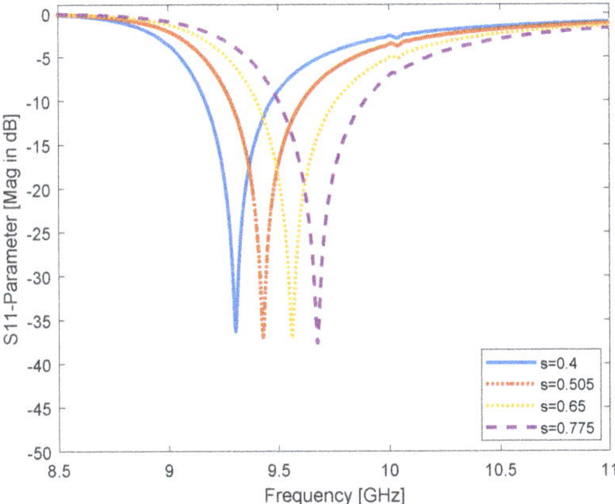

Figure 7. Engraving the center width of both CSRR rings, the s parameter for both loops and its changing s = 0.4, 0.505, 0.65, and 0.775 mm; the resonant frequency is shifted to the right as a result of changing capacitance.

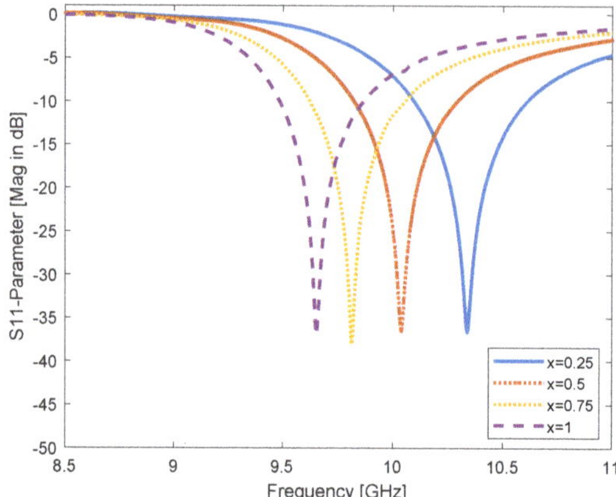

Figure 8. Alteration of both CSRR rings (upper and lower loops) vertical separation, x parameter for x = 0.25, 0.5, 0.75, and 1 mm, and resonant frequency is shifted to the left according to the coupling effect.

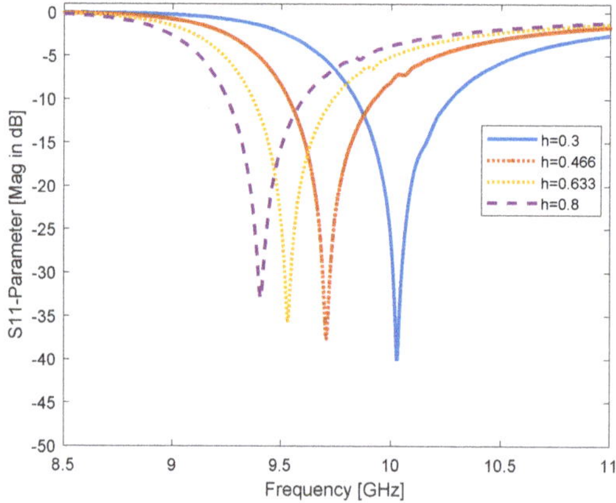

Figure 9. Alteration of CSRR height, h parameter for h = 0.3, 0.466, 0.633, and 0.8 mm, and the resonant frequency is shifted to the left.

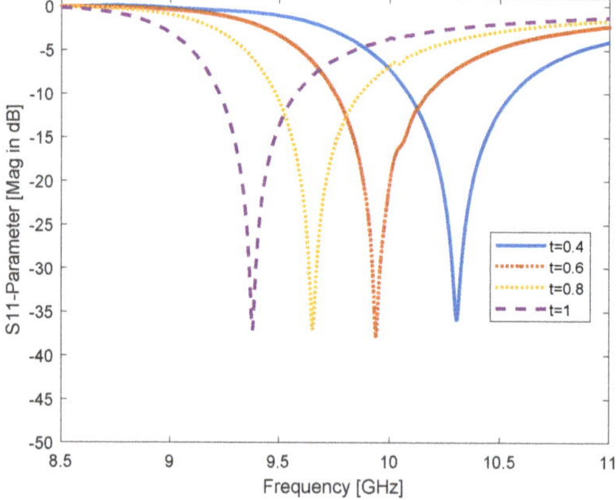

Figure 10. Changing the line width t parameter of both CSRR rings, for t = 0.4, 0.6, 0.8, and 1 mm, the resonant frequency is shifted to the left and the changing is high in frequency according to this parameter.

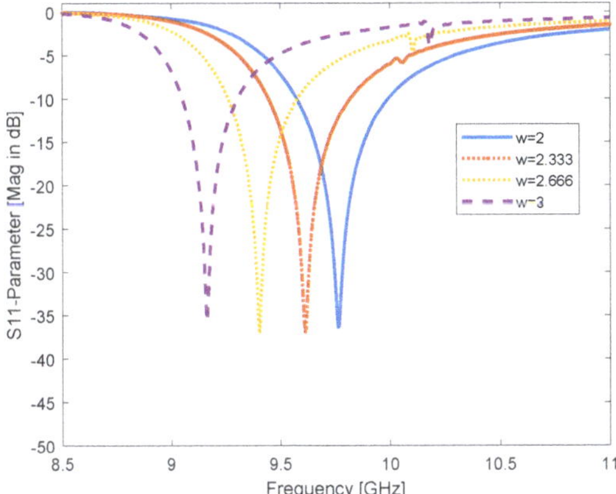

Figure 11. Alteration of CSRR rings width w parameter and its efffect on resonant frequency w = 2, 2.333, 2.666, and 3 mm; resonant frequency is shifted to the left as well as the bandwidth being decreased.

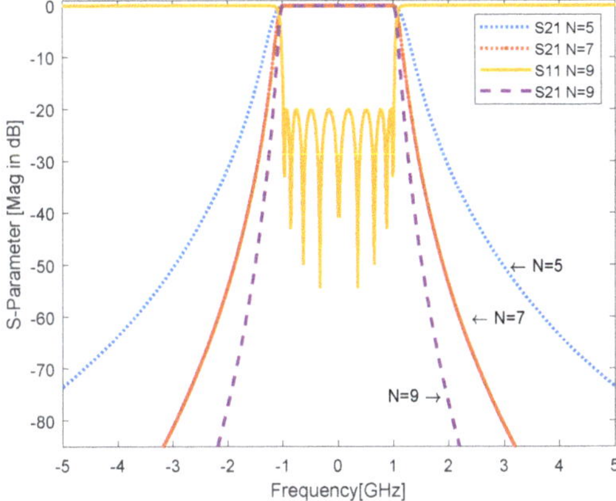

Figure 12. The alternation of the Chebyshev response as well as changing the bandwidth and the sharpness of a filter could be affected based on the number of its resonators, meaning that an increase of filter order N will give more sharpness and enhance bandwidth for $N = 5, 7$ and 9.

Table 1. The influence of changing parameters of the meta-resonator and its effect on cutoff frequency and bandwidth of a bandpass filter, $\leftarrow$ response shifts to the left, $\rightarrow$ response shifts to the right, and $\approx$ response doesn't change.

Parameters	Resonant Frequency	Bandwidth
w ↑	$\leftarrow$	decreases
s ↑	$\rightarrow$	increases
h ↑	$\leftarrow$	decreases
t ↑	$\approx$	increases
x ↑	$\leftarrow$	decreases
sh ↑	$\leftarrow$	$\approx$

With increasing value of modes (propagation patterns) in the waveguide, attenuation increases as well (evanescent); particularly, the waveguide acts as a high pass filter and supports all possible modes higher than cutoff frequency, but the attenuation losses are minimum in the lower order mode. Thus, assuming that the dominant pattern of propagation is TE_{101} for the cavity, this means that transverse electric or there are no components of electric field in the direction of propagation $E_z = 0$, and magnetic components exist in the the same direction of propagation $H_z \neq 0$. The proposed waveguide filter has the symmetry modeling of resonators, so this feature reduces the number of variables of design (physical dimensions). The structure of waveguide filter, initially used quarter wavelength $\frac{\lambda}{4}$ distance between adjacent resonators. The gap between the inserted metamaterial and the metallic walls of waveguide is less than 30 μm. The procedure used here to design the waveguide bandpass filter is proposed in [38], which depends on the coupling between cascaded meta-resonators. The first waveguide resonator is run and optimized at 10 GHz as shown in Figure 4, the initial length of the waveguide resonator without inserting metamaterial is 40 mm and with inserting metamaterial is 17 mm. After that, another resonator has been inserted within the waveguide, or cascaded with the previous one and so on, as shown in Figure 13. Once the overall waveguide filter is totally designed, the initial response is optimized by using an Interpolated Quasi Newton method to reach the desired filter specifications. The flowchart of the filter design procedure is shown in Figure 14. It is noticed that, at 10 GHz, the return loss S_{11} is the lesser of -12 dB with insertion loss S_{21} being close to 0.04 dB, cutoff at -70 dB, and Q factor is 7563 as shown in Figures 15 and 16.

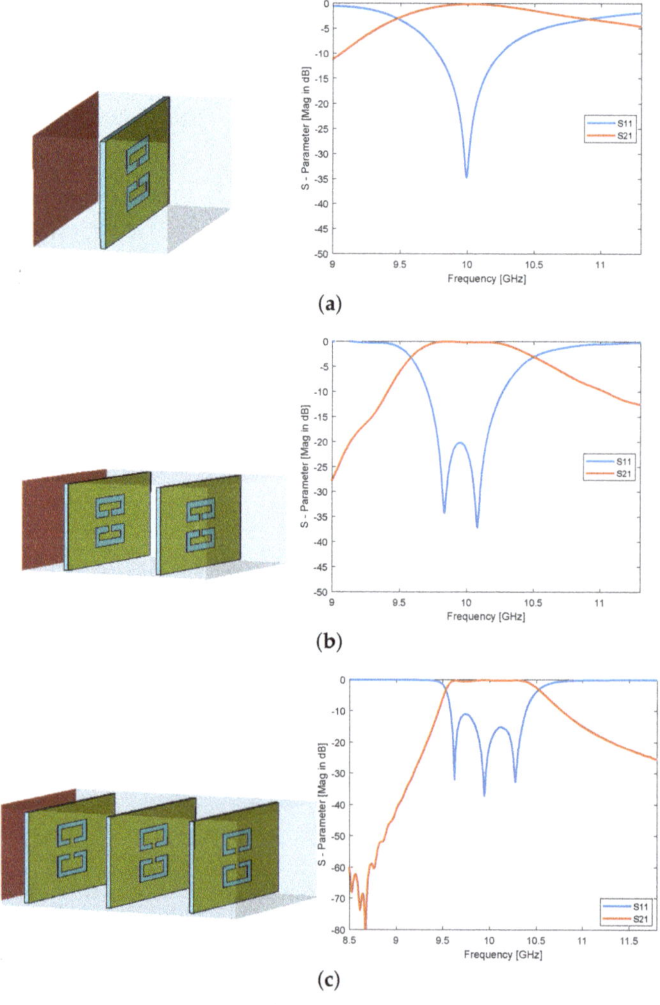

(a)

(b)

(c)

Figure 13. *Cont.*

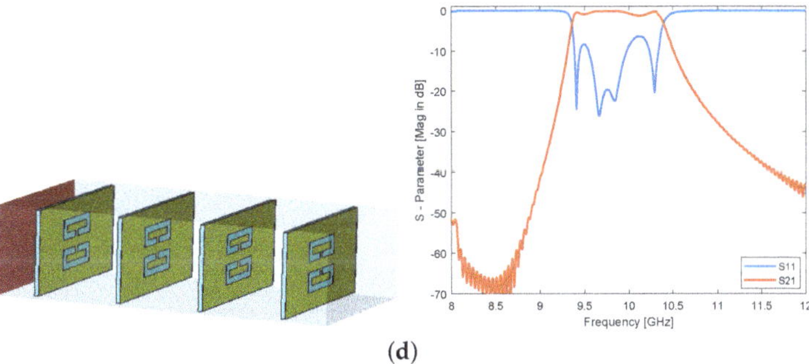

(d)

Figure 13. Design steps of bandpass waveguide filter based on rectangular meta-resonator and its S-parameters for (**a**) first resonator and its S-parameter; (**b**) second resonator is added to the first one and its S-parameter; (**c**) the third resonator is added to both resonators and its S-parameter; (**d**) the fourth resonator is added in, cascaded to three resonators and its S-parameter.

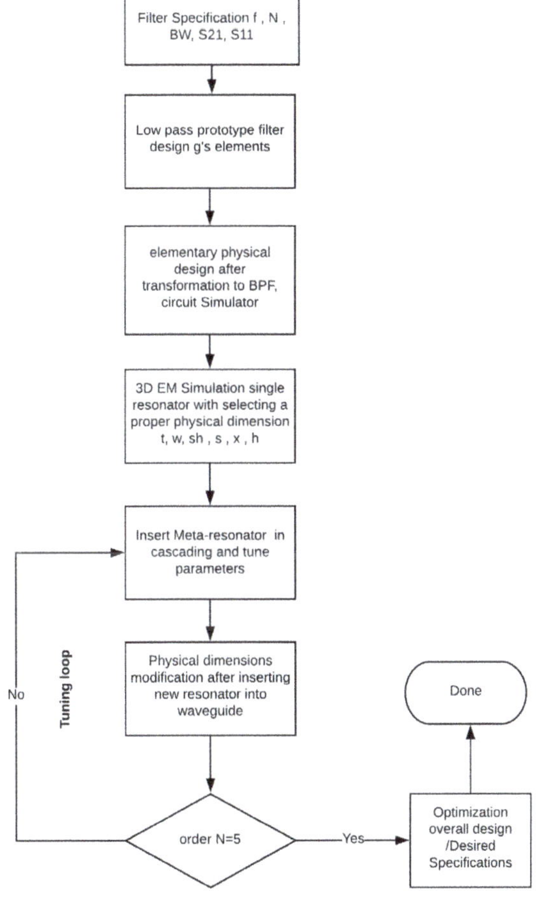

Figure 14. Flowchart of the waveguide filter design methodology.

Figure 15. Final design of the proposed waveguide meta-resonator filter $N = 5$ for X-band at resonant frequency 10 GHz.

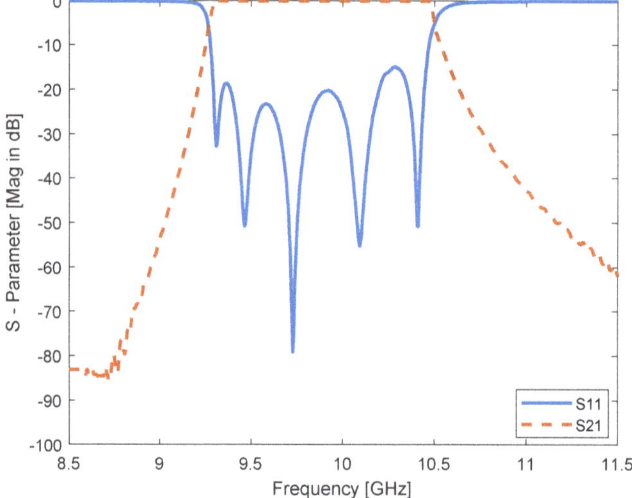

Figure 16. Optimized final response of the proposed waveguide meta-resonator filter and its S-parameters S_{11} at -20 dB and S_{21} 0.045 dB at center frequency 10 GHz, $N = 5$.

4. Discussion

Based on the initial design specifications of the bandpass filter H-plane, resonant frequency f_0, number of resonators (cavities) N, relative bandwidth BW, low pass prototype with element values g_i, and insertion loss of filter being assumed, then the conventional resonators are designed to get the desired BPF. To obtain the bandpass filter from a low pass prototype filter with g parameters $g_0 = 1$, $g_1 = 0.9714$, $g_2 = 1.3721$, $g_3 = 1.8014$, $g_4 = 1.3721$, $g_5 = 0.9714$, and $g_6 = 1$, the frequency transformation method is used here [37] as shown in Figure 17. If the conventional bandpass waveguide filter for X-band application as shown in Figure 18a, with its design method (insertion loss method) as has been proposed in [9,38] that consists of five resonators with inductive apertures (discontinuities), and the proposed metamaterial bandpass waveguide filter (cascaded meta-resonators) as shown in Figure 15 are compared with each other, the use of metamaterial technology shortening the overall physical length of the waveguide filter by 31% and enhancing the bandwidth up to 1.22 GHz with increasing 37.5% at 10 GHz will be noticed. In Table 2, a quick comparison between two waveguide filters has been summarized. In Figure 19, both responses S_{21} of the conventional and the proposed bandpass filters could be seen.

Table 2. Comparison between conventional waveguide bandpass filter and the metamaterial waveguide filter for the X-band at 10 GHz.

Filter Type	f_0 (GHz)	L (mm)	s11 (dB)	Z_w (Ω)	BW (GHz)
Proposed filter	10	51	-20	499	1.22
Conventional filter	10	97.1	-19	497	0.550

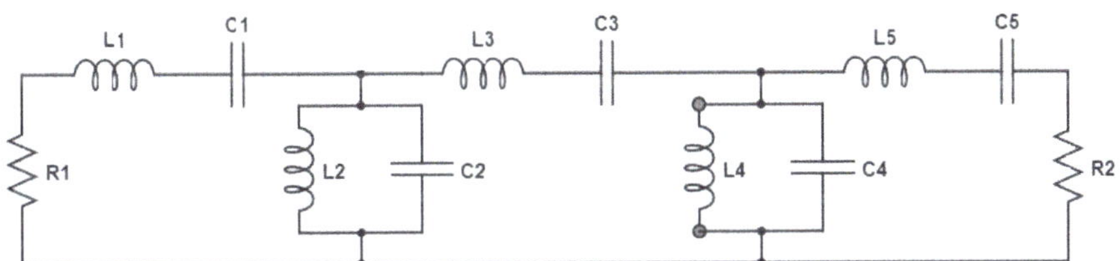

Figure 17. Circuit model of a symmetrical bandpass filter for the Chebyshev response after transformation from low pass prototype filter, $R_1 = R_2 = 498\ \Omega$, $C_1 = C_5 = 0.002644$ pF , $C_2 = C_4 = 0.6932$ pF, $C_3 = 0.001113$ pF, $L_1 = L_5 = 122.7$ nH, $L_2 = L_4 = 0.3653$ nH, and $L_3 = 227.54$ nH [37].

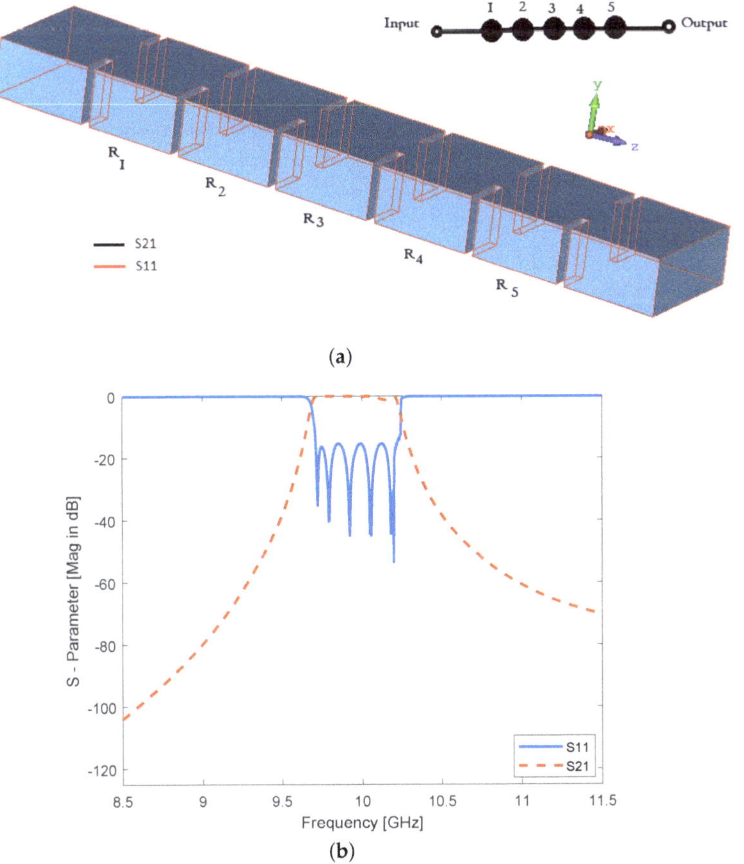

(**a**)

(**b**)

Figure 18. (**a**) conventional Bandpass waveguide filter with inductive irises fifth order $N = 5$ for WR- 90 standard, $a > b$, $a = 22.86$ mm, $b = 10.16$ mm, and (**b**) illustration of S-parameters S11 and S21.

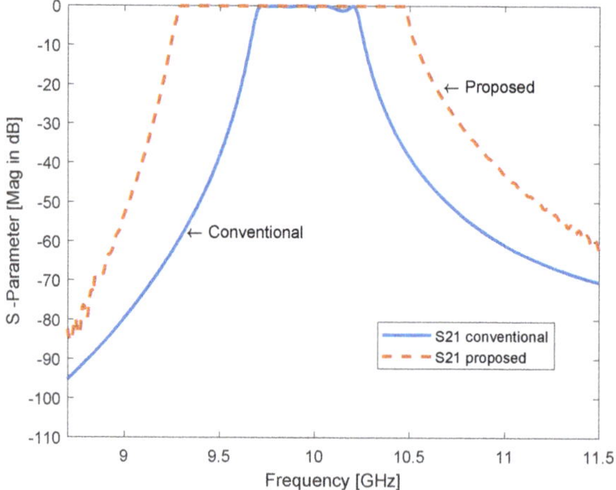

Figure 19. S-parameter comparison between conventional rectangular waveguide filter with direct waveguide junctions and meta-resonator waveguide filter without waveguide junctions for X-band, $N = 5$. The proposed filter has enhanced the bandwidth to 37.5% as seen in the figure.

5. Conclusions

In this paper, a simple methodology to control bandwidth and cutoff frequency of waveguide filter is proposed and simulated. Inserting CSRRs in the same plane of transverse in the waveguide filter for X-band applications such as the RADAR communication system and LNA is presented. A CSRR-loaded bandpass waveguide filter with a Chebyshev response using a symmetry structure with a shifted upper or lower ring of CSRR is designed and discussed here. Adding a meta-resonator to another at the same time in the structure, with an initial distance $\left(\frac{\lambda}{4}\right)$ between them, is simulated. A fifth order CSRR-loaded bandpass waveguide filter is designed and simulated by computer simulation technology (CST) along with a circuit simulator as well. The filter's dimensions are optimized and tuned toward the desired center frequency at 10 GHz, as well as the shifting of upper and/or lower rings from each other. The bandwidth of the proposed bandpass filter increases up to 1.22 GHz, meaning that the bandwidth of the filter is increased by 37.5%, and the filter's size reduced by 31% as well if it was to be compared with the conventional X-band bandpass filter, which is designed by using the insertion loss method at the same center frequency. For future work, the proposed waveguide filter will be employed to design a waveguide diplexer and multiplexer based on the metamaterial technology.

Author Contributions: Conceptualization, M.A.; methodology, M.A.; software, M.A.; Design, M.A., U.C.H; validation, M.A., writing–review and editing, U.C.H.; original draft preparation, M.A.; supervision, U.C.H.; visualization, M.A.; optimization and tuning, M.A. All authors have read and agree to the published version of the manuscript.

Funding: This research received no external funding.

Conflicts of Interest: The authors declare no conflict of interest.

Abbreviations

The following abbreviations are used in this manuscript:

MM	Metamaterial
SRR	Split Ring Resonator
CSRR	Complementary Split Ring Resonator
CST	Computer Simulation Technology

References

1. Pozar, D.M. *Microwave Engineering*; Wiley: Hoboken, NJ, USA, 2012.
2. Guo, C.; Shang, X.; Lancaster, M.J.; Xu, J. A 3D printed lightweight X-band waveguide filter based on spherical resonators. *IEEE Microw. Wirel. Compon. Lett.* **2015**, *25*, 442–444. [CrossRef]
3. Atia, A.E.; Williams, A.E. Narrow-bandpass waveguide filters. *IEEE Trans. Microw. Theory Tech.* **1972**, *20*, 258–265. [CrossRef]
4. Bonetti, R.R.; Williams, A.E. Application of dual TM modesto triple-and quadruple-mode filters. *IEEE Trans. Microw. Theory Tech.* **1987**, *35*, 1143–1149. [CrossRef]
5. Levy, R.; Cohn, S.B. A history of microwave filter research, design, and development. *IEEE Trans. Microw. Theory Tech.* **1984**, *32*, 1055–1067. [CrossRef]
6. Levy, R.; Snyder, R.V.; Matthaei, G. Design of microwave filters. *IEEE Trans. Microw. Theory Tech.* **2002**, *50*, 783–793. [CrossRef]
7. Hunter, I.C.; Billonet, L.; Jarry, B.; Guillon, P. Microwave filters-applications and technology. *IEEE Trans. Microw. Theory Tech.* **2002**, *50*, 794–805. [CrossRef]
8. Zou, X.; Tong, C.M.; Yu, D.W. Design of an X-band symmetrical window bandpass filter based on substrate integrated waveguide. In Proceedings of the 2011 Cross Strait Quad-Regional Radio Science and Wireless Technology Conference, Harbin, China, 26–30 July 2011; Volume 1, pp. 571–574.
9. Damou, M.; Nouri, K.; Feham, M.; Chetioui, M. Design and Optimization of Rectangular Waveguide Filter based on Direct Coupled Resonators. *Int. J. Electron. Telecommun.* **2017**, *63*, 375–380. [CrossRef]
10. Morelli, M.; Hunter, I.; Parry, R.; Postoyalko, V. Stopband performance improvement of rectangular waveguide filters using stepped-impedance resonators. *IEEE Trans. Microw. Theory Tech.* **2002**, *50*, 1657–1664. [CrossRef]
11. Liao, S.S.; Sun, P.T.; Chen, H.K.; Liao, X.Y. Compact-size coplanar waveguide bandpass filter. *IEEE Microw. Wirel. Compon. Lett.* **2003**, *13*, 241–243. [CrossRef]
12. Goussetis, G.; Budimir, D. Novel periodically loaded E-plane filters. *IEEE Microw. Wirel. Compon. Lett.* **2003**, *13*, 193–195. [CrossRef]
13. Mohottige, N.; Glubokov, O.; Jankovic, U.; Budimir, D. Ultra Compact Inline *E*-Plane Waveguide Bandpass Filters Using Cross Coupling. *IEEE Trans. Microw. Theory Tech.* **2016**, *64*, 2561–2571. [CrossRef]
14. Mohottige, N.; Glubokov, O.; Budimir, D. Ultra Compact Inline *E*-Plane Waveguide Extracted Pole Bandpass Filters. *IEEE Microw. Wirel. Compon. Lett.* **2013**, *23*, 409–411. [CrossRef]
15. Budimir, D.; Glubokov, O.; Potrebić, M. Waveguide filters using T-shaped resonators. *Electron. Lett.* **2011**, *47*, 38–40. [CrossRef]
16. Chen, X.P.; Wu, K. Substrate integrated waveguide filters: Design techniques and structure innovations. *IEEE Microw. Mag.* **2014**, *15*, 121–133. [CrossRef]
17. Zhou, S.; Wang, Z.; Xu, R.; Shen, D.; Zhan, M. A novel X-band half mode substrate integrated waveguide (HMSIW) bandpass filter. In Proceedings of the 2009 Asia Pacific Microwave Conference, Singapore, 7–10 December 2009; pp. 1387–1389.
18. Jin, C.; Chen, J.X.; Chu, H.; Bao, Z.H. X-band differential bandpass filter with high common-mode suppression using substrate integrated waveguide cavity. *Electron. Lett.* **2014**, *50*, 88–89. [CrossRef]
19. Wong, S.W.; Chen, R.S.; Wang, K.; Chen, Z.N.; Chu, Q.X. U-shape slots structure on substrate integrated waveguide for 40-GHz bandpass filter using LTCC technology. *IEEE Trans. Compon. Packag. Manuf. Technol.* **2015**, *5*, 128–134. [CrossRef]
20. Dahle, R.; Laforge, P.; Kuhling, J. 3-D Printed Customizable Inserts for Waveguide Filter Design at X-Band. *IEEE Microw. Wirel. Compon. Lett.* **2017**, *27*, 1080–1082. [CrossRef]
21. Liu, L.; Zhang, P.; Weng, M.H.; Tsai, C.Y.; Yang, R.Y. A Miniaturized Wideband Bandpass Filter Using Quarter-Wavelength Stepped-Impedance Resonators. *Electronics* **2019**, *8*, 1540. [CrossRef]
22. Pendry, J.B.; Holden, A.J.; Robbins, D.J.; Stewart, W. Magnetism from conductors and enhanced nonlinear phenomena. *IEEE Trans. Microw. Theory Tech.* **1999**, *47*, 2075–2084. [CrossRef]
23. Veselago, V. Electrodynamics of substances with simultaneously negative electrical and magnetic permeabilities. *Physics-Uspekhi* **1968**, *10*, 504–509.
24. Hrabar, S.; Bartolic, J.; Sipus, Z. Waveguide miniaturization using uniaxial negative permeability metamaterial. *IEEE Trans. Antennas Propag.* **2005**, *53*, 110–119. [CrossRef]

25. Ortiz, N.; Baena, J.; Beruete, M.; Falcone, F.; Laso, M.; Lopetegi, T.; Marques, R.; Martin, F.; Garcia-Garcia, J.; Sorolla, M. Complementary split-ring resonator for compact waveguide filter design. *Microw. Opt. Technol. Lett.* **2005**, *46*, 88–92. [CrossRef]

26. Stefanovski, S.L.; Potrebić, M.M.; Tošić, D.V. Design and analysis of bandpass waveguide filters using novel complementary split ring resonators. In Proceedings of the 2013 11th International Conference on Telecommunications in Modern Satellite, Cable and Broadcasting Services (TELSIKS), Nis, Serbia, 16–19 October 2013; Volume 1, pp. 257–260.

27. Fallahzadeh, S.; Bahrami, H.; Tayarani, M. A novel dual-band bandstop waveguide filter using split ring resonators. *Prog. Electromagn. Res.* **2009**, *12*, 133–139. [CrossRef]

28. Fathnan, A.; Amrullah, Y.; Wijayanto, Y.; Mahmudin, D.; Daud, P. A compact X-Band bandpass filter using rectangular split ring resonators for radar applications. In Proceedings of the 2015 International Conference on Radar, Antenna, Microwave, Electronics and Telecommunications (ICRAMET), Bandung, Indonesia, 5–7 October 2015; pp. 60–63.

29. Odabasi, H.; Teixeira, F. Electric-field-coupled resonators as metamaterial loadings for waveguide miniaturization. *J. Appl. Phys.* **2013**, *114*, 214901. [CrossRef]

30. Bage, A.; Das, S. Stopband Performance Improvement of CSRR-Loaded Waveguide Bandpass Filters Using Asymmetric Slot Structures. *IEEE Microw. Wirel. Compon. Lett.* **2017**, *27*, 697–699. [CrossRef]

31. Torabi, Y.; Dadashzadeh, G.; Oraizi, H. Miniaturized sharp band-pass filter based on complementary electric-LC resonator. *Appl. Phys. A* **2016**, *122*, 273. [CrossRef]

32. Dong, Y.; Itoh, T. Miniaturized dual-band substrate integrated waveguide filters using complementary split-ring resonators. In Proceedings of the 2011 IEEE MTT-S International Microwave Symposium, Baltimore, MD, USA, 5–10 June 2011; pp. 1–4.

33. ul Haq, T.; Khan, M.; Siddiqui, O. Design and implementation of waveguide bandpass filter using complementary metaresonator. *Appl. Phys. A* **2016**, *122*, 34. [CrossRef]

34. Bahrami, H.; Hakkak, M.; Pirhadi, A. Using complementary split ring resonators (CSRR) to design bandpass waveguide filters. In Proceedings of the 2007 Asia-Pacific Microwave Conference, Bangkok, Thailand, 11–14 December 2007; pp. 1–4.

35. Yoon, K.C.; Kim, S.C.; Lee, J.C. Compact low noise amplifier with high stable gain using distributed CRLH metamaterial. *Microw. Opt. Technol. Lett.* **2016**, *58*, 1715–1720. [CrossRef]

36. Chen, R.; Li, R.; Zhou, S.; Chen, S.; Huang, J.; Wang, Z. An X-Band 40 W Power Amplifier GaN MMIC Design by Using Equivalent Output Impedance Model. *Electronics* **2019**, *8*, 99. [CrossRef]

37. Hong, J.S.G.; Lancaster, M.J. *Microstrip Filters for RF/Microwave Applications*; John Wiley & Sons: Hoboken, NJ, USA, 2004; Volume 167.

38. Shang, X.; Xia, W.; Lancaster, M.J. The design of waveguide filters based on cross-coupled resonators. *Microw. Opt. Technol. Lett.* **2014**, *56*, 3–8. [CrossRef]

 *electronics*

Article

An Ultra-Wideband Bandpass Filter with a Notch Band and Wide Upper Bandstop Performances

Min-Hang Weng [1], Che-Wei Hsu [2], Siang-Wen Lan [3] and Ru-Yuan Yang [2,*]

[1] School of Information Engineering, Putian University, Putian 351100, China; hcwweng@gmail.com
[2] Graduate Institute of Materials Engineering, National Pingtung University of Science and Technology, Pingtung County 912, Taiwan; namiemayday@gmail.com
[3] Microelectronics and Department of Electrical Engineering, National Cheng Kung University, Tainan 701, Taiwan; s907952271@gmail.com
* Correspondence: ryyang@mail.npust.edu.tw; Tel.: +886-8-770-3202 (ext. 7555); Fax: +886-8-774-0552

Received: 20 October 2019; Accepted: 4 November 2019; Published: 8 November 2019

Abstract: This paper presents an ultra-wideband bandpass filter (UWB-BPF) with a notch band and a wide upper stopband. Two pairs of half-wavelength high-impedance line resonators tightly and strongly coupled to the input/output lines are used to provide the wideband responses. The first UWB responses of 3.4–5.0 GHz and the second UWB of 6.0–10.0 GHz are designed independently first and then combined together for the application of a direct-sequence ultra-wideband bandpass (DS-UWB) system. Without using any extra bandstop structure, a notch band at 5.2 GHz can be obtained. The fabricated UWB-BPF with a compact circuit size exhibits good passband performances including insertion losses of 1 ± 0.3 and 2 ± 0.4 dB for first and second passbands, respectively, a high isolation at 5.2 GHz with an attenuation level of 22.7 dB, and wide upper stopband responses from 11 GHz to 19 GHz, simultaneously. The measured results also exhibit good agreement with the simulated results.

Keywords: ultra-wideband; bandpass filter; half-wavelength resonator; insertion loss

---

1. Introduction

In recent years, multi-functional mobile wireless communication systems have been developed rapidly. For example, since February 2002, the U.S. Federal Communications Commission (FCC) has established the ultra-wideband (UWB) range from 3.1 to 10.6 GHz, used for imaging systems, vehicular radar systems, communication and measurement systems, etc. [1]. Due to advantages of low power consumption and high data transmission rate, UWB applications attract great attention. The bandpass filters (BPFs) are crucial devices in the communication system. Recently, both academia and industry have conducted many significant research activities exploring various UWB components and devices applied to a direct-sequence ultra-wideband (DS-UWB) system from 3.1 to 10.6 GHz with a notch band at 5.7 GHz [2,3].

In the past, several UWB-BPF design methods have been proposed [4–16]. In [4], the stepped-impedance resonator (SIR) was used to effectively obtain the wideband performance by controlling the spurious frequencies. However, the structure of the SIR suffered from design complexity, especially in the analysis of resonant frequencies. In [5], the multiple-mode resonator (MMR) was used for a UWB filter with good in-band and out-of-band performances, but revealed a large circuit size and insertion loss. In [6], the asymmetric stepped-impedance resonator (ASIR) was first presented to achieve the UWB performance with a wide stopband and high selectivity. However, the design of ASIRs still suffered from design complexity. In [7,8], the defect ground structure (DGS) was used in the UWB design; however, the DGS would usually destroy the signal integrity. In [9–11], wideband BPFs with notched bands were designed using different open/shorted stub-loaded resonators. In [12], a wideband bandpass filter was proposed by means of split circular rings and rectangular stubs. However, the bandwidth was not large enough and the notched band was not provided. In [13], a wideband bandpass filter was reported using a multi-layer structure; however, the design procedure and the device structure were complex. In [14], a wideband BPF was directly designed by using composite series and shunt resonators. In [15,16], a wideband BPF was realized using multiple-mode split-ring resonators in a waveguide cavity. However, it is still a challenge to design a UWB-BPF without using a complex design process and structure to achieve a notch band, good band isolation and wide stopband response, simultaneously.

In this paper, we developed a simple technique and structure to achieve a compact and high-performance UWB filter by using two half-wavelength high-impedance line resonators. The bandwidth is decided by estimating the image impedance Zi of a one-stage coupled line. A notch band at 5.2 GHz can be provided, without using any extra bandstop structure, to avoid interference with the signals of the wireless local area network (WLAN) system. Moreover, with two open stubs, the designed filter achieved the characteristics of a wide stopband and a high band-edge attenuation rate, simultaneously. The designed BPF was fabricated and measured to verify the design concept.

2. Design Procedure

Figure 1 is the structure of the proposed UWB-BPF filter. This filter is mainly formed by using two pairs of half-wavelength high-impedance line resonators and the input/output lines. Two arms of the half-wavelength high-impedance line resonators are tightly parallel, coupled to the input/output lines, formed as the coupled lines, to obtain a strong coupling for creating a wideband response. The resonator at the right side is designed for the first UWB responses of 3.4–5.0 GHz and the resonator at the left side is designed for the second UWB of 6.1–10.0 GHz. In this study, the proposed UWB-BPF is implemented on the 0.787 mm-thick Duroid 5880 substrate having a dielectric constant εr of 2.2, and a loss tangent of 0.0009. The full-wave electromagnetic (EM) simulation tool (IE3D) is used in the design [17]. The design procedure is illustrated in detail as the following.

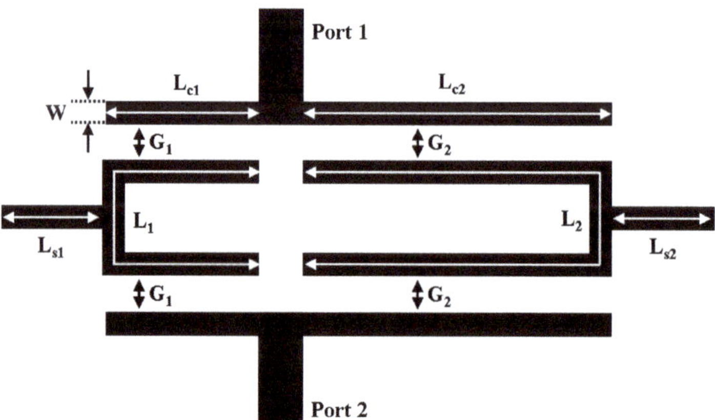

Figure 1. Structure of the proposed bandpass filter. (W = 0.2 mm, G_1 = G_2 = 0.2 mm, L_1 = 16.4 mm, L_2 = 29.2 mm, L_{c1} = 6.9 mm, L_{c2} = 13.7 mm, L_{s1} = 4.9 mm, and L_{s2} = 4.4 mm).

2.1. Step 1: Determining Centered Frequency of the Passband

Firstly, the centered frequencies of the two passbands should be determined in accordance with the application of direct-sequence ultra-wideband bandpass (DS-UWB). The resonator is simply constructed by using a half guided-wavelength ($\lambda g/2$) high-impedance line, where λg is the guided-wavelength at the specific frequency. The line with high impedance, 158 ohms in this case, is needed to have a strong coupling with a suitable gap to the input/output ports. To simplify the design, the impedance as well as the width of two resonators is the same. Figure 2 displays the resonant frequencies of the half-wavelength high-impedance line with different physical lengths. In the design, the required center frequencies of the first and second passbands are around 4.2 and 7.8 GHz, respectively. It is found that the resonator with a physical length of 29.2 mm excites a resonant peak at 4 GHz, responsible for the first passband at 3.4–5.0 GHz, and the resonator with the physical length of 14.8 mm excites a resonant peak at 8 GHz, responsible for the second passband at 6.0–10.0 GHz. Thus, L_1 of 14.8 mm and L_2 of 29.2 mm are firstly chosen for the first and second resonators, respectively.

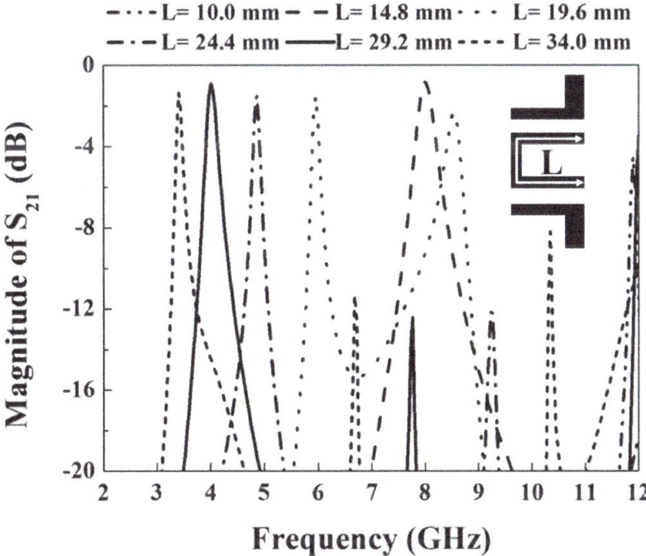

Figure 2. The resonant frequency of the half-wavelength high-impedance line with different physical lengths.

2.2. Step 2: Forming the Wideband Response

In general, the gap between the resonators and the input/output lines shall be as close as possible to obtain a wideband response. The resonators with high-impedance lines can avoid the gap smaller than the fabrication limit of the circuit board engraving machine. In previous research [18], the image impedance method was used to design a wideband bandpass filter, and the bandwidth of the passbands could be controlled by choosing different gaps of the coupled lines. The image impedance Zi is used to estimate the bandwidth of the coupled line and is expressed as [18]:

$$Z_i = \sqrt{(Z_{0e} - Z_{0o})^2 \csc^2 \theta_1 - (Z_{0e} + Z_{0o})^2 \cot^2 \theta_1} \tag{1}$$

where Z_{0e} and Z_{0o} are the even and odd mode characteristic impedances, respectively, and θ_1 is the electrical length of the coupled line. Moreover, by using the conformal mapping method, the Z_{0e} and Z_{0o} can be calculated as functions of the line width and coupling gap of the parallel coupled microstrip line [19]. Namely, as the dielectric constant (εr) and thickness (h) of the used substrate are identified, the Z_{0e} and Z_{0o} can be plotted as the design graphs versus different strip width (W/h) and coupling gaps (G/h), as discussed and shown in Figure 2 [18]. Therefore, the required bandwidth of the coupled line can be calculated by selecting the image impedance Z_i, and then the Z_{0e} and Z_{0o} can be selected to achieve the desired image impedance Z_i. In this study, without using the design curve, a transmission line calculator of the full-wave EM simulator of Zeland IE3D software can be used to extract the Z_{0e} and Z_{0o} of the coupled line to match the dimension values [20].

Figure 3 shows the band responses with different coupling gaps (G_1 and G_2) between the resonators and the input/output lines. It is noted that the harmonics mode at 7.8 GHz of the first resonator, as shown in Figure 2, would also be combined with the resonant mode of the second resonator to form the second UWB passband response. Thus, the physical length of the second resonator shall be slightly tuned to satisfy the requirement of the second UWB passband response. After optimized simulation, the physical length L_1 is selected as 16.4 mm for the second passband at 6.0–10.0 GHz. The gaps G_1 and G_2 of 0.2 mm can achieve the desired bandwidths for the first and second passbands and are larger than the minimum operating range of the circuit board engraving machine.

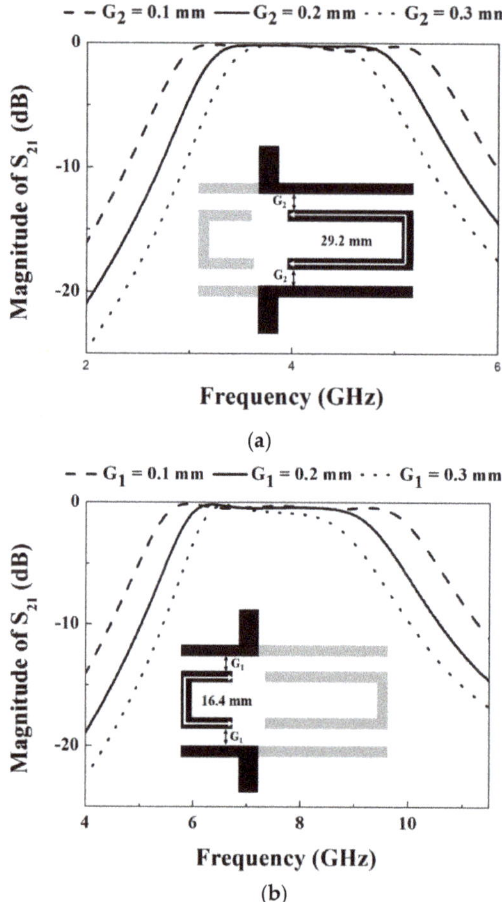

Figure 3. The band responses with different gaps between the resonators and the input/output lines. (a) first passband of $f_1 = 4$ GHz with L_2 of 19.2 mm (b) second passband of $f_2 = 8$ GHz with L_1 of 16.4 mm.

2.3. Step 3: Combining Two UWB Responses

In this step, the two UWB responses are combined to form the desired band requirements for the DS-UWB performance. Figure 4 shows (a) first passband and (b) second passband responses simulated by tuning the coupling gap of (a) G_1 (b) G_2, respectively. It is clearly observed that after combination, two UWB responses are not affected and are the same as those in Figure 3. Varying the coupling gap of G_1 and G_2 only slightly affects the second and first UWB response, respectively. Thus, the coupling gaps G_1 and G_2 are obtained as 0.2 mm for the proposed filter design. Moreover, a notch band, namely, a stopband at 5.2 GHz can be obtained naturally without using any extra bandstop structure, since the two wide passbands are individually formed and controlled. Namely, the frequency and attenuation of this stopband can be determined easily by controlling the upper band edge of the first wide band and the lower band edge of the second wide band in this design. Due to this notch band, the interference with the signals of the WLAN system can be avoided.

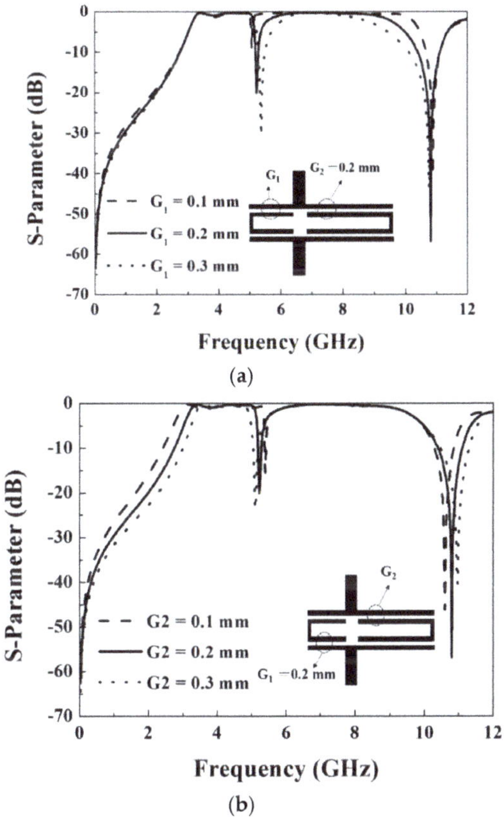

Figure 4. (**a**) First passband and (**b**) second passband responses simulated by tuning the coupling gap of (**a**) G_1 and (**b**) G_2, respectively (W = 0.2 mm, L_1 = 16.4 mm, L_2 = 29.2 mm, L_{c1} = 6.9 mm, L_{c2} = 13.7 mm, L_{s1} = 4.9 mm, and L_{s2} = 4.4 mm).

2.4. Step 4: Forming the Transmission Zeros

As shown in Figure 4, the higher band of the designed UWB filter is poor and shall be improved. In order to extend the stopband region, two uniform open stubs are simply added to suppress the undesired spurious responses. It is known that the symmetrical plane of the half-wavelength high-impedance line shows a virtual ground characteristic [14]. Therefore, a wide stopband region could be achieved by tapping open stub loads at the symmetrical plane of the two half-wavelength high-impedance lines to form the transmission zeros. With the adoption of two tapped open stubs, the attenuation poles are located near the passband and the two passband performances are not affected, as shown in Figure 5. These two tapped open stubs with L_{s1} = 4.9 mm and L_{s2} = 4.4 mm are regarded as the quarter-guided wavelength resonators at 10.85 GHz and 12.42 GHz to suppress the undesired spurious responses, respectively.

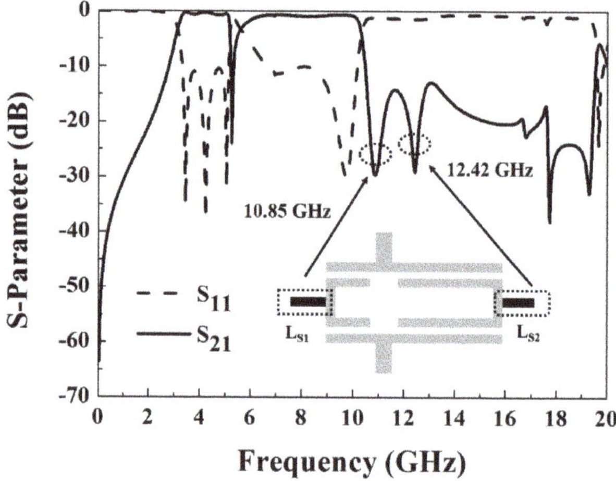

Figure 5. Frequency responses of the designed filter with two added open stubs.

Current distributions are used to confirm the resonant conditions of this UWB filter [1]. Figure 6a shows the simulated current distributions at the center frequencies of two passband responses, namely at 4 and 8 GHz. As discussed in Step 2, the strong coupling effect of the half-wavelength high-impedance line with narrow coupling gap (G_1 and G_2) is used to excite the two ultra-wideband responses. Therefore, most current energy is absorbed by these resonators, thus creating a two-coupling path to excite two passband responses. Figure 6b shows the simulated current distributions of the filter at the frequencies of the transmission zeros. As discussed in Step 4, when the tapped open stubs are regarded as the quarter-wavelength resonator, the transmission zeros are also created. Obviously, two tapped open stubs absorb, relatively, a lot the current energy, and the energy of the two half-wavelength high-impedance lines is not enough to excite a response, even with narrow coupling gaps (G_1 and G_2), thus forming transmission zeros at 10.85 GHz and 12.42 GHz.

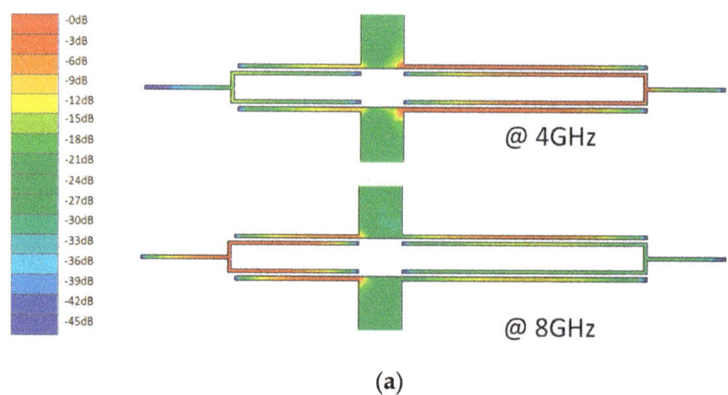

(**a**)

Figure 6. *Cont.*

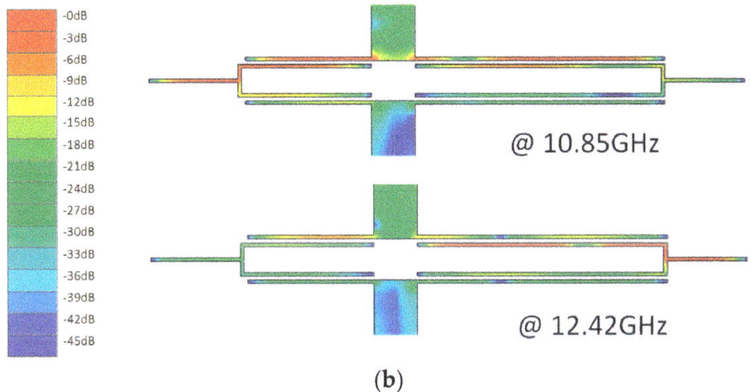

(b)

Figure 6. Current distributions of (**a**) the center frequencies of two wide bands, and (**b**) the frequencies of the transmission zeros.

3. Experimental Results

The designed UWB-BPF was fabricated by using an engraving machine, which can achieve the smallest line width and line gap of 0.15 mm. The fabricated filter is measured by the HP8510C Network Analyzer after the standard short/open/load/through calibration. Figure 7 is the simulated and measured frequency responses of the fabricated UWB-BPF, with a picture of the fabricated filter inserted. The proposed UWB-BPF has a whole size of 32.3×13 mm^2, namely, around $0.70\ \lambda g \times 0.28\ \lambda g$, where λg is the guided-wavelength at 5 GHz. For the first passband, a low insertion loss of 1 ± 0.3 dB and a fractional bandwidth (FBW) of about 42% are obtained. For the second passband, a low insertion loss of 2 ± 0.4 dB and a fractional bandwidth of about 58% are obtained. The transmission zeros appear at 10.85 GHz and 12.42 GHz to form a wide stopband region from 11 GHz to 19 GHz, with a rejection level larger than 12 dB. Also, due to the appearance of the transmission zero around 10.85 GHz, the upper band edge shows a high attenuation rate of 60 dB/GHz. Moreover, due to individual excitation of the two wide passbands, a notch band at 5.2 GHz is obtained, thus achieving a high passband isolation with an isolation level of 22.7 dB without using extra bandstop structure and avoiding the interference with the signals of the WLAN system. The measured results are slightly different than the simulated results in the higher band, resulting from the material variation of the substrate.

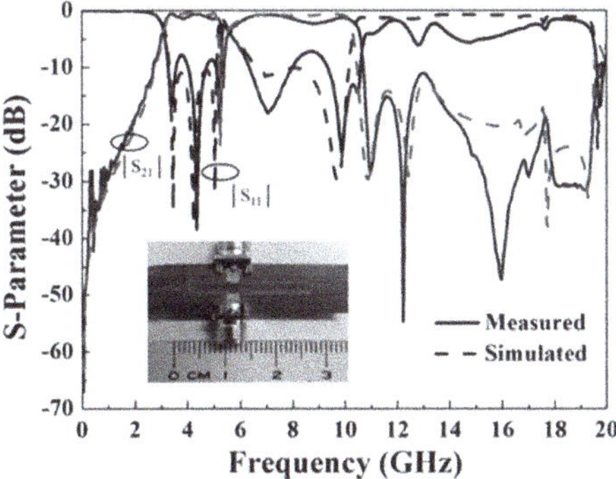

Figure 7. Simulated and measured frequency responses of the fabricated UWB-BPF. Inserted is the photograph of the filter.

Measuring the deviation in the group delays across the two passbands of the proposed filter is necessary to quantify the level of distortion introduced by the designed filter [5]. Figure 8 shows the group delay of the measured results in the (a) first passband and (b) second passband. The group delays for two passbands are similar and exhibit a small varying range from 0.15 to 0.75 ns. The measured results actually verify the possibility of the proposed design procedure.

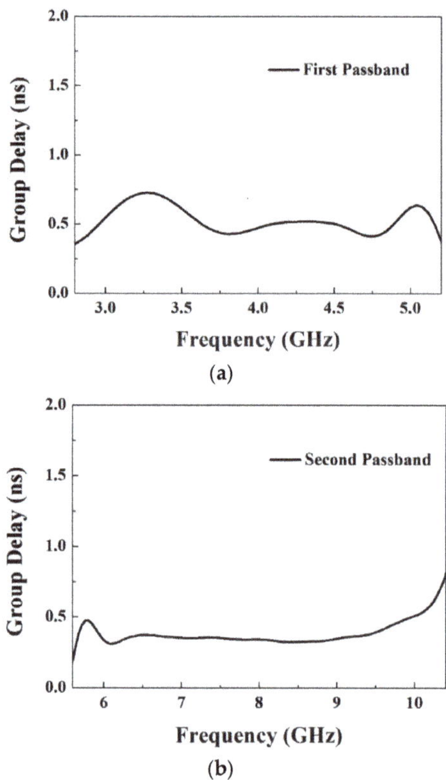

Figure 8. Group delays of the measured results in the (a) first passband and (b) second passband.

Table 1 shows the comparison of the designed filter with some published wideband BPFs. It is clear that the proposed design offers several advantages such as the appearance of the notched band and the wide stopband as well as the higher attenuation rate near the upper band edge compared to the previous reported work. Moreover, the proposed design has a very simple structure without requiring any defected ground area. Thus, the proposed UWB-BPF shows a good potential for the application of DS-UWB system.

Table 1. Comparison of filter performances of the proposed filter with the previous published works.

	Ref. [8]	Ref. [9]	Ref. [10]	Ref. [11]	Ref. [12]	Ref. [13]	This Work
Center frequency (GHz)	5	1.1	4	3	2.3	5	5
$|S11|$ (dB)	12	13	14	11.7	>13	15	10
$|S21|$ (dB)	1.2	0.2	1.4	2.1	0.35	2	2
3-dB FBW (%)	100	110	45	107	80	100	100
Circuit Size ($\lambda g \times \lambda g$)	Unknown	0.15×0.12	0.30×0.10	0.89×0.46	0.53×0.43	0.74×0.42	0.70×0.28
Number of notched band	1	0	0	2	0	1	1
Wide stopband	No	Yes	No	No	No	Yes	Yes
Defected ground	Yes	No	No	Yes	No	Yes	No

4. Conclusions

In this paper, a UWB-BPF with a low insertion loss, high band isolation and a wide upper stopband was reported. A simple method and structure were presented to realize the proposed UWB filter by using two half-wavelength high-impedance line resonators. The two UWB performances in the 3.4–5.0 and 6.1–10.0 GHz ranges are controlled individually and thus a notch band at 5.2 GHz is achieved to avoid signal interference with the WLAN signals. With the adoption of two tapped open stubs, the attenuation poles are located near the passband to suppress undesired spurious responses, causing a wide upper stopband. The measured results agree with the simulated results, and actually verify the design concept.

Author Contributions: M.-H.W. contributed the design concept and wrote the paper, C.-W.H. and S.-W.L. simulated, fabricated and measured the filter, and R.-Y.Y. provided the design resource and advised the design procedure.

Funding: This research received no external funding.

Conflicts of Interest: The authors declare no conflict of interest.

References

1. Federal Communication Commission. *Revision of Part 15 of the Commission's Rules Regarding Ultra-Wideband Transmission Systems*; ET-Docket 98-153, FCC02-48; Federal Communication Commission: Washington, DC, USA, 2002.
2. Kim, C.H.; Chang, K. Ultra-wideband (UWB) ring resonator bandpass filter with a notched band. *IEEE Microw. Wirel. Compon. Lett.* **2011**, *21*, 206–208. [CrossRef]
3. Hao, Z.C.; Hong, J.S.; Parry, J.P.; Hand, D.P. Ultra-wideband bandpass filter with multiple notch bands using nonuniform periodical slotted ground structure. *IEEE Trans. Microw. Theory Technol.* **2009**, *57*, 3080–3088.
4. Hung, C.Y.; Weng, M.H.; Su, Y.K. Design of compact and sharp rejection UWB BPFs using interdigital stepped-impedance resonators. *IEICE Electron. Lett.* **2007**, *90*, 1652–1654. [CrossRef]
5. Wong, S.W.; Zhu, L. Implementation of compact UWB bandpass filter with a notch-band. *IEEE Microw. Wirel. Compon. Lett.* **2008**, *18*, 10–12. [CrossRef]
6. Chang, Y.C.; Kao, C.H.; Weng, M.H.; Yang, R.Y. Design of the compact wideband bandpass filter with low loss, high selectivity and wide stopband. *IEEE Microw. Wirel. Compon. Lett.* **2008**, *18*, 187–189. [CrossRef]
7. Song, Y.; Yang, G.M.; Geyi, W. Compact UWB bandpass filter with dual notched bands using defected ground structures. *IEEE Microw. Wirel. Compon. Lett.* **2014**, *24*, 230–232. [CrossRef]
8. Liu, J.B.; Ding, W.H.; Chen, J.H.; Zhang, A. New ultra-wideband filter with sharp notched band using defected ground structure. *Prog. Electromagn. Res. Lett.* **2019**, *83*, 99–105. [CrossRef]
9. Deng, K.; Feng, W. Wideband bandpass filter with multiple transmission zeros and compact size. *Microw. Opt. Technol. Lett.* **2016**, *58*, 2452–2455. [CrossRef]
10. Zhang, Z.C.; Liu, H. A ultra compact wideband bandpass filter using a quadmode stub-loaded resonator. *Prog. Electromagn. Res. Lett.* **2018**, *77*, 35–40. [CrossRef]
11. Li, Y.; Choi, W.W.; Tam, K.W.; Zhu, L. Novel wideband bandpass filter with dual notched bands using stub-loaded resonators. *IEEE Microw. Wirel. Compon. Lett.* **2017**, *27*, 25–27.
12. Choudhary, D.K.; Chaudhary, R.K. A compact via-less metamaterial wideband bandpass filter using split circular rings and rectangular stub. *Prog. Electromagn. Res. Lett.* **2018**, *72*, 99–106. [CrossRef]
13. Ji, X.C.; Ji, W.S.; Feng, L.Y.; Tong, Y.Y.; Zhang, Z.Y. Design of a novel multi-layer wideband bandpass filter with a notched band. *Prog. Electromagn. Res. Lett.* **2019**, *82*, 9–16. [CrossRef]
14. Li, Z.; Wu, K.L. Direct synthesis and design of wideband bandpass filter with composite series and shunt resonators. *IEEE Trans. Microw. Theory Technol.* **2017**, *65*, 3789–3800. [CrossRef]
15. Hameed, M.; Xiao, G.; Qiu, L.; Xiong, C.; Hameed, T. Multiple-mode wideband bandpass filter using split ring resonators in a rectangular waveguide cavity. *Electronics* **2018**, *7*, 356. [CrossRef]
16. Hameed, M.; Xiao, G.; Najam, A.I.; Qiu, L.; Hameed, T. Quadruple-mode wideband bandpass filter with improved out-of-band rejection. *Electronics* **2019**, *8*, 300. [CrossRef]
17. Zeland Software, Inc. *IE3D Simulator*; Zeland Software, Inc.: Fremont, CA, USA, 2002.

18.   Ye, C.S.; Su, Y.K.; Weng, M.H.; Hung, C.Y.; Tang, R.Y. Design of the compact parallel-coupled lines wideband bandpass filters using image parameter method. *Prog. Electromagn. Res.* **2010**, *100*, 153–173. [CrossRef]
19.   Hong, J.-S.; Lancaster, M.J. *Microstrip Filters for RF/Microwave Applications*; Wiley: New York, NY, USA, 2001.
20.   Pozar, D.M. *Microwave Engineering*, 4th ed.; Wiley: New York, NY, USA, 2012; ISBN 978-0-470-63155-3.

Article

Polarization-Independent Tunable Ultra-Wideband Meta-Absorber in Terahertz Regime

Shuxiang Liu, Li Deng, Meijun Qu and Shufang Li *

Beijing Key Laboratory of Network System Architecture and Convergence, Beijing University of Posts and Telecommunications, Beijing 100876, China
* Correspondence: bupt_paper@126.com

Received: 23 May 2019; Accepted: 17 July 2019; Published: 26 July 2019

Abstract: In this paper, we demonstrate an ultra-broadband terahertz bilayer graphene-based absorption structure. It has two stacking graphene layers sandwiched by an Au cylinders array, backed by a metallic ground plane. Au cylinders are used to adjust the input impedance to be closely matched to the free space, enabling an ultra-broadband absorption. The absorption spectrum of the bilayer graphene-based absorption structure with Au cylinder arrays shows a bandwidth of 7.1 THz, with the absorption exceeding 80%. The achieved ultra-wideband THz meta-absorber has high absorption, independence of polarization property, simultaneously, illustrating to be a promising candidate for teraherz broadband absorption application.

Keywords: ultra-wideband; THz absorber; high absorption; polarization independent

---

1. Introduction

The terahertz (THz) range of the electromagnetic spectrum, located between microwave and infrared band, has received fast growing research interest in the past two decades, due to extensive applications such as safety inspection, imaging, sensing, material detection, secure communication and so on in virtue of its unique properties [1–5]. The THz absorbing structure has been part of important elements in these THz applications. Limited by the lack of natural absorbing materials in the terahertz gap, metamaterial absorbers (MAs), where the periodic elements are much smaller than the working wavelength of THz waves, have been considered to be an effective solution [6–10]. Therefore, the emergence of terahertz metamaterials brings new opportunities for the development and application of terahertz absorbing technology. MAs are generally realized by using lossy materials in periodic patterns. According to the equivalent medium theory, by reducing transmission and reflection, the energy of the electromagnetic wave is confined in the material, thus illustrating absorption characteristics.

However, THz MAs studied to date still exhibit compromised performance between broadband operation, high absorption, polarization dependence, angle insensitivity and structural complexity. Long Ju and Yongzhi Cheng proposed the polarization independent tunable absorber, however, the mechanism of metamaterial resonance absorption still makes the absorption bandwidth of metamaterials narrow and limit the practical application [11,12]. Yannan Jiang et al. studied a wide-angle incidence Tunable Terahertz Absorber, but the tunable range is only about 1 THz [13]. In addition, dual-band, multi-band and broadband terahertz metamaterial absorber has attracted more attention from engineering researchers [14–17]. To achieve broadband absorption, one way is tantamount to merge multiple resonators in one cell [18–21]. Another strategy is used by stacking structure or cascading the multi-layered sheet with different frequency responses separated by dielectric with different thicknesses [22–24]. Delin Jia designed a broadband THz bi-metasurfaces absorber composed of two stacking metasurfaces [25]. The absorber combines the above two methods by horizontally arranging multiplexed resonators on each planar layer and then vertically stacking the two layers. However, the absorption rate is lower than 60% at the middle part of the spectrum, and lack of the wide-angle incidence characteristics.

Motivated by the methods and structures of researches to achieve broad angle [26,27], and high absorption [28,29], and driven by the essential need for polarization-independent ultra-wideband meta-absorber, we theoretically investigate a bilayer graphene-based absorption structure with Au cylinder arrays on parylene substrate. We use the cascading bilayer graphene structure with two different frequency resonances. The first graphene layer achieves wave absorption in low frequency band, the second graphene layer achieves wave absorption in high frequency band. Due to the two frequency bands are close to each other, we put the two graphene layers together with cascading structure. To achieve as high absorption as possible, we need to have the input impedance of the whole structure to be as close to the impedance of free space. However, the cascading structure suffers an unsatisfactory impedance matching, then the metallic cylinders are attached to improve the impedance matching of the whole structure. Finally, we have the ultra-broadband Meta-absorber about 7.1 THz with absorption higher than 80% with polarization insensitive for both TE and TM polarizations.

2. Models and Design

The schematic of the proposed absorption structure is depicted in Figure 1. The whole structure consists of cylinders array with the first graphene on the top, and substrate with second graphene on the top and the metallic layer on the bottom. Both the first graphene layer and the second graphene layer use the gradient diameter of the graphene sheet as the resonance unit cell. Aiming to obtain absorption peaks at low frequency, the period of the top graphene is 75 μm, employing one outer annulus with outer radius $R_1 = 35.5$ μm, inner annulus with outer radius $R_2 = 23.5$ μm, and inner circle disk with radius $R_3 = 17.5$ μm. The gap between outer annulus with inner annulus and inner annulus with inner circle disk are $g_1 = 1$ μm and $g_2 = 1$ μm, respectively. To get an absorption targeting at high spectrum, the second graphene has an 18.75 μm periodicity of resonators array with a 25% fill factor, thus making the bilayer graphene-based absorber has a periodicity P of 75 μm. The bottom graphene layer consists of 16 identical circle disks, each one has one outer annulus with outer radius $R_4 = 8.47$ μm, and inner annulus with outer radius $R_5 = 5.83$ μm, and inner circle disk with radius $R_6 = 2.84$ μm. The gap between outer annulus with inner annulus and inner annulus with inner circle disk are $g_4 = 0.4$ μm and $g_3 = 1.1$ μm, respectively. As shown in Figure 1, four Au cylinders as resonators are periodically arranged in both the x- and y-directions. The radius of the Au cylinder, and the height of the Au cylinder are denoted as $R = 3.75$ μm and $DH = 16.8$ μm, respectively. At the bottom layer, Au is also selected as the ground plate, whose conductivity is described by Drude model as $\epsilon(\omega) = \epsilon_\infty - \omega_p^2/(\omega^2 + i\rho\omega)$, with $\epsilon_\infty = 1.0$, $\omega_p = 4.35\pi \times 10^{15}$ s^{-1} and $\rho = 8.17 \times 10^{13}$ s^{-1} [30]. The low loss parylene with thickness $UH = 8.75$ μm is chosen as the substrate. A permittivity $\epsilon = 2.60$ and loss $tan\delta = 0.04$ are used to model the parylene film. The electrical vector of plane wave is parallel to x-direction in simulation. The bilayer graphene can be modeled as a 2-D surface, with surface conductivity of top layer δ_{g1} for top graphene layer and δ_{g2} for second graphene layer. In the terahertz range, for both δ_{g1} and δ_{g2}, governed by the Kubo formula including the interband and intraband transition contributions [11]. The contribution of the interband δ_{inter} is negligible compared with the intraband δ_{intra}, which can be expressed as:

$$\delta_{intra1}(\omega_1, \mu_{c1}, \Gamma_1, T) \approx \frac{-je^2 k_B T}{\pi \hbar^2 (\omega_1 - j2\Gamma_1)} \left[\frac{\mu_{c1}}{k_B T} + 2ln(e^{-\mu_{c1}/(k_B T)} + 1) \right]. \tag{1}$$

$$\delta_{intra2}(\omega_2, \mu_{c2}, \Gamma_2, T) \approx \frac{-je^2 k_B T}{\pi \hbar^2 (\omega_2 - j2\Gamma_2)} \left[\frac{\mu_{c2}}{k_B T} + 2ln(e^{-\mu_{c2}/(k_B T)} + 1) \right]. \tag{2}$$

where k_B is the Boltzmann's constant, T is the temperature in Kelvin, e is the electron charge, and μ_{c1}, μ_{c2} is the chemical potential. Moreover, ω_1, ω_2 are the angular frequency, $\hbar$ is the reduced Planck's constant, τ_1, τ_2 is the electron relaxation time, and $\Gamma_1 = 1/(2\tau_1)$, $\Gamma_2 = 1/(2\tau_2)$ are the electron scattering rate. Here assuming $T = 300\ K$ and $\tau_1 = 0.09\ ps$, $\tau_2 = 0.06\ ps$. The whole structure, including the bilayer graphene and four Au cylinder arrays, is objected to possess the symmetry principle, ensuring the insensitivity to the polarization of electromagnetic waves. The dimensions of the whole structure are optimized with HFSS software. The absorption rate is described as $A(\omega) = 1 - R(\omega) - T(\omega)$, where $A(\omega)$, $R(\omega)$, and $T(\omega)$ represent frequency dependent absorptivity, reflectivity, and transmission rate, respectively. Moreover, the transmissivity of the structure $T(\omega)$ is equal to 0 due to the gold-backed slab ($n = 200\ nm$), leading to the absorption rate $A(\omega) = 1 - R(\omega) = 1 - |S_{11}|^2$.

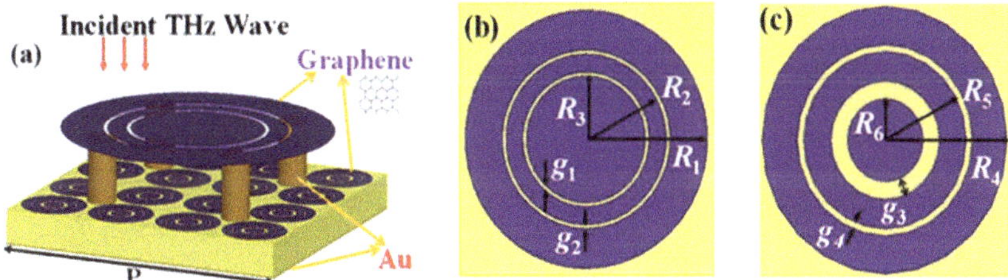

Figure 1. Proposed meta-absorber with bilayer graphene and Au cylinders. (**a**) Schematic of the proposed absorber. (**b**) Top view of the unit cell of top layer, (**c**) Top view of the unit cell of second graphene layer.

3. Simulation and Results

In the simulation, the first graphene layer is set as impedance boundary1 with resistance1: $1/coff_1/\tau_1$, and reactance1: $2\pi Freq/coff_1$. In addition, the second graphene layer is set as impedance boundary2 with resistance2: $1/coff_2/\tau_2$, and reactance2: $2\pi Freq/coff_2$. Where, $coff_1 = e_0^2 k_B T/\pi/\hbar^2(\mu_{c1}/k_B/T + 2ln(e^{(-\mu_{c1}/k_B/T)} + 1))$, $coff_2 = e_0^2 k_B T/\pi/\hbar^2(\mu_{c2}/k_B/T + 2ln(e^{(-\mu_{c2}/k_B/T)} + 1))$.

Master-Slave boundary conditions are used to represent the periodicity of the structure. As shown in Figure 2a, the simulation model of one unit cell is based on two FloquetPorts air-filled waveguide with master–slave boundary conditions for numerically computing the absorption rate of the absorber. FloquetPorts are set for the two-sides of the z-direction, while periodic boundary condition is used in x/y-direction.

The simulated absorption spectrum of only the first graphene absorber is plotted in Line 1 of Figure 2b, the absorption peaks are at 1.9 THz and 3.3 THz. The simulation results show that the absorption peaks are 93% and 97%, respectively. The simulated absorption peak of the second graphene layer is at 6 THz with an absorption of 81%, as shown in Line 2 of Figure 2b. When we combine the first and second graphene layers sandwiched by parylene, a low ebb occurs, the absorber appears to have dual-band characteristics, as presented in Line 4 of Figure 2b. It may be caused by the mismatch of the impedance. Then, the diagrammatic sketch of an equivalent circuit model of the absorber is drawn in Figure 3. The principle can be started with considering the bottom metallic ground with acceptable approximation as a short circuit. The dielectric layer can be characterized as an ideal transmission line, which represents a capacitance or inductive element. The impedance of the periodic graphene is composed of an infinite number of RLC circuits, each representing a mode of graphene. In this equivalent circuit model, Z_0 represents free space impedance and Z_g1, Z_g2 represent the impedance of the first and second graphene layer in the structure. The goal of the following design is to set the real part of the input impedance approximate to Z_0 through optimizing the parameters of bilayer graphene and cylinders array. It is easy to find that the impedance matching improves a lot when the bilayer structure are sandwiched by the cylinders array, as depicted in Figure 4a. To validate the influence of Au cylinders array, various γ, the ratio of period P and the radius R of Au cylinders, are simulated.

As plotted in Figure 4b, that brings obvious affections to the absorption performance, when γ is chosen as 0.2, the proposed structure obtains the optimum, with absorption higher than 80% in about 7.1 THz.

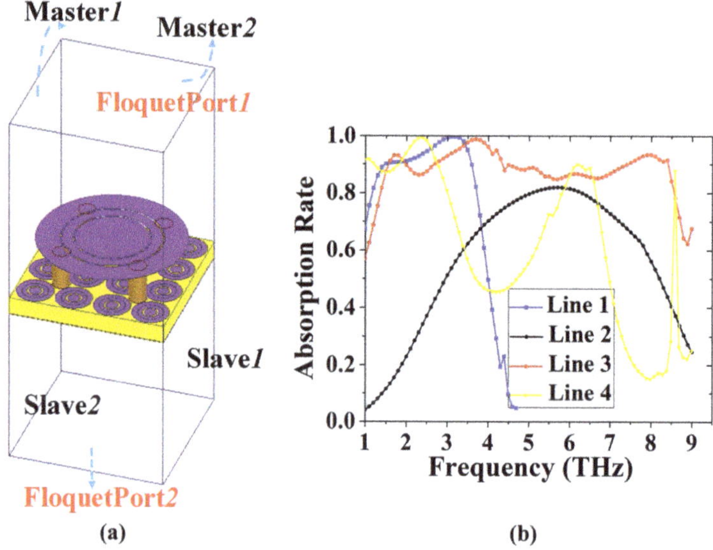

Figure 2. (a) Boundary and Excitation setting for the unit cell of the proposed absorber, (b) The simulated frequency responses for Line 1: with the first graphene layer, Line 2: with the second graphene layer, Line 3: the proposed absorber, Line 4: without the cylinders array.

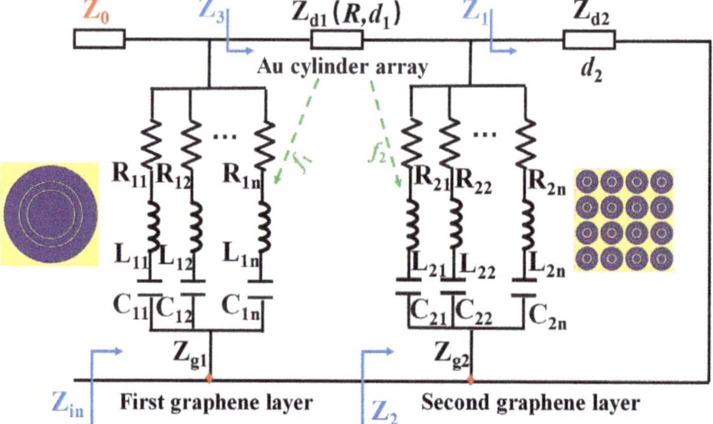

Figure 3. The diagrammatic sketch of an equivalent circuit model of the proposed absorber.

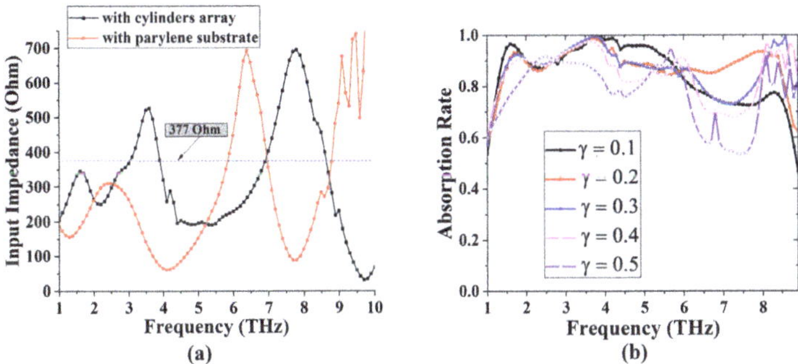

Figure 4. (a) Comparison of input impedance with cylinders array and with parylene substrate. (b) The simulated frequency responses with different γ.

4. Discussion

To reveal the mechanism inducing such high absorption, the electric field distribution and surface current distribution of the absorber will be provided. Simulated electric field distribution, surface current density of the first graphene layer and surface current density of the backed Au ground at $f_1 = 3.8$ THz are plotted in Figure 5. Apparently, electric resonance occurs in the edge of both the outer Annulus and the inner Annulus, and the transient current directions of the first graphene layer and the ground plane are just opposite, forming a magnetic dipole response at 3.8 THz. As for the second graphene layer, the absorption peak is obtained at $f_3 = 6$ THz with the absorption of 81.73%, as shown in Figure 2b. The surface current distribution on the second graphene layer and backed Au ground layer are next investigated to further explore the high spectrum absorption. Obviously, there exists an electric resonance on each super unit cell, as depicted in Figures 6 and 7 offers that the directions of transient current of the second graphene layer and the ground plane are also reversed, emerging the magnetic dipole response at 6 THz.

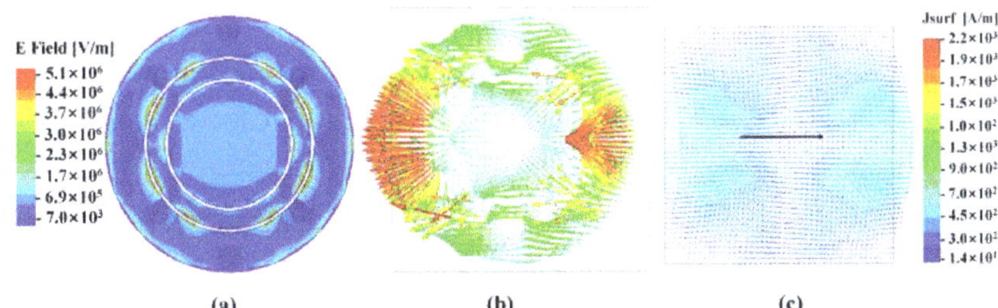

Figure 5. The simulated (a) electric field distribution and (b) surface current density of the first graphene layer and (c) surface current density of the backed Au ground at $f_1 = 3.8$ THz.

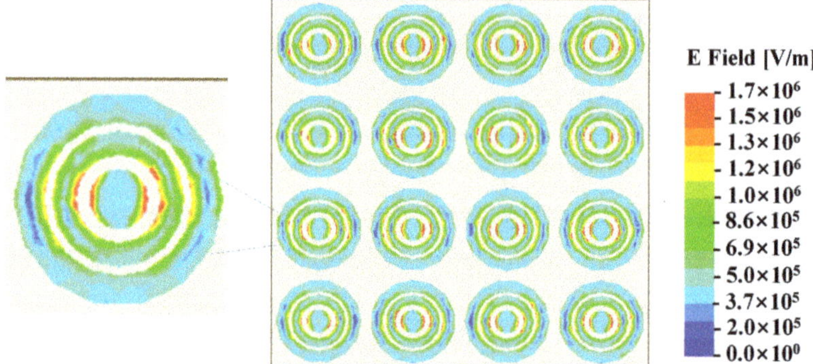

Figure 6. The simulated electric field distribution of the second graphene layer at $f_2 = 6$ THz.

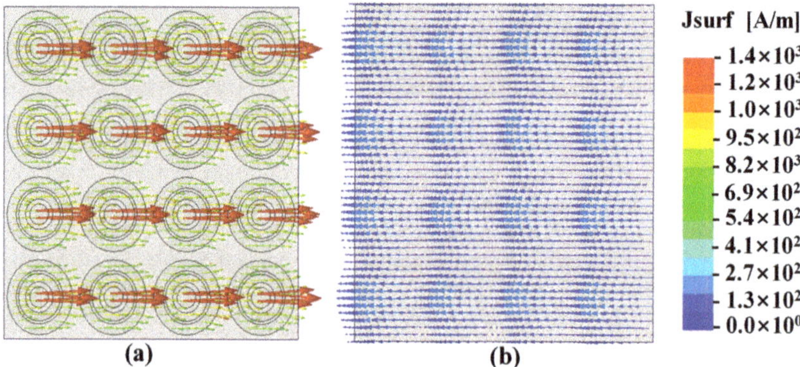

(a)   **(b)**

Figure 7. The simulated surface current density of (**a**) the second graphene layer and (**b**) the backed ground at the frequency of $f_2 = 6$ THz.

To further explain the mechanism of improved absorption, electromagnetic field profile for the structures with and without graphene are delineated, as shown in Figure 8. At 3.7 THz, it is easy to find that few changes appear, the absorption holds normal at lower spectrum. At 6.5 THz, we can catch the sight of local electric field enhancement, which could explain the ascendant absorption at higher spectrum. In addition, at 7.6 THz, upon the basis of cylinders' resonance, interaction within graphene and metal columns is perceived, with improved absorption rate of high spectrum. Compared with the recently reported THz broadband absorber with a bandwidth of 3.31 THz while absorption exceeding 50% [25], the bandwidth of our proposed absorber has wider bandwidth of 7.1 THz and higher absorption of 80%. Furthermore, the electrically tunable property of the ultra-broadband absorber is investigated. The surface conductivity of graphene sheet relates largely to its Fermi energy, which can be controlled by electrostatic doping or applying bias voltage. Through varying the Fermi level of the graphene sheet located on the separate layer, the absorption amplitude of low spectrum and high spectrum can be independently controlled with almost unchanging influence of adjacent band, as shown in Figure 9. It should be noted that through changing the chemical potential of graphene, the conductivity of graphene can be changed correspondingly. Thus by tuning the chemical potential of graphene can indirectly change impedance matching. Trough changing the impedance matching, we can flexibly obtain the demanding absorption bandwidth. When the chemical potential of the first layer u_{c1} goes down, the absorption for the simulated frequency response of the lower spectrum goes down too, vice versa. In addition, when the chemical potential of the second layer u_{c2} goes down, the absorption for the simulated frequency response of the upper spectrum goes down too. Only when we choose the optimized value of combination of the two variables, u_{c1}, u_{c2}, can we get the optimal bandwidth.

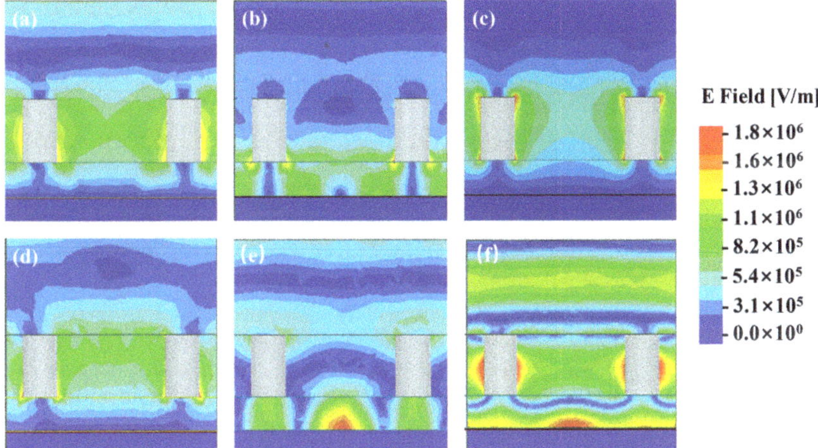

Figure 8. The electromagnetic field profile for the structures, (**a**) without graphene at 3.7 THz, (**b**) without graphene at 6.5 THz, (**c**) without graphene at 7.6 THz, (**d**) with graphene at 3.7 THz, (**e**) with graphene at 6.5 THz, (**f**) with graphene at 7.6 THz.

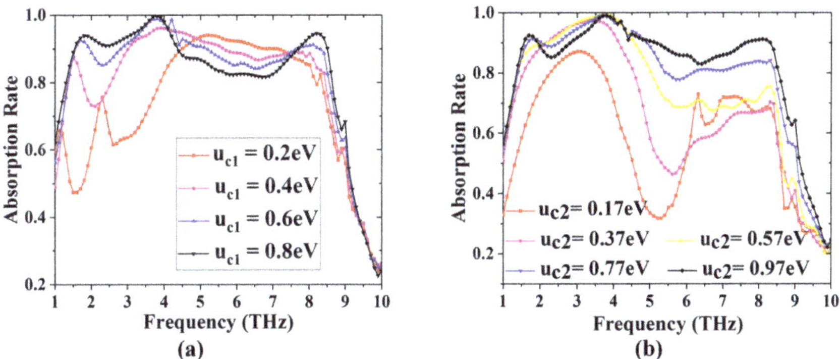

Figure 9. The simulated absorption spectrum with changing graphene loading voltage of (**a**) the first graphene layer, (**b**) the second graphene layer.

Next, we will discuss the characteristics of independence of polarization angle. The polarization angle is defined as included angle between x-axis and electric vector of plane wave. In the above mentioned analysis, we discussed the characteristics only under normal incidence scenario. We have investigated the absorption consistency of variational trend under the changes of the polarization angle ranging from 0° to 90° in steps of 15°, at varying incidence angle from 0° to 60°, with a step width of 15°. As shown in Figure 10, little variance of absorbance emerges for the structure, illustrating highly consistent absorption performance under various polarization angles of incident plane wave. In other words, the absorption performance of the structure is independent of polarization. It has a great advantage in many THz applications where non-polarized source is preferred to achieve high absorption efficiency.

Additionally, the absorption spectrum under varied oblique incidence in both the TE and TM modes are studied, as depicted in Figure 11a,b, respectively. In the simulation, the incidence angles vary from $0°$ to $80°$ with the step width of $10°$. It can be presented that the peak absorption keeps larger than 70% up to $70°$ incidence angles for both the TE mode and the TM mode, respectively.

In addition, regarding the angular stability study, both TE- and TM-polarized incident plane waves are considered. The angular stability of the proposed absorber under oblique incidence is determined by the thickness of the metallic cylinders and the substrate, the geometry and the size of the unit-cell (periodicity) compared to the whole thickness of the structure [31]. However, it should be aware that these factors not only influence the angular stability but also the absorbing bandwidth. It can be explained from Figure 3, the impedance of Z_{d1} is related to the thickness of the metallic cylinders and Z_{d2} has relation with the thickness of the substrate. Changing d_1 and d_2 can affect the input impedance of the whole structure, further influencing the absorbing bandwidth. Thus, a trade-off solution has to be optimized for the values of the thickness and the absorption bandwidth. Commercial Software HFSS of Ansoft based on finite elements methods (FEM) and Quasi Newton algorithm are chosen to achieve the optimization process. First, we set the incidence angle (theta) varied from $0°$ to $70°$ in steps of $10°$, at varying polarization angles from $0°$ to $80°$, with a step width of $20°$. The comparison is implemented in terms of the absolute bandwidth, comprising the frequencies for which the absorption of the structure keeps higher than 70%, no matter the phi and theta angles in the considered range. As shown in Tables 1 and 2, when theta is lower than $50°$, the bandwidth has wide bandwidth with absorption higher than 70%. In addition, when the theta angle is higher than $50°$, it is obvious that the absorption bandwidth reduces.

Table 1. Absorption bandwidth under the changes of the incident angle theta for TE incident wave.

Bandwidth (THz)	phi = 0°	phi = 20°	phi = 40°	phi = 60°	phi = 80°
theta = 0°	7.3	7.4	7.3	7.4	7.4
theta = 10°	7.4	7.5	7.4	7.4	7.5
theta = 20°	7.6	7.7	7.6	7.6	7.7
theta = 30°	7.9	7.9	7.9	7.9	7.9
theta = 40°	8.3	8.0	8.3	8.1	6.7
theta = 50°	6.2	4.3	6.2	4.9	4.7
theta = 60°	2.7	2.4	2.7	2.5	1.8
theta = 70°	1.3	1.0	1.3	1.2	0.9

Table 2. Absorption bandwidth under the changes of the incident angle theta for TM incident wave.

Bandwidth (THz)	phi = 0°	phi = 20°	phi = 40°	phi = 60°	phi = 80°
theta = 0°	7.3	7.4	7.3	7.4	7.4
theta = 10°	7.4	7.5	7.4	7.4	7.5
theta = 20°	7.6	7.8	7.6	7.6	7.7
theta = 30°	8.2	8.0	8.2	8.2	7.9
theta = 40°	8.4	8.6	8.4	8.6	7.8
theta = 50°	8.6	8.3	8.6	6.9	8.1
theta = 60°	7.4	8.0	7.4	7.5	7.6
theta = 70°	3.9	3.4	3.9	3.5	3.5

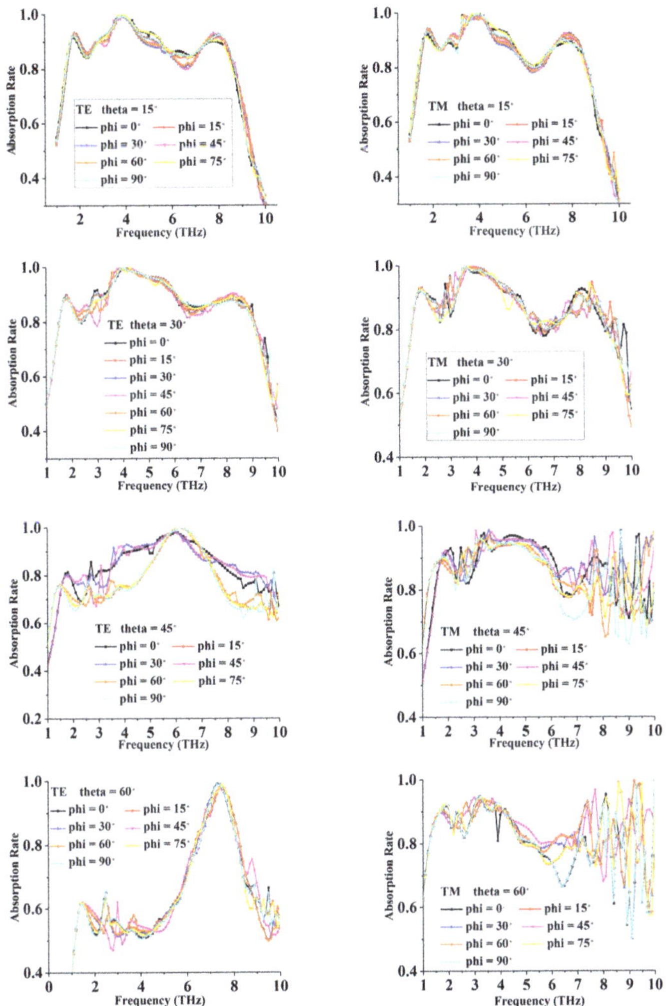

Figure 10. Simulated absorption spectra of the absorber for different polarization angles under various incidence.

As shown in Figure 10, for different oblique illuminations, we can clearly find that when the incident angle is above 60°, the bandwidth with absorption higher than 70% reduced a lot. It may be explained that the sheet impedance of graphene does not change with the incident angle and polarization states, but as the incident angle of plane wave changes, projection impedance of incident wave varies, the impedance mismatch becomes worse. So the meta-absorber has an incident angle about 45° with broadband absorption higher than 70%.

Finally, we have achieved a bilayer graphene-based absorption structure with Au cylinder arrays on parylene substrate. It can achieve high absorption rate of 80% within 7.1 THz with polarization-insensitive both TE and TM polarizations, and an incident-angle about 45° with broadband absorption higher than 70%.

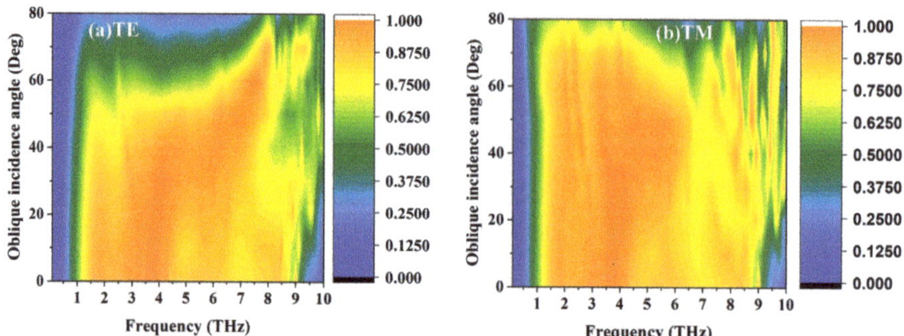

Figure 11. Absorption contour map of the absorber under various incident angles (**a**) TE mode and (**b**) TM mode.

5. Conclusions

In summary, a bilayer graphene-based absorption structure with Au cylinder arrays on parylene substrate is proposed and numerically simulated at THz wavelengths. We use the cascading bilayer graphene structure with two different frequency resonances. The first graphene layer achieves wave absorption in low frequency band, the second graphene layer achieves wave absorption in high frequency band. Ultra-wideband is realized when the guided resonance is excited in the Au cylinder array, promoting input impedance of the whole structure to closely match with the free space, connecting the low frequency and high frequency absorption bands and broadening the whole spectrum. The bandwidth and resonance position of the absorption resonance can be independently tuned by varying the chemical potential of separate graphene layers. Simulation results demonstrate that the absorption of the proposed structure can be as high as more than 80% from 2.9 THz to 7.1 THz with variation of the graphene's chemical potential from 0.2 eV to 0.97 eV with polarization insensitive for both TE and TM polarizations stable against frequency.

Author Contributions: Conceptualization, S.L. (Shuxiang Liu) and L.D.; methodology, S.L. (Shuxiang Liu); software, S.L. (Shuxiang Liu); validation, S.L. (Shuxiang Liu); formal analysis, S.L. (Shuxiang Liu); investigation, S.L. (Shuxiang Liu); resources, L.D. and S.L. (Shufang Li); data curation, S.L. (Shuxiang Liu); writing—original draft preparation, S.L. (Shuxiang Liu); writing—review and editing, S.L. (Shuxiang Liu), M.Q. and S.L. (Shufang Li); visualization, S.L. (Shuxiang Liu); supervision, L.D.; project administration, L.D. and S.L. (Shufang Li); funding acquisition, L.D. and S.L. (Shufang Li).

Funding: This research was funded by National Nature Science Foundation of China No. 61601040 and 61427801.

Acknowledgments: This work was supported by National Nature Science Foundation of China (No. 61601040 and 61427801), 111 project (No. B17007) and Director Funds of Beijing Key Laboratory of Network System Architecture and Convergence (No. 2017BKL-NSAC-ZJ-01), Beijing Nova Program (No. Z181100006218039).

Conflicts of Interest: The authors declare no conflict of interest.

References

1. Zhang, C.; Huang, C.; Pu, M.; Song, J.; Zhao, Z.; Wu, X.; Luo, X. Dual-band wide-angle metamaterial perfect absorber based on the combination of localized surface plasmon resonance and Helmholtz resonance. *Sci. Rep.* **2017**, *7*, 5652. [CrossRef] [PubMed]

2. Stantchev, R.I.; Phillips, D.B.; Hobson, P.; Hornett, S.M.; Padgett, M.J.; Hendry, E. Compressed sensing with near-field THz radiation. *Optica* **2017**, *4*, 989–992. [CrossRef]

3. Chen, M.; Singh, L.; Xu, N.; Singh, R.; Zhang, W.; Xie, L. Terahertz sensing of highly absorptive water-methanol mixtures with multiple resonances in metamaterials. *Opt. Express* **2017**, *25*, 14089–14097. [CrossRef] [PubMed]

4. Tao, H.; Landy, N.I.; Bingham, C.M.; Zhang, X.; Averitt, R.D.; Padilla, W.J. A metamaterial absorber for the terahertz regime: Design, fabrication and characterization. *Opt. Express* **2008**, *16*, 7181–7188. [CrossRef] [PubMed]

5. Cong, L.; Xu, N.; Zhang, W.; Singh, R. Polarization Control in Terahertz Metasurfaces with the Lowest Order Rotational Symmetry. *Adv. Opt. Mater.* **2015**, *3*, 1176–1183. [CrossRef]

6. Yen, T.J.; Padilla, W.J.; Fang, N.; Vier, D.C.; Smith, D.R.; Pendry, J.B.; Basov, D.N.; Zhang, X. Terahertz magnetic response from artificial materials. *Science* **2004**, *303*, 1494–1496. [CrossRef] [PubMed]

7. Zhang, S.; Park, Y.S.; Li, J.; Lu, X.; Zhang, W.; Zhang, X. Negative refractive index in chiral metamaterials. *Phys. Rev. Lett.* **2009**, *102*, 023901. [CrossRef] [PubMed]

8. Singh, R.; Plum, E.; Zhang, W.; Zheludev, N.I. Highly tunable optical activity in planar achiral terahertz metamaterials. *Opt. Express* **2010**, *18*, 13425–13430. [CrossRef] [PubMed]

9. Bahk, Y.M.; Kang, B.J.; Kim, Y.S.; Kim, J.Y.; Kim, W.T.; Kim, T.Y.; Kang, T.; Rhie, J.; Han, S.; Park, C.H.; et al. Electromagnetic saturation of angstrom-sized quantum barriers at terahertz frequencies. *Phys. Rev. Lett.* **2015**, *115*, 125501. [CrossRef]

10. Yoshida, K.; Shibata, K.; Hirakawa, K., Terahertz field enhancement and photon-assisted tunneling in single-molecule transistors. *Phys. Rev. Lett.* **2015**, *115*, 138302. [CrossRef]

11. Ju, L.; Geng, B.; Horng, J.; Girit, C.; Martin, M.; Hao, Z.; Bechtel, H.A.; Liang, X.; Zettl, A.; Shen, Y.R.; et al. Graphene plasmonics for tunable terahertz metamaterials. *Nat. Nanotechnol.* **2011**, *6*, 630. [CrossRef] [PubMed]

12. Cheng, Y.; Gong, R.; Cheng, Z. A photoexcited broadband switchable metamaterial absorber with polarization-insensitive and wide-angle absorption for terahertz waves. *Opt. Commun.* **2016**, *361*, 41–46. [CrossRef]

13. Jiang, Y.; Wan, X.; Wang, J.; Wang, J. Tunable Terahertz Absorber Based on Bulk-Dirac-Semimetal Metasurface. *IEEE Photonics J.* **2018**, *10*, 1–7. [CrossRef]

14. Zhang, J.; Tian, J.; Li, L. A Dual-Band Tunable Metamaterial Near-Unity Absorber Composed of Periodic Cross and Disk Graphene Arrays. *IEEE Photonics J.* **2018**, *10*, 1–12. [CrossRef]

15. Deng, G. Triple-band polarisation-independent metamaterial absorber at mm wave frequency band. *IET Microw. Antennas Propag.* **2018**, *12*, 1120–1125. [CrossRef]

16. Huang, X.; Lu, C.; Rong, C.; Hu, Z.; Liu, M. Multiband Ultrathin Polarization-Insensitive Terahertz Perfect Absorbers With Complementary Metamaterial and Resonator Based on High-Order Electric and Magnetic Resonances. *IEEE Photonics J.* **2018**, *10*, 1–11. [CrossRef]

17. Dorche, A.E.; Abdollahramezani, S.; Chizari, A.; Khavasi, A. Broadband, Polarization-Insensitive, and Wide-Angle Optical Absorber Based on Fractal Plasmonics. *IEEE Photonics Technol. Lett.* **2016**, *28*, 2545–2548. [CrossRef]

18. Zhang, Y.; Li, Y.; Cao, Y.; Liu, Y.; Zhang, H. Graphene induced tunable and polarization-insensitive broadband metamaterial absorber. *Opt. Commun.* **2017**, *382*, 281–287. [CrossRef]

19. Guo, W.; Liu, Y.; Han, T. Ultra-broadband infrared metasurface absorber. *Opt. Express* **2016**, *24*, 20586–20592. [CrossRef]

20. Xiong, H.; Wu, Y.B.; Dong, J.; Tang, M.C.; Jiang, Y.N.; Zeng, X.P. Ultra-thin and broadband tunable metamaterial graphene absorber. *Opt. Express* **2018**, *26*, 1681–1688. [CrossRef]

21. Torabi, E.S.; Fallahi, A.; Yahaghi, A. Evolutionary Optimization of Graphene-Metal Metasurfaces for Tunable Broadband Terahertz Absorption. *IEEE Trans. Antennas Prog.* **2017**, *65*, 1464–1467. [CrossRef]

22. Dong Y.; Liu, P.; Yu, D.; Li, G.; Yang, L. A Tunable Ultrabroadband Ultrathin Terahertz Absorber Using Graphene Stacks. *IEEE Antennas Wirel. Propag. Lett.* **2017**, *16*, 1115–1118. [CrossRef]

23. Fardoost, A.; Vanani, F.G.; Safian, R. Design of a Multilayer Graphene-Based Ultrawideband Terahertz Absorber. *IEEE Trans. Nanotechnol.* **2017**, *16*, 68–73.

24. Pan, W.; Yu, X.; Zhang, J.; Zeng, W. A Novel Design of Broadband Terahertz Metamaterial Absorber Based on Nested Circle Rings. *IEEE Photonics Technol. Lett.* **2016**, *28*, 2335–2338. [CrossRef]

25. Jia, D.; Xu, J.; Yu, X. Ultra-broadband terahertz absorption using bi-metasurfaces based multiplexed resonances. *Opt. Express* **2018**, *26*, 26227. [CrossRef] [PubMed]

26. Sayem, A.A.; Mahdy, M.R.C.; Rahman, M.S. Broad angle negative refraction in lossless all dielectric or semiconductor based asymmetric anisotropic metamaterial. *J. Opt.* **2016**, *18*, 015101. [CrossRef]

27. Cong, L.; Pitchappa, P.; Wu, Y.; Ke, L.; Lee, C.; Singh, N.; Yang, H.; Singh, R. Active Multifunctional Microelectromechanical System Metadevices: Applications in Polarization Control, Wavefront Deflection, and Holograms. *Adv. Opt. Mat.* **2016**, *5*, 1600716. [CrossRef]
28. Cong, L.; Tan, S.; Yahiaoui, R.; Yan, F.; Zhang, W.; Singh, R. Experimental demonstration of ultrasensitive sensing with terahertz metamaterial absorbers: A comparison with the metasurfaces. *Appl. Phys. Lett.* **2015**, *106*, 031107. [CrossRef]
29. Cong, L.; Pitchappa, P.; Lee, C.; Singh, R. Active Phase Transition via Loss Engineering in a Terahertz MEMS Metamaterial. *Adv. Mater.* **2017**, *29*, 1700733. [CrossRef]
30. Liu, N.; Liu, H.; Zhu, S.; Giessen, H. Stereometamaterials. *Nat. Photonics* **2009**, *3*, 157–162. [CrossRef]
31. De Cos, M.E.; Las-Heras, F. On the advantages of loop-based unit-cell's metallization regarding the angular stability of artificial magnetic conductors. *Appl. Phys. A* **2015**, *118*, 699–708. [CrossRef]

 **electronics**

Article

Impact of IT Devices Production Quality on the Level of Protection of Processed Information against the Electromagnetic Infiltration Process

Ireneusz Kubiak

Electromagnetic Compatibility Department and Laboratory, Military Communication Institute; 05-130 Zegrze Poludniowe, Poland; i.kubiak@wil.waw.pl

Received: 14 August 2019; Accepted: 16 September 2019; Published: 19 September 2019

Abstract: Due to the variety and multiplicity of electronic devices, the issue of electromagnetic environment protection is becoming more and more important. We often hear about how necessary it is for electronic devices to meet appropriate requirements. Meeting these requirements determines whether a device can be marketed. Unfortunately, the electrical parameters of electronic components have a very wide range of tolerances. For this reason, measured values of electromagnetic disturbances generated by devices of the same type are not always identical. Differences between those values may reach up to several dB. This problem also concerns electromagnetic emissions correlated with the processed information, which are very sensitive to electromagnetic infiltration process. Issues related to the problem of protection of electromagnetic environment are shown on the basis of research results obtained for several devices of the same type. Mentioned level differences of electromagnetic emissions can decide about a classification of device from viewpoint of protection of information against electromagnetic penetration process. These differences may be a treat to information security. Higher levels of valuable emissions force an application of additional methods limiting an effectiveness of electromagnetic penetration process. This particularly applies to IT devices with a wide range of applications, e.g., laptops and desktop computers. In this paper, this phenomenon was presented on the basis of tests of several devices of the same type. Also, there were carried out analyses of influence of increase levels of valuable emissions on a security zone radius.

Keywords: electromagnetic compatibility; protection of information; electromagnetic emissions; computers and information processing; data acquisition; image recognition

---

1. Introduction

A technological development and an availability of various electronic devices constantly increase the number of potential sources of electromagnetic disturbances. This phenomenon directly attracts interest in a protection of the electromagnetic spectrum [1,2]. This is particularly important in cases of very sensitive electronic devices and commonly used wireless communication. In addition, the range of frequencies in which electromagnetic disturbances may occur covers almost the full spectrum of radio frequencies [3,4].

In the article electromagnetic disturbances are analyzed (valuable emissions), generated by typical and commonly used IT devices such as personal computers (desktops) and laptops from viewpoint of possibilities of processing classified information. The issue is very important because a lot of protected information is processed on such devices. The essence of the problem applies to different effectiveness of IT devices radiation as sources of valuable emissions. Of course, special devices (TEMPEST devices classified according to, e.g., SDIP-27/2 document) could be used. However, they are very expensive because such devices have special solutions, limiting levels of unintentional electromagnetic emissions. Appropriate tests were conducted in an anechoic chamber using a specialized measuring equipment and an antenna. The analysis shows a threat connected with a lack of technical parameters; repeatability of electronic components used for construction of IT devices. The lack of proper parameters can result in very different values of measured electromagnetic disturbances. These differences can reach even 20 dB. This is unacceptable in cases of sensitivity of electronic devices to external electromagnetic fields.

Worrying phenomena related to lack of repeatability of measured values of electromagnetic disturbances within acceptable limits include:

- a fulfilment of appropriate EMC requirements (e.g., MIL-STD-461G, EMC Directive) by all electronic devices of a given type;
- a possible correlation of electromagnetic emissions with processed information (SDIP-27/2, SDIP-28/2) and increase of risk connected with electromagnetic infiltration process [5–7].

In case of the electromagnetic infiltration process, an increase of value of measured signal level by even a few dB may cause loss of protected information [8,9]. Therefore, we have to closely look into the observed phenomenon presented in the next sections of the article. It is also necessary to reflect on possibilities to ensure repeatability of manufactured devices in terms of levels of electromagnetic disturbances.

Such huge differences of levels of electromagnetic emissions are related to the quality of electronic components. This is especially important in the case of the electromagnetic protection of electronically processed information. The degree of susceptibility to the electromagnetic infiltration process of some types of devices is higher, and lower in the case of others [10,11].

To protect processed information against electromagnetic penetration process, different methods are used. However, existing phenomenon of different levels of the electromagnetic emissions makes impossible to apply one effective solution. For example, Tajima [12,13] proposed to apply a noise generator (Figure 1). The generator is a source of electromagnetic disturbances. The levels of these disturbances should be higher than levels of sensitive emissions correlated with processed information. The solution can have negative influence on other electronic devices, especially when the levels of valuable emissions are very high. Then the levels of additional disturbances couldn't meet requirements of electromagnetic compatibility (e.g., EMC Directive).

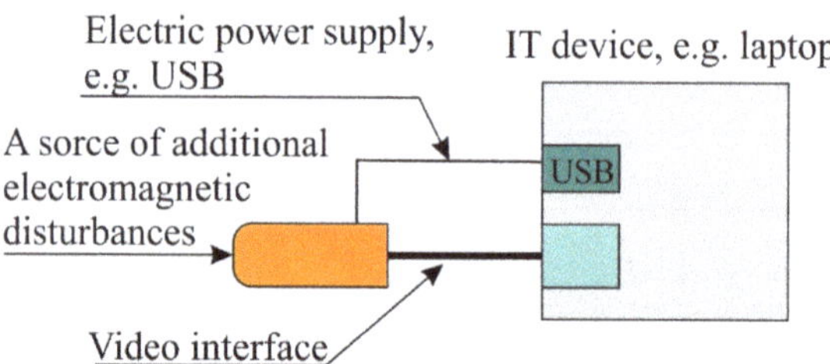

Figure 1. A photograph of the countermeasure device connected to a mobile PC.

Kuhn [14] proposed to apply filtering of video signals (Figure 2a). The use of the filtering limits spectrum of the valuable emission (Figure 2b, 20%, 30%, 40%, 50% of signal spectrum is cut from the top). It is possible in the case of personal computers where we have an access to video interfaces. However, graphic elements displayed on a screen become unclear (edges, e.g., of letters, are diffuse). Another problem appears in the case of mobile computers. Video interface (e.g., LVDS—Low-Voltage Differential Signaling) is not accessible. In this case, there can be a selected device, which is a source of lower levels of electromagnetic emissions.

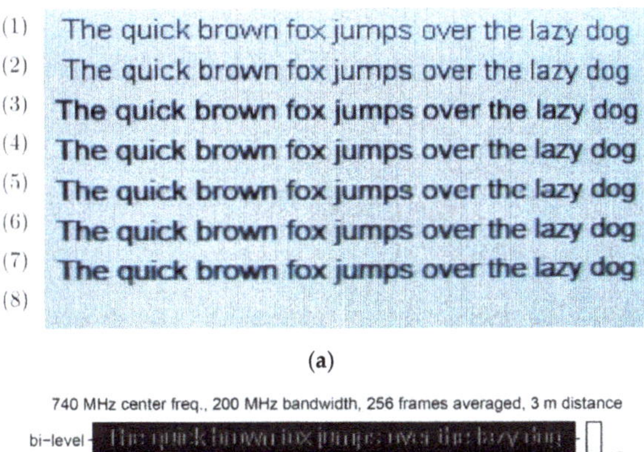

(a)

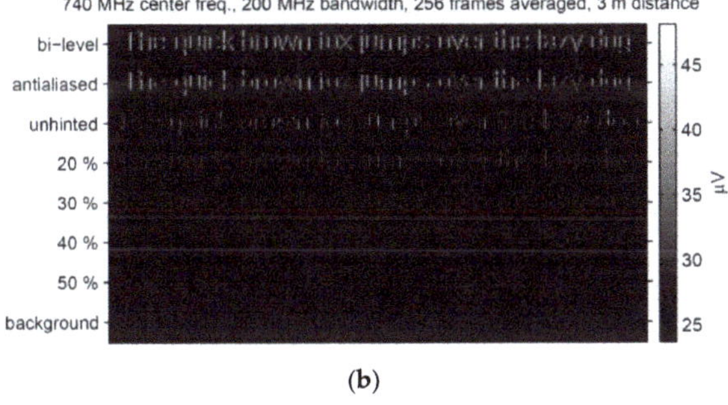

(b)

Figure 2. (a) A picture of the filtered fonts [14], (b) a picture for signal with limited spectrum [14].

A lot of producers of special devices (devices which could process classified information) use an electromagnetic shielding. The solution also limits levels of electromagnetic emissions. However, sometimes inside such devices is not the place for application of additional elements.

Therefore, above methods could be replaced by a selection of device, which is characterized by lower levels of electromagnetic emissions. This is possible because devices of the same type are sources of sensitive emissions about different levels appearing on the same frequency. Additionally, weaker emissions can protect processed information against electromagnetic infiltration process. In this paper the electromagnetic emissions were analyzed from viewpoint of existing risk connected with the loss of processed information.

At the same time, it was shown that commercial devices cannot be used uncritically to process classified data. Thus, attention was drawn to the existing quality problem of manufactured IT devices.

2. Materials and Methods

2.1. Test Devices

Five types of devices (15 measured devices of the same type) were selected for the study: three types of laptops and two types of desktops, which for the purposes of the analyzes were marked as laptops *A*, laptops *B*, laptops *C*, and desktops *A* and desktops *B*. A device configuration, including:

- mother board;
- graphic card;
- CD/DVD;
- HDD (SSD);
- LCD display;
- keyboard;

within a framework of given type was the same. During the tests the laptops were battery-powered.

2.2. Test Conditions

The tests of the devices were carried out in an anechoic chamber (Figure 3, Laboratory of Military Communication Institute, Zegrze Poludniowe, Poland). The technical parameters of the chamber (20 m × 16 m × 8 m, length × width × height), as well as internal conditions of electromagnetic environment ensured repeatability of conducted tests. An external electromagnetic environment did not have impact on the internal electromagnetic environment. The anechoic chamber is characterized by required shielding effectiveness, which is at least 100 dB in the frequencies range from 10 kHz to 18 GHz (IEC 61000–5–7, IEEE STD 299).

Figure 3. An anechoic chamber.

DSI-1550-A TEMPEST test system (up to 2 GHz), Microwave Downconverter DSI-1580-A (up to 22 GHz), as well as Rhode and Schwarz antenna (dipole antenna HE527 (200 MHz – 1 GHz)) were used for the tests. The distance of the measuring antenna from the tested device was equal 1 m (Figure 4). The tests were carried out for the frequency range of 200 MHz to 1 GHz.

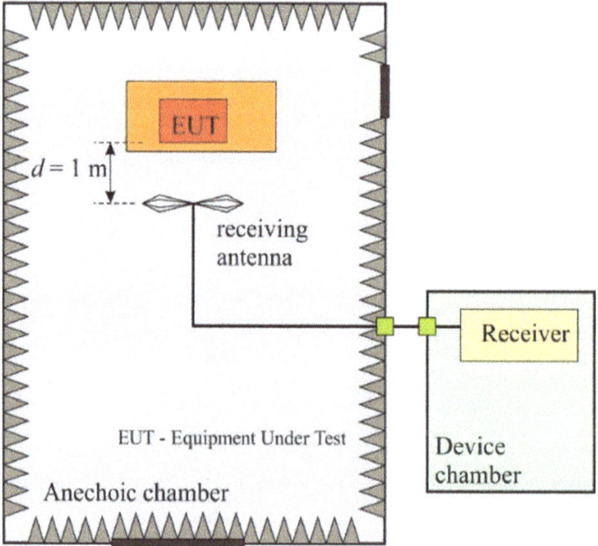

Figure 4. Test system.

During the tests, a special computer program was used. The program ensured simultaneous operation of all components of the tested device:

- data reading from CD/DVD;
- data reading and recording on HDD (SSD);
- data reading and recording in RAM;
- recording of a selected sign from a keyboard in a text editor;
- displaying of a video on LCD display.

3. Results

Test results for each type of device were placed on a single graph (Figures 5–9). This enabled easy analysis of levels of electromagnetic disturbances, as well as valuable emissions and assessment of observed risk [15–17]. Maximal E_{max} and minimal E_{min} values of electromagnetic disturbances:

$$E_{max}(f_n) \ = \ max[E_k(f_n)] \tag{1}$$

$$E_{min}(f_n) \ = \ min[E_k(f_n)] \tag{2}$$

where E_k—electromagnetic disturbances for k device (k = 1, 2, ..., K, K = 15), f—frequency, were marked with wider lines.

Laptop A Type

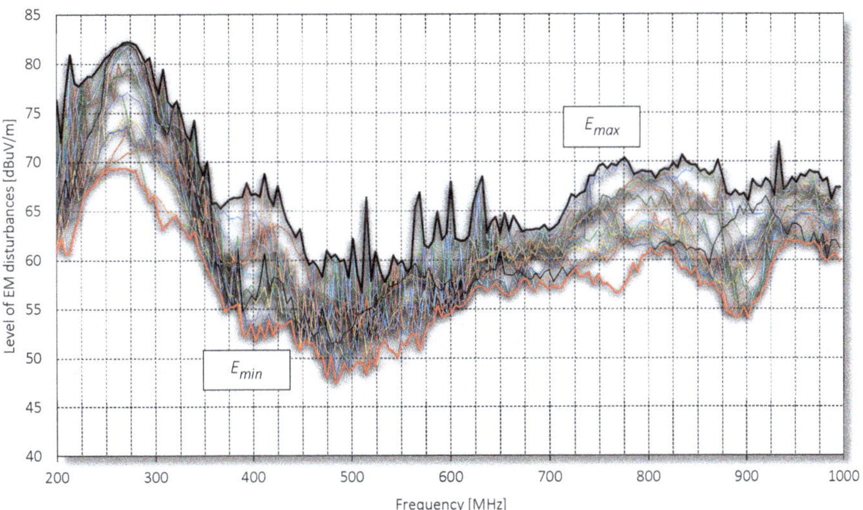

Figure 5. Levels of electromagnetic disturbances (valuable emissions, 15 measured devices of the same type) measured for laptop A in the range of frequencies from 200 MHz to 1 GHz, BW = 50 MHz, measurement distance: 1 m.

Laptop B Type

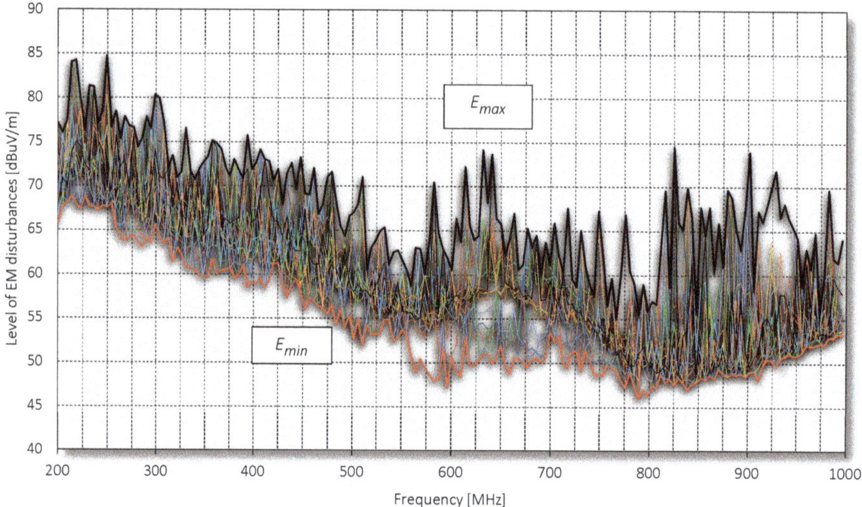

Figure 6. Levels of electromagnetic disturbances (valuable emissions, 15 measured devices of the same type) measured for laptops B in the range of frequencies from 200 MHz to 1 GHz, BW = 50 MHz, measurement distance: 1 m.

Laptop C Type

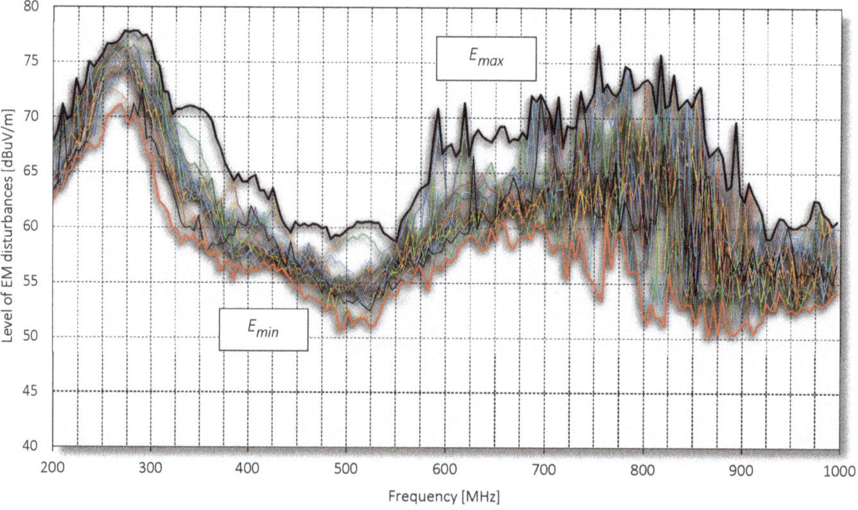

Figure 7. Levels of electromagnetic disturbances (valuable emissions, 15 measured devices of the same type) measured for laptops C in the range of frequencies from 200 MHz to 1 GHz, BW = 50 MHz, measurement distance: 1 m.

Desktop A Type

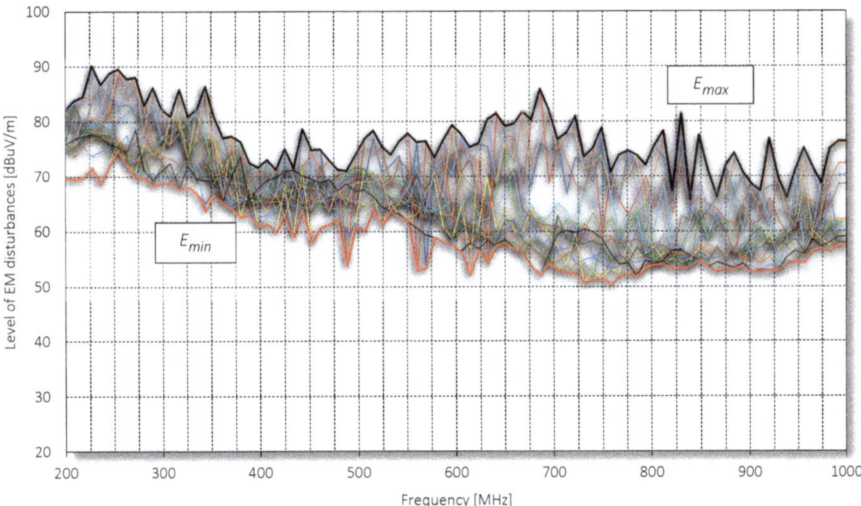

Figure 8. Levels of electromagnetic disturbances (valuable emissions, 15 measured devices of the same type) measured for desktops A in the range of frequencies from 200 MHz to 1 GHz, $BW = 50$ MHz, measurement distance: 1 m.

Desktop B Type

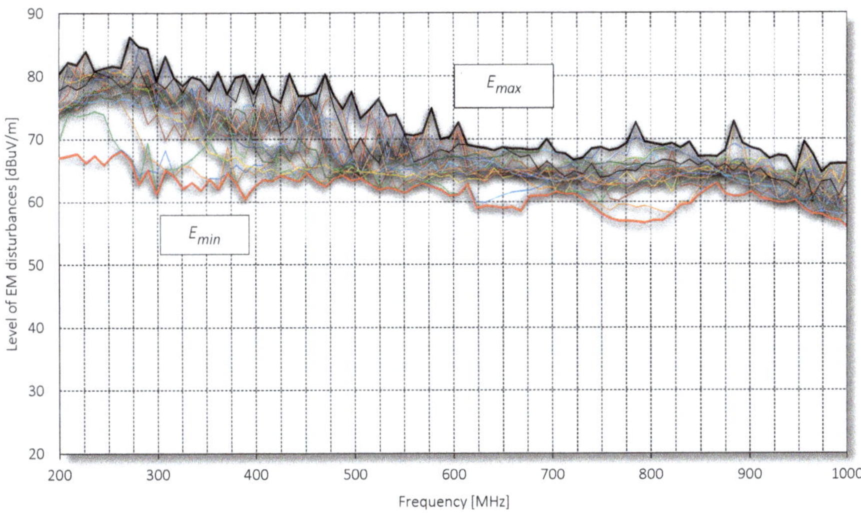

Figure 9. Levels of electromagnetic disturbances (valuable emissions, 15 measured devices of the same type) measured for desktops B in the range of frequencies from 200 MHz to 1 GHz, $BW = 50$ MHz, measurement distance: 1 m.

Differences and Average Values of Levels of Electromagnetic Disturbances

The analysis of the test results of the levels of electromagnetic disturbances was conducted based on differences E_{Diff} between maximum E_{max} and minimum E_{min} values for a given type of device (Figures 10 and 11):

$$E_{Diff}(f_n) \; = \; E_{max}(f_n) - E_{min}(f_n),\tag{3}$$

As well as on the basis of average values E_{Avr} for different types of devices (Figures 12 and 13):

$$E_{Avr}(f_n) \; = \; \frac{\sum_{k=1}^{K} E_k(f_n)}{K}\tag{4}$$

Based on obtained results, the potential impact of the quality of electronic components of the devices on the levels of EM disturbances, as well as the electromagnetic protection level of processed information, can be noticed.

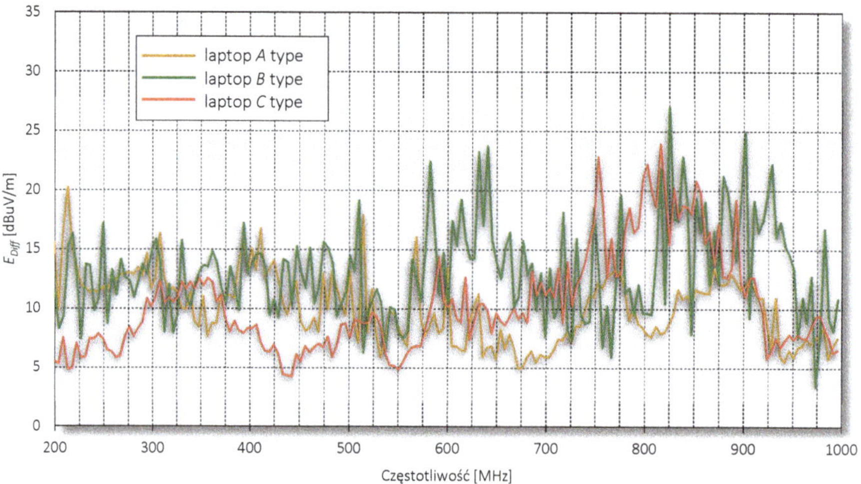

Figure 10. Differences of measured levels of electromagnetic (EM) disturbances (valuable emissions) for three types of laptops in the range of frequencies from 200 MHz to 1 GHz.

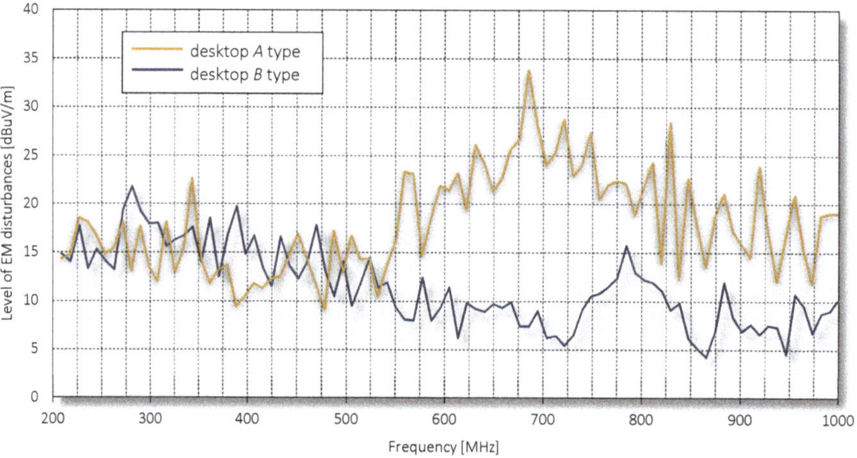

Figure 11. Differences of measured levels of EM disturbances (valuable emissions) for two types of desktop computers in the range of frequencies from 200 MHz to 1 GHz.

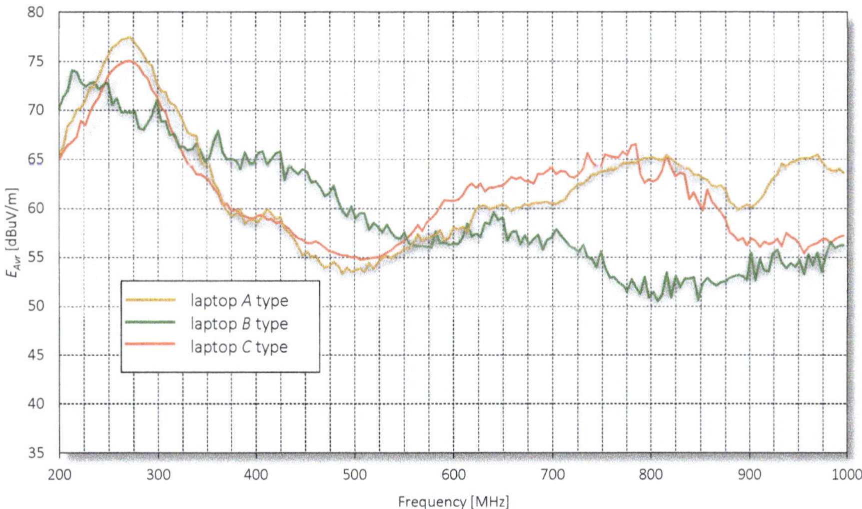

Figure 12. Average values of EM disturbances (valuable emissions) for three measured laptops in the range of frequencies from 200 MHz to 1 GHz.

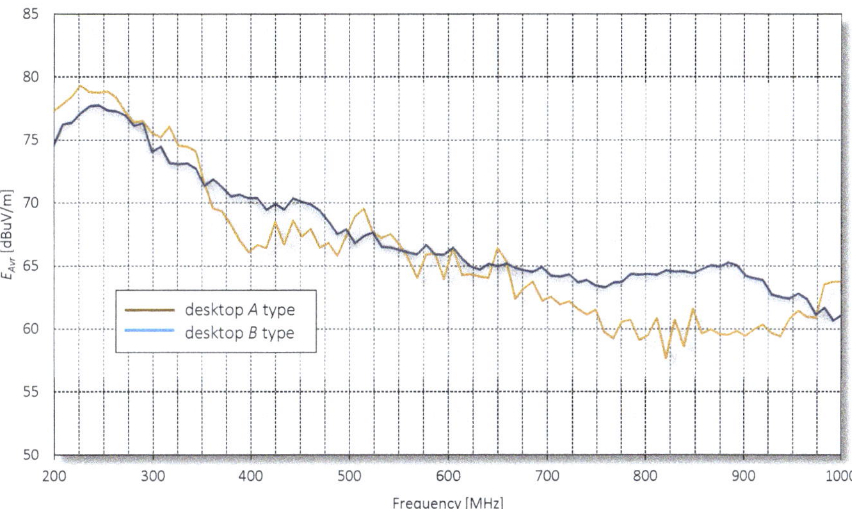

Figure 13. Average values of EM disturbances (valuable emissions) for two measured desktop computers in the range of frequencies from 200 MHz to 1 GHz.

4. Discussions

4.1. Valuable Emissions

The electromagnetic disturbances can have characteristics of processed information [18,19]. Such disturbances are called "valuable (sensitive) emissions".

In this case, an increase of levels of measured emission by approximately several dB could enable the possibility of non-invasive acquisition of information (Figure 14) [20,21].

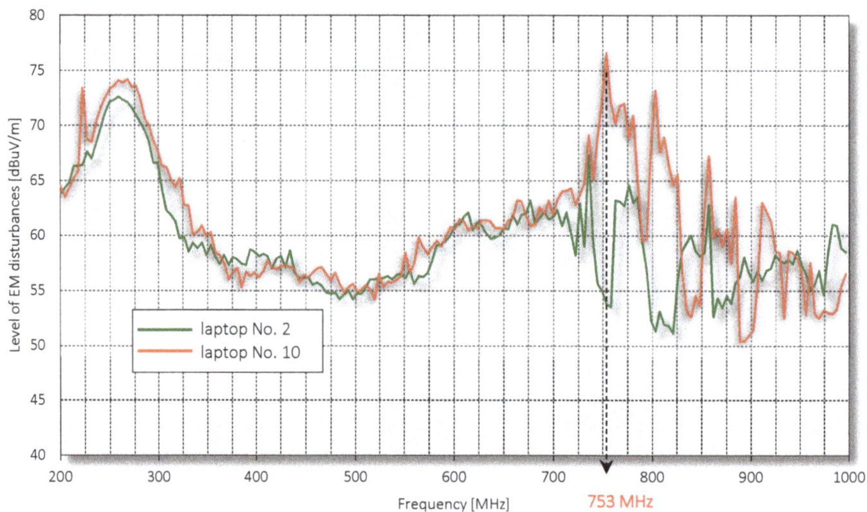

Figure 14. Levels of electromagnetic emissions from two selected laptops C and frequency of valuable emissions in the range of frequencies from 200 MHz to 1 GHz.

Examples of reconstructed images using recorded sensitive emissions on frequency 753 MHz for two different laptops of *C* type are shown in Figure 15.

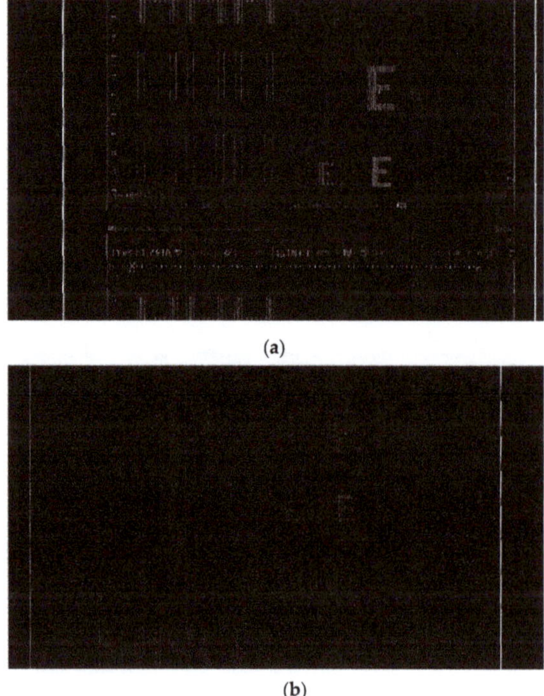

(a)

(b)

Figure 15. Reconstructed images for valuable emissions measured on frequency 753 MHz for two laptops type *C*: (**a**) laptop No. 10, (**b**) laptop No. 2.

4.2. Attenuation of Valuable Emissions

Analyzing Figure 14, it can be noticed that the difference of levels of electromagnetic emission, which appear on frequency 753 MHz, is very high (22.8 dB). This phenomenon allows reconstructing of primary information, and the information is legible. There can be selected other devices of the same type, which are characterized by lower levels of valuable emission. In this case, the emission does not allow reconstructing of primary information (Figure 16a). Of course, the first device can be used to process classified information. Then, many requirements have to be met, which decrease levels of sensitive emission (Figure 16b (zoning) and Figure 16c (zoning and shielding)). It is not easy. Different solutions have to be connected, which could make difficult the use of such a device.

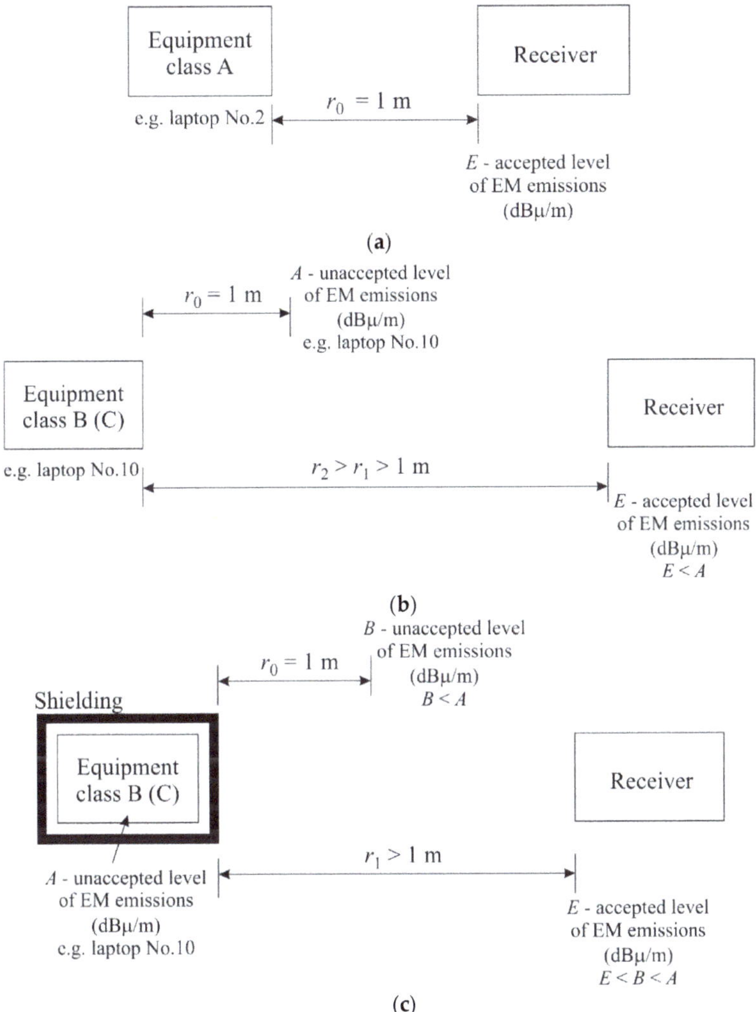

Figure 16. Attenuation methods of emission from electronic device: (**a**) equipment class A and protection zone about radius $r_0 = 1$ m, (**b**) equipment class B and protection zone about radius $r_2 > r_1 > r_0$, (**c**) equipment class B inside of shielding and protection zone about radius r_1.

It is connected with special zones and the shielding which have to ensure an appropriate attenuation of electromagnetic emission, according to the formula (Figure 17, [22]):

$$E = A - B - D(r_1); \tag{5}$$

$$E = A - D(r_2); \tag{6}$$

where:

$$D(r_x) = 20 \log\left(\frac{r_x}{r_0}\right)[dB]; \tag{7}$$

and

r_x—distance r_1 or r_2;
A—unaccepted level of EM emission;
B—attenuation inserted by shielding;
D—attenuation inserted by distance r_x.
The value of D depends on the shielding material.

In case of emission measured on frequency 753 MHz, and without additional shielding, a protection zone has to have a minimal radius according to a formula:

$$22.8 = 20 \log(r_2); \tag{8}$$

$$\log(r_2) = \frac{22.8}{20} = 1.14; \tag{9}$$

$$r_2 = 10^{1.14} = 13.8038 \ [m]. \tag{10}$$

In practice, a radius equaling more than 13 m is very big. It means that the protection zone on the radius, 13.8038 m, has to be organized around our device, which processes classified information.

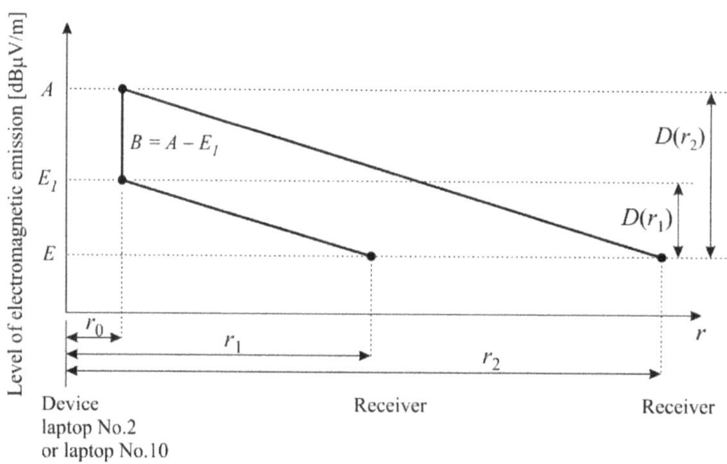

Figure 17. Attenuation of emission from electronic device

5. Conclusions

This paper analyzes levels of electromagnetic disturbances from different types of electronic devices. Obtained results were connected, with possibilities of existence of sensitive emissions correlating with processed data. Each device of a given type was measured in the same conditions.

The results were largely scattered. Differences between levels of electromagnetic disturbances reached nearly 25 dB, especially in the frequency range from 600 MHz to 900 MHz. The phenomenon is very disquieting. There are frequencies in which valuable emissions most often exist [23]. Of course, the emissions come from sources in a shape of graphic lines (e.g., cables) and circuits (e.g., graphic cards). Experience shows that increasing the level of sensitive emissions by approximately 5 dB could determine classification of these emissions. In this context, we can consider the impact of the quality of components of electronic devices.

The quality of components additionally impacts the levels of electromagnetic disturbances. The differences between levels of electromagnetic disturbances of approximately 20–25 dB can decide whether a device meets the requirements of the EMC Directive. Not every device meets the limit values of EMC standardizations.

Another phenomenon considered in the article is the threat of information loss. Devices of the same type can be classified as meeting the requirements of electromagnetic protection of processed information or not. For devices for which the levels of sensitive emissions exceed the permissible values, it is necessary to use additional organizational solutions for electromagnetic safety. Such a solution may be a physical protection zone with a very large radius that is not always possible to create, especially when the device processing classified data must be used in offices. As shown in the analysis carried out, it may be necessary to use protection zones with a radius of more than 10 m. Therefore, the best solution is to choose one device out of several that is characterized by lower levels of sensitive emissions. As such, the device will allow the processing of classified information by meeting both the requirements of the SDIP-27/2 document (NATO TEMPEST requirements and Evaluation procedures) and the EMC directive.

Funding: This research received no external funding.

Conflicts of Interest: The author declare no conflict of interest.

References

1. Lee, H.K.; Kim, J.H.; Kim, S.C. Emission Security Limits for Compromising Emanations Using Electromagnetic Emanation Security Channel Analysis. *IEICE Trans. Commun.* **2639**, *10*, 2639–2649. [CrossRef]
2. Xie, C.h.; Wang, T.; Hao, X.; Yang, M.; Zhu, Y.; Li, Y. Localization and Frequency Identification of Large-Range Wide-Band Electromagnetic Interference Sources in Electromagnetic Imaging System. *Electronics* **2019**, *8*, 499. [CrossRef]
3. Zhang, N.; Lu, Y.H.; Cui, Q.; Wang, Y.Y. Investigation of Unintentional Video Emanations from a VGA Connector in the Desktop Computers. *IEEE Trans. Electromagn. Compat.* **2017**, *59*, 1826–1834. [CrossRef]
4. Glowacz, A.; Glowacz, Z. Diagnostics of stator faults of the single-phase induction motor using thermal images, MoASoS and selected classifiers. *Measurement* **2016**, *93*, 86–93. [CrossRef]
5. Yuan, K.; Grassi, F.; Spadacini, G.; Pignari, S.A. Crosstalk-Sensitive Loops and Reconstruction Algorithms to Eavesdrop Digital Signals Transmitted Along Differential Interconnects. *IEEE Trans. Electromagn. Compat.* **2017**, *59*, 256–265. [CrossRef]
6. Kubiak, I. Video signal level (colour intensity) and effectiveness of electromagnetic infiltration. *Bull. Pol. Acad. Sci. – Tech. Sci.* **2016**, *64*, 2007–2018. [CrossRef]
7. Loughry, J.; Umphress, D.A. Information Leakage from Optical Emanations. *ACM Trans. Inf. Syst. Secur.* **2002**, *5*, 262–289. [CrossRef]
8. Kubiak, I. Laser printer as a source of sensitive emissions. *Turk. J. Electr. Eng. Comput. Sci.* **1354**, *26*, 1354–1366.
9. Cihan, U.; Aşık, U.; Cantürk, K. Analysis of Information Leakages on Laser Printers in the Media of Electromagnetic Radiation and Line Conductions. In Proceedings of the International Conference on Information Security and Cryptology, Ankara, Turkey, 30–31 October 2015.
10. Wu, C.; Gao, F.; Dai, H.; Wang, Z.A. Topology-Based Approach to Improve Vehicle-Level Electromagnetic Radiation. *Electronics* **2019**, *8*, 364. [CrossRef]

11. Jurič, I.; Nedeljković, U.; Novaković, D.; Pinćjer, I. Visual experience of noise in digital images. *Tehnicki Vjesnik-Technical Gazette* **2016**, *23*. [CrossRef]

12. Tajima, K.; Ishikawa, R.; Mori, T.; Suzuki, Y.; Takaya, K. A study on risk evaluation of countermeasure technique for preventing electromagnetic information leakage from ITE. In Proceedings of the International Symposium on Electromagnetic Compatibility EMC Europe, Angers, France, 4–7 September 2017. [CrossRef]

13. Telecommunication Standardization Sector of ITU. K.115: Mitigation methods against electromagnetic security threats. Available online: https://www.itu.int/rec/T-REC-K.115-201511-I (accessed on 16 September 2019).

14. Kuhn, M.G. Compromising Emanations: Eavesdropping Risks of Computer Displays. Technical Report. 2003. Available online: http://www.cl.cam.ac.uk/TechReports/ (accessed on 16 September 2019).

15. Cakir, G.; Cakir, M.; Sevgi, L. Electromagnetic radiation from multilayer printed circuit boards: a 3D FDTD-based virtual emission predictor. *Turk. J. Electr. Eng. Comput. Sci.* **2009**, *17*, 315–326.

16. Litao, W.; Bin, Y. Analysis and Measurement on the Electromagnetic Compromising Emanations of Computer Keyboards. In Proceedings of the Seventh International Conference on Computational Intelligence and Security, Sanya, Hainan, China, 3–4 December 2011; pp. 640–643.

17. Ko, W.L. Time Domain Solution of Electromagnetic Problems. *Electromagnetics* **1992**, *12*, 403–433.

18. Kubiak, I. The Influence of the Structure of Useful Signal on the Efficacy of Sensitive Emission of Laser Printers. *Measurement* **2018**, *119*, 63–76. [CrossRef]

19. Chervyakov, N.; Lyakhov, P.; Kaplun, D.; Butusov, D.; Nagornov, N. Analysis of the Quantization Noise in Discrete Wavelet Transform Filters for Image Processing. *Electronics* **2018**, *7*, 135. [CrossRef]

20. Kubiak, I. LED printers and safe fonts as an effective protection against the formation of unwanted emission. *Turk. J. Electr. Eng. Comput. Sci.* **2017**, *25*, 4268–4279. [CrossRef]

21. Kubiak, I. TEMPEST font counteracting a non-invasive acquisition of text data. *Turk. J. Electr. Eng. Comput. Sci.* **2018**, *26*, 582–592. [CrossRef]

22. Telecommunication Standardization Sector of ITU. Test methods and guide against information leaks through unintentional electromagnetic emissions. Available online: https://www.itu.int/rec/T-REC-K.84-201101-I/en (accessed on 16 September 2019).

23. Kubiak, I. Influence of the method of colors on levels of electromagnetic emissions from video standards. *IEEE Trans. Electromagn. Compat.* **2018**, *61*, 1–9. [CrossRef]

electronics

Article

Transversely Compact Single-Ended and Balanced Bandpass Filters with Source–Load-Coupled Spurlines

Fang Yan [1,*]**, Yong Mao Huang** [2]**, Tao Huang** [2]**, Shuai Ding** [3] **and Maurizio Bozzi** [4]

[1] Center of Airworthiness, Civil Aviation University of China, Tianjin 300300, China; wangkenian@126.com
[2] School of Electrical and Electronic Information, Xihua University, Chengdu 610039, China; ymhuang128@gmail.com (Y.M.H.); huang1987tao@yeah.net (T.H.)
[3] Institute of Applied Physics, and School of Physical Electronics, University of Electronic Science and Technology of China, Chengdu 610054, China; uestcding@hotmail.com
[4] Department of Electrical, Computers and Biomedical Engineering, University of Pavia, 27100 Pavia, Italy; maurizio.bozzi@unipv.it
* Correspondence: fyan@cauc.edu.cn; Tel.: +86-158-229-69896

Received: 18 March 2019; Accepted: 6 April 2019; Published: 10 April 2019

Abstract: Multi-function wireless systems demand multi-channel transmit/receive (TR) modules, particularly as multiple functions are required to operate simultaneously. In each channel, passive components, including bandpass filters, must be compact, or at least transversely compact; thus, the entire circuitry of the channel will be slender, and consequently multiple channels can be parallel-arranged conveniently. In this work, single-ended and balanced bandpass filters for multi-channel applications are presented. As a unique resonator, the U-shaped stepped impedance resonator (USIR) can achieve size miniaturization compared with its corresponding uniform impedance resonator (UIR) counterpart. Hence, with the utilization of USIRs, the proposed bandpass filters are able to acquire compact transverse sizes. Moreover, by using the source–load coupling scheme, two transmission zeros (TZs) are respectively generated at the lower and upper sides of the passbands, which is useful for improvement of the selectivity performance. In addition, spurlines are introduced at the input and output ports to produce another TZ to further enhance the stopband performance, which cannot be acquired by the UIR or stepped impedance resonator (SIR). To verify the aforementioned idea, one single-ended and one balanced bandpass filter are implemented, with experimental results in good agreement with the corresponding simulations. Meanwhile, as compared with some similar works, the proposed balanced filter achieves compact transverse size, sharp selectivity skirt, and wide stopbands up to the fourth-order harmonic with suppression over 20 dB, which illustrates its suitability for differential signal transmission application in microwave circuits and systems.

Keywords: balanced bandpass filter; common mode suppression; spurline; source–load coupling; stepped impedance resonator (SIR)

---

1. Introduction

Recently, different kinds of radio frequency (RF) and microwave systems for various wireless applications like commutation, radars, and sensors have been widely investigated. As applications become increasingly concrete and specific, more and more functions are expected to be integrated into a single system, and the system circuitry eventually gets more and more complex. Meanwhile, there is also an increasing demand on high-performance passive components, to ensure that multiple functions in the single system can all operate well. Passive components including filters with excellent performance in amplitude noise, phase jitter, spurs, and harmonics are highly desired, particularly for applications such as fifth generation (5G) mobile access and transmission, high-resolution imaging radar, automotive anti-collision sensors, and many other specific applications. For instance, Figure 1 gives the diagram of a typical multi-function wireless system with multi-channel receivers. It can be easily obtained that various kinds of active and passive components, including antennae, low noise amplifiers (LNAs), mixers, power amplifiers (PAs), and filters, are contained in each channel (frequency synthesizers for carriers are omitted). On one hand, as depicted in [1], active components are mostly integrated circuits with extremely compact sizes. Hence, the passive components of a multi-channel radio typically occupy about 65% of its circuitry size. Therefore, size reduction of passive components is important to miniaturize the overall size of the system. On the other hand, for the receiver of each channel in Figure 1, all components, especially the passive ones, are supposed to be compact (or at least transversely compact) therefore the entire channel circuitry is slender. In this way, all channels can be parallel-arranged easily and the circuit board of the multi-channel system will eventually be compact. Thus, passive components with compact transverse size are quite useful for multi-channel systems.

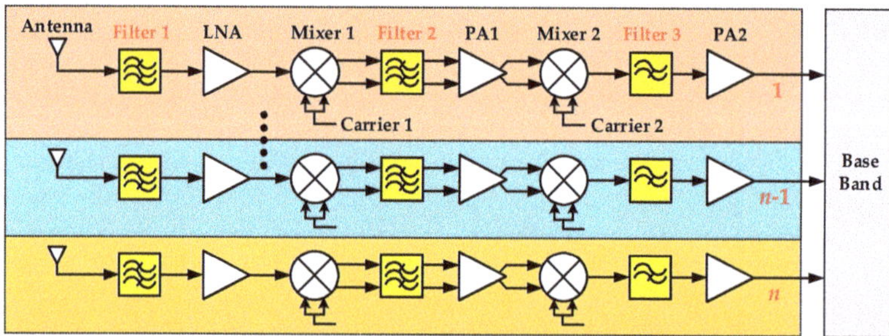

Figure 1. Diagram of a typical multi-function wireless system with multi-channel receivers.

Meanwhile, it can be inferred from Figure 1 that most of the passive components in the multi-channel system are filters. For instance, as shown in Figure 1, Filter 1 is a single-ended bandpass filter, Filter 2 is a balanced one (which is also called a differential filter) and Filter 3 is a single-ended lowpass filter. Actually, passband selecting by filter seems be one of the most effective approaches for suppression of the unwanted spurs or harmonics in the received signals. In the literature, guided-wave structure of different forms, planar and nonplanar, have been developed to implement high-performance filters. It is known that microstrip technology is conventional but has the advantage of being able to integrate with other planar components and circuits due of its simple and compact structure [2]. Compared to the waveguide, substrate-integrated waveguide, and coplanar waveguide technologies, it is very easy to design a variety of filters with complex structures and coupling mechanisms as well as miniature size by using microstrip line. In past decades, microstrip uniform impedance resonators (UIRs) of various electrical lengths have been widely utilized to design filters [2]. To improve the design flexibility of UIRs and miniaturize their size, the stepped impedance resonator (SIR) has been developed [3]. Its modified forms—folded SIR (FSIR) and U-shaped SIR (USIR)—can achieve further size reduction [4]. On the other hand, some interesting techniques, such as defected ground structure (DGS), defected microstrip structure (DMS), spurline, and source–load coupling, have been proposed to improve the performance of microstrip filters [5–7]. In [8], both the SIR and spurline are utilized to develop a balanced filter with wide stopband. However, its superconducting process limits its application in practical RF systems. Later, the spurline-stub resonator is used to suppress the even-mode noise of differential filter. However, it needs to undergo micromachining fabrication and wire-bonding processes, which are complex [9]. A dual-band balanced filter based on open-ended half-wavelength UIR resonator and interdigital capacitor is proposed in [10], while another case was made of a short-ended half-wavelength UIR resonator with multilayer structure [11]. In [12], the aggressive space mapping method is utilized to synthesize a balanced filter incorporating the SIR, which achieves an excellent common mode rejection ratio (CMRR) about 50 dB. However, since the filter contains seven SIRs, its physical size is bulky. Here, it is essential to declare that the CMRR in this work is defined as usual, as CMRR = $|S_{dd21}| / |S_{cc21}|$. On the other hand, the matrix transformation method is employed to synthesize the balanced filter with common mode (CM) noise absorption, which seems to be an effective strategy [13]. On the other hand, selectivity and stopband performance of the synthesized filter are not good owing to its impedance–transformation-based structure. The dumbbell-shaped resonator is etched on the ground plane of the differential filter to improve its CMRR to 57 dB [14]. Recently, both the SIRs and interdigital capacitors are utilized together to produce two TZs for the balanced filter in [15]. In addition, some multi-function balanced filters are also investigated. In [16], a SIR-based diplexer with high performance is reported. The multi-stub-loaded transmission line was included to realize quasi-reflectionless single- and dual-band balanced filters [17]. Another work even integrates power dividing function in the single-ended-to-balanced filter with wideband CM suppression [18].

Most of the aforementioned works exhibit high performance in CMRR, stopband, or selectivity. However, these high performances are mainly achieved at the cost of large physical size, which is not proper for system integration and packaging applications. It is known that for system integration and packaging applications, both the physical size and stopband performance, as well as selectivity, are significant. In this work, a single-ended and a balanced filter are investigated, taking account of size, selectivity, and stopband simultaneously. The USIR is utilized as basic resonator for size reduction while the source–load coupling topology is able to produce two TZs for selectivity performance, and spurlines included at the source and load node can generate an extra TZ to enhance the stopband performance. This work is organized as below: Section 2 introduces the USIR and spurline in detail. Sections 3 and 4 investigate implementation of the single-ended and balanced filters, respectively. In Section 5, a brief discussion is sketched, with a conclusion given at the end.

2. Basic Elements of the Proposed Filters

2.1. Half-Wavelength U-Shaped Stepped Impedance Resonator

Figure 2a gives geometries and corresponding equivalent-circuit models of the half-wavelength UIR, SIR, and USIR. As is shown, the UIR with the characteristic impedance of Z_0 can be modeled as a parallel network of inductance L_0 and capacitance C_0. The SIR consists of one high-impedance section in the center and two low-impedance sections on both sides. The high-impedance section has characteristic impedance of Z_H and electrical length of θ_H, while values are Z_L and θ_L for the low-impedance case. The effective inductance and capacitance of the SIR can be represented by L_s and C_s. For the USIR, effective inductance and capacitance are denoted by L_u and C_u. Additionally, as listed in Figure 2a, folding the straight SIR into a U shape will generate extra strip-to-strip capacitive effect between the two parallel sections and can be modeled as C_e. Firstly, as the half-wavelength USIR operates at the dominant resonant mode, the electric field will distribute symmetrically along the transverse centerline. Hence, under the dominant resonant mode, the transverse centerline can be regarded as the short-circuited plane [3,19]. Then, the half-wavelength USIR can be equivalent to the quarter-wavelength USIR, as shown in Figure 2a. Then, resonance properties of the half-wavelength USIR can be captured from the corresponding properties of the quarter-wavelength USIR. Therefore, the input admittance of the USIR in Figure 2a (i.e., Y_{in}) is given as:

$$Y_{in} = \frac{1}{Z_{in}} = \frac{Z_L - Z_H \tan\theta_L \tan\frac{\theta_H}{2}}{jZ_L\left(Z_H \tan\frac{\theta_H}{2} + Z_L \tan\theta_L\right)} \tag{1}$$

The USIR operates at the dominant resonant frequency, $Y_{in} = 0$, which means:

$$Z_L - Z_H \tan\theta_L \frac{\theta_H}{2} = 0. \tag{2}$$

Thus, it can be captured that:

$$\theta_{\mathrm{H}} = 2\arctan\frac{Z_{\mathrm{L}}}{Z_{\mathrm{H}}\tan\theta_{\mathrm{L}}}. \tag{3}$$

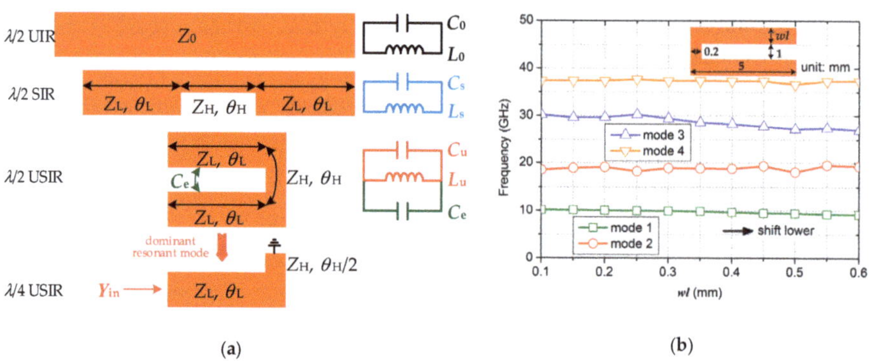

(a)

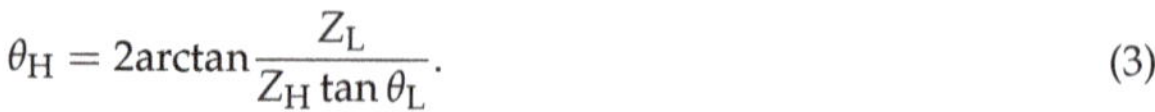

(b)

Figure 2. Geometry and resonant frequency of the U-shaped stepped impedance resonator (USIR): (**a**) Geometries and equivalent-circuit models of the uniform impedance resonator (UIR), stepped impedance resonator (SIR), and USIR; (**b**) resonant frequencies of the lowest four modes in the USIR versus width of the low-impedance section.

Then, electrical lengths of the half-wavelength USIR, θ_{half}, can be calculated as:

$$\theta_{\mathrm{half}} = 2\theta_{\mathrm{L}} + \theta_{\mathrm{H}} = 2\theta_{\mathrm{L}} + 2\arctan\frac{Z_{\mathrm{L}}}{Z_{\mathrm{H}}\tan\theta_{\mathrm{L}}} = 2\arctan[\tan\theta_{\mathrm{L}}] + 2\arctan\frac{Z_{\mathrm{L}}}{Z_{\mathrm{H}}\tan\theta_{\mathrm{L}}}. \tag{4}$$

By considering the basic properties of the arctangent function and Equation (4) together, the following can be derived: as $Z_{\mathrm{L}} = Z_{\mathrm{H}}$, θ_{half} is equal to π, which is the very electrical length of the conventional half-wavelength UIR under the dominant resonant mode; as $Z_{\mathrm{L}} > Z_{\mathrm{H}}$, θ_{half} is able to larger than π; as $Z_{\mathrm{L}} < Z_{\mathrm{H}}$, θ_{half} can be smaller than π. Therefore, by selecting the value of Z_{L} to be smaller than that of Z_{H}, the electrical length of USIR can be smaller than that of the conventional UIR, which means the physical length of USIR is able to be shorter than that of the conventional UIR, and eventually size reduction of the resonator will be achieved.

Secondly, as depicted in [2,3], the dominant resonant frequencies of UIR, SIR, and USIR in Figure 2a can be expressed as:

$$f_0 = \frac{1}{2\pi\sqrt{L_0 C_0}}, \tag{5}$$

$$f_{\mathrm{s}} = \frac{1}{2\pi\sqrt{L_{\mathrm{s}} C_{\mathrm{s}}}}, \tag{6}$$

$$f_{\mathrm{u}} = \frac{1}{2\pi\sqrt{L_{\mathrm{u}}(C_{\mathrm{u}} + C_{\mathrm{e}})}}. \tag{7}$$

Generally, since the extra strip-to-strip capacitive effect produced by the folding is much weaker than the inherent inner capacitive effect of USIR, it can be obtained that $C_{\mathrm{e}} \ll C_{\mathrm{u}}$. Hence, for the simplification of analyses, the parasitic strip-to-strip capacitive effect of the USIR can be neglected and its dominant resonant frequency can be calculated as:

$$f_{\mathrm{u}} \approx \frac{1}{2\pi\sqrt{L_{\mathrm{u}} C_{\mathrm{u}}}}. \tag{8}$$

Subsequently, distribution of the higher-order modes is important for the transmission property of the resonator. To study the property of higher-order modes of the USIR directly, some numerical simulations are carried out using a commercial three-dimensional full-wave simulator. The substrate for simulation is set with a thickness of 0.508 mm, a relative permittivity of 2.2, a loss tangent of 0.001, and a copper cover of 0.035 mm. Figure 2b lists the simulated resonant frequencies of the lowest four modes in the USIR versus width of the low-impedance section, with some fixed geometrical parameters of the USIR being given in the inset as well. From Figure 2b, it can be obtained that the resonant frequencies of modes 2, 3, and 4 are about the second-, third-, and fourth-order harmonics of mode 1 (the dominant resonant mode) of the USIR. Thus, USIR is a relatively simple resonator and can be utilized easily, as no complex parasitic modes exist. Furthermore, as the width of the low-impedance section *wl* gets wider, the dominant resonant frequency of the USIR becomes lower. That is, the smaller the impedance ratio Z_L/Z_H is, the lower the dominant resonant frequency will be. Therefore, a smaller impedance ratio can be helpful for size reduction of the resonator and other related components. Meanwhile, Figure 3 shows electric field distributions of the lowest four modes in the USIR. For these simulations, the value of *wl* is 0.4 mm.

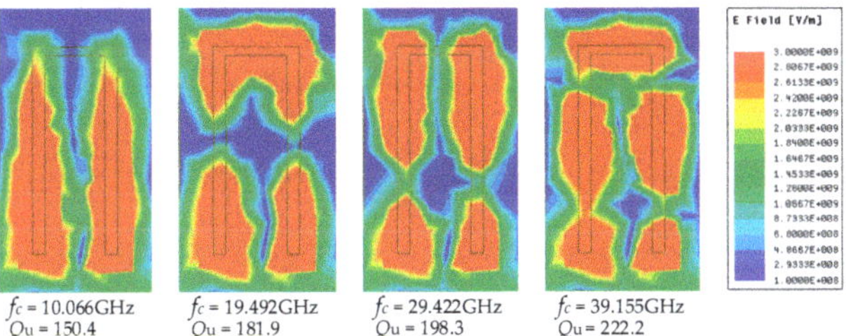

Figure 3. Electric field distributions of the lowest four modes in the USIR.

2.2. T-Shaped Spurline

As mentioned in Section 1, defected ground structure (DGS), defected microstrip structure (DMS), and spurline are advanced techniques for improvement of the stopband performance of microstrip filters. Figure 4 gives derivation from the conventional microstrip transmission line to the DGS, DMS, and various spurlines. As depicted in [5,6], the DGS, DMS, and quarter-wavelength spurline can all be equivalent to a parallel network of resistance, inductance, and capacitance (RLC). In practical design, this parallel network is cascaded into the component and can produce a TZ at a specific frequency. Location of the TZ is mainly determined by physical size of the defected structure in DGS, DMS, and spurline. In the literature, various modified spurlines, including the mender spurline and multi-spurline, have been developed. As discussed in [6], the mender has a longer slot compared with the conventional spurline (which has straight slots), which will provide a larger product of equivalent capacitance and inductance to generate a TZ at lower frequency. For filter cases, generating a TZ at lower frequency of the upper stopband usually means improving the selectivity performance. On the other hand, it can be seen from Figure 4 that the multi-spurline has multiple slots of different lengths, which will produce multiple TZs at different frequencies and be helpful for improvement of stopband performance of filters.

In this work, the T-shaped spurline is proposed, with its geometry shown in Figure 4. Obviously, compared with the conventional spurline, the proposed one contains more flexible dimensions, since the T-shaped slot includes three sections of various widths and lengths. The proposed T-shaped spurline can be equivalent to a parallel RLC network as well, with the lumped elements being given as [6]:

$$R_T = 2Z_0 \left(\frac{1}{|S_{21}|} - 1 \right)\Bigg|_{f=f_0}, \tag{9}$$

$$C_T = \frac{\sqrt{0.5(R_T + 2Z_0)^2 - 4Z_0^2}}{2.83\pi Z_0 R_T \Delta f}, \tag{10}$$

$$L_{\mathrm{T}} = \frac{1}{(4\pi f_0)^2 C_{\mathrm{T}}},\tag{11}$$

where L_{T}, C_{T}, and R_{T} are the equivalent inductance, capacitance, and resistance, respectively, of the proposed T-shaped spurline. $|S_{21}|$ is the insertion loss. f_0 is the resonant frequency (frequency of the TZ) and Δf is the 3 dB bandwidth of $|S_{21}|$. Z_0 is the characteristic impedance of the transmission line, which is typically set as 50 ohm. In the specific implementation, values of L_{T}, C_{T}, and R_{T} can be calculated by considering Equations (9)–(11), incorporating the specified design frequency of the TZ and numerical simulation results of the T-shaped spurline.

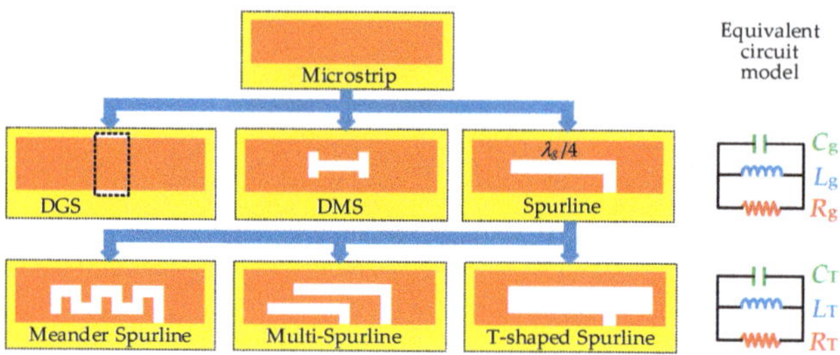

Figure 4. Derivation from the conventional microstrip to the defected ground structure (DGS), defected microstrip structure (DMS), and various spurlines.

3. Single-Ended Bandpass Filter with Source–Load-Coupled Spurlines

To demonstrate the effectiveness of the USIR and proposed T-shaped spurline, a single-ended bandpass filter is designed. First of all, as mentioned above, the source-load coupling scheme is able to improve performance of the filter notably. Hence, the source–load coupling scheme is employed in the proposed single-ended filter design as well. Figure 5 sketches the detailed configuration and coupling scheme of the proposed single-ended filter based on the USIR and T-shaped spurline. Firstly, as shown in Figure 5a, the proposed single-ended bandpass filter contains two USIRs, two spurlines, two 50 ohm microstrip lines at the input and output ports, and two tapered transitions between the spurlines and 50 ohm microstrips. Secondly, as shown in Figure 5b, the filter can generate a major coupling path as S-R1-R2-L. Here, S and L represent the source and load nodes, respectively. R1 and R2 denote resonator nodes. Meanwhile, in the proposed single-ended filter, there is another coupling path (i.e., S-L) which is mainly contributed to by the source–load coupling scheme. With the two aforementioned coupling paths operating together, the cross-coupling will be generated, so that two TZs, separately located at the lower and upper sides of the passband, can be generated, which will help to enhance the selectivity of the filter remarkably. Thirdly, the source and load nodes mainly consist of the proposed T-shaped spurlines, thereby an extra transmission zero can be produced at the upper stopband to extend the bandwidth of the stopband of the proposed single-ended filter.

Thereafter, as introduced in [2], operation frequency of the filter is mainly determined by the sizes of the employed resonators, whereas its bandwidth is mainly controlled by the coupling coefficients between adjacent resonators. Furthermore, tapered transitions will also influence both the central frequency and the bandwidth, since they can affect the value of the external quality factor notably. To study the influence from the USIR on the operation frequency of the filter more clearly, some numerical simulations have been done on the same aforementioned substrate by using the same commercial electromagnetic simulator used above. In these simulations, some fixed geometrical parameters are: $ws = 1.52$, $lt = 1.5$, $a0 = 1.2$, $b0 = 2.0$, $a1 = 0.6$, $b1 = 6.0$, $c1 = 2.0$, $v1 = 0.5$, $s1 = 0.5$, $s2 = 0.3$, $s3 = 0.3$ (units: mm). Figure 6 shows simulated $|S21|$ of the proposed single-ended filter with the variation of some geometrical parameters. For Figure 6a, $b3 = 2.2$ mm, $a2 = 0.4$ mm, $a3 = 0.2$ mm. As $b2$ increases from 3.9 mm to 4.5 mm, both the passband and the first two TZs shift lower, while the highest TZ only shifts slightly lower. For Figure 6b, $b2 = 4.2$ mm, $a2 = 0.4$ mm, $a3 = 0.2$ mm. In this case, with a longer $b3$, the passband, TZ1, and TZ2 all shift lower, whereas TZ3 remains nearly unchanged. In Figure 6c, $b2 = 4.2$ mm, $b3 = 2.2$ mm, $a3 = 0.2$ mm. For this case, the passband, TZ1, and TZ2 all shift lower with the increasing of $a2$, while TZ3 is also nearly unmoved. Finally, for Figure 6d, $b2 = 4.2$ mm, $b3 = 2.2$ mm, $a2 = 0.3$ mm. From Figure 6d, the wider $a3$ is, the lower the passband is. Moreover, TZ1 and TZ2 will shift much lower than TZ3 as $a3$ gets wider.

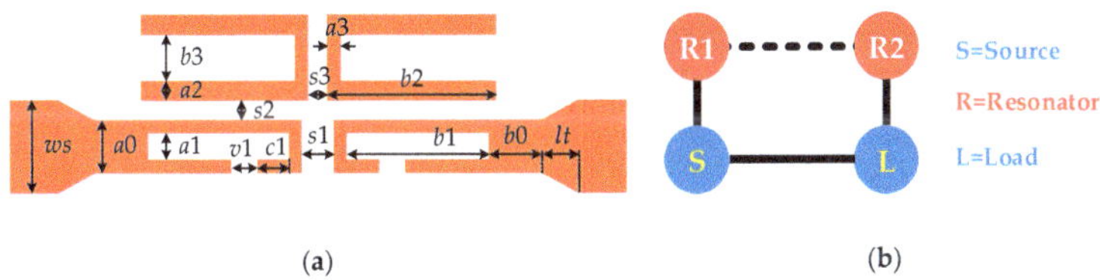

(a)

(b)

Figure 5. Configuration and coupling scheme of the proposed single-ended bandpass filter: (**a**) Detailed configuration; (**b**) coupling scheme.

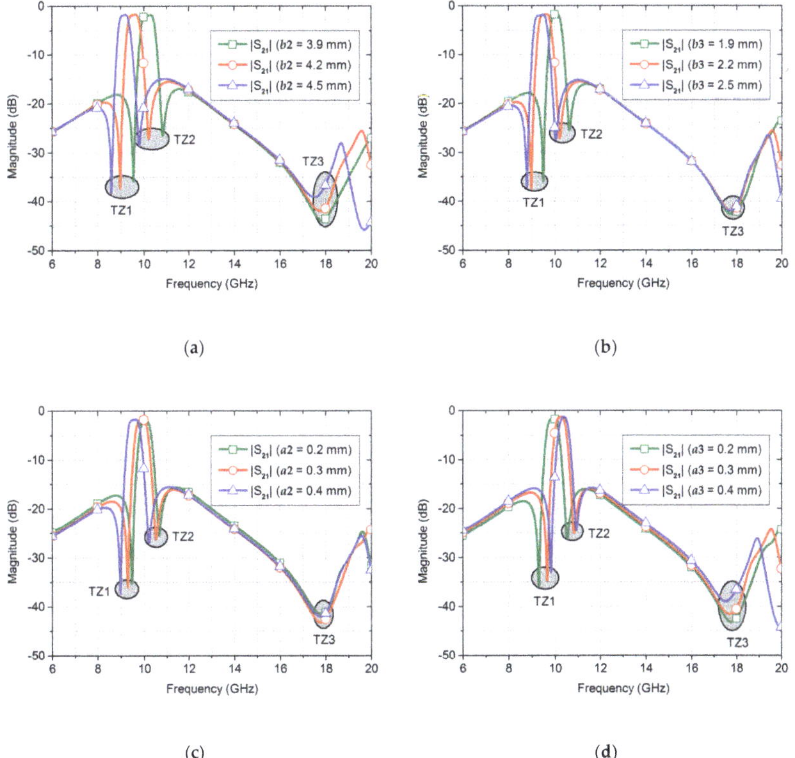

(a)

(b)

(c)

(d)

Figure 6. Simulated |S21| of the proposed single-ended bandpass filter with the variation of some geometrical parameters: (**a**) $b2$; (**b**) $b3$; (**c**) $a2$; (**d**) $a3$.

The proposed single-ended filter in Figure 5a is designed with a central frequency of 10 GHz, a bandwidth of 70 MHz (corresponding to a fractional bandwidth (FBW) of 7%) and a ripple in passband of 0.1 dB. With the aforementioned analyses, the filter is modeled and simulated by using the simulator mentioned above, and the optimized dimensions are listed as below: $ws = 1.52$, $lt = 1.8$, $a0 = 1.0$, $b0 = 2.0$, $a1 = 0.4$, $b1 = 6.0$, $c1 = 2.0$, $v1 = 0.4$, $s1 = 0.54$, $s2 = 0.25$, $s3 = 0.35$, $a2 = 0.36$, $a3 = 0.2$, $b2 = 4.2$, $b3 = 2.1$ (units: mm). Afterwards, the optimized single-ended filter is fabricated on a

RT/Duroid 5880 substrate with a thickness of 0.508 mm, a relative permittivity of 2.2 ± 0.2, and a loss tangent of 0.0009. The fabrication is carried out by utilizing the standard printed circuits board (PCB) process, with a 2 µm-thick gold cover being electroplated on the copper metal after fabrication. The fabricated single-ended filter is measured by using a Keysight N5245A vector network analyzer and a test fixture. Figure 7 shows photographs and measured results of the fabricated single-ended filter. It can be seen in Figure 7a that the fabricated single-ended filter has a physical size of 15 mm × 3.7 mm, corresponding to an electrical size of 0.135 λ_g^2, where λ_g is the guided wavelength of the microstrip line at the central frequency. According to Figure 7b, the fabricated single-ended filter achieves a central frequency of 10.12 GHz, a FBW of 6.5%, an in-band insertion loss of 1.4 dB, and a return loss better than 22 dB. Moreover, three TZs, at 8.8 GHz, 11.2 GHz, and 19 GHz, respectively, are achieved as well. Furthermore, as shown in Figure 7c, the fabricated single-ended filter exhibits a stopband covering 11 GHz to 35 GHz with rejection over 21 dBc, illustrating good stopband performance. Finally, from Figure 7d, the simulated group delay of S21 is less than 0.75 ns, with a small in-band variation less than 0.2 ns. Meanwhile, the measured group delay of S21 is less than 1 ns, and the variation of group delay in the passband is less than 0.35 ns, which is suitable for linear phase applications. Above all, the fabricated single-ended filter achieves selectivity, stopband and phase performance, and its measured results are in good agreement with the simulated ones.

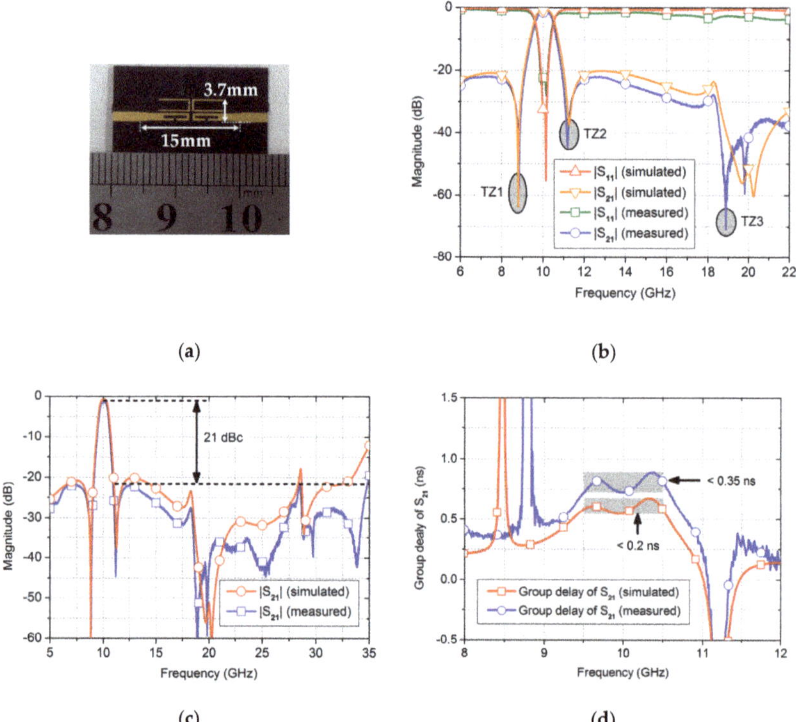

Figure 7. Photograph and measured results of the fabricated single-ended filter: (**a**) Photograph; (**b**) measured and simulated results; (**c**) wideband response of |S₂₁|; (**d**) measured group delay.

4. Balanced Bandpass Filters with Source–Load-Coupled Spurlines

It is well known that the differential transmission is better than the single-ended case, since the balanced signals are able to exhibit much better performance, including better immunity to environmental noise, weaker electromagnetic interference, better electromagnetic compatibility, and larger dynamic range [20]. Due to these advantages, balanced circuits and components have been widely used in wireless communication and sensing applications. For instance, in the fifth generation (5G) mobile communication system, the operating frequency becomes higher and its corresponding bandwidth gets wider as well. Consequently, circuits and components in the 5G system will be more sensitive to environmental noises. Hence, balanced components including filters are in urgent demand in wireless communication and sensing communities.

In order to further explore the availability of the USIR and the proposed T-shaped spurline in complex or multi-port microwave components, a balanced bandpass filter is designed. Figure 8 shows the detailed configuration and coupling scheme of the proposed balanced filter. As shown in Figure 8a, the proposed balanced filter consists of two USIRs, four T-shaped spurlines, and four 50 ohm microstrip lines at the input and output ports. Moreover, it can be seen from Figure 8b that there are four individual coupling paths (i.e., S-R1-R2-L, S-L, S'-R1-R2-L', and S'-L') in the proposed balanced filter. Among these coupling paths, S-R1-R2-L and S'-R1-R2-L' are the major paths for signal transmission, while S-L and S'-L' are the source–load coupling paths for TZ generation. More specifically, once the proposed balanced filter is in operation, coupling paths S-R1-R2-L and S-L can cross-couple to generate two TZs at the lower and upper sides of the passband. Meanwhile, the other two coupling paths, S'-R1-R2-L' and S'-L', will function in a similar way. Nevertheless, this does not mean that these four coupling paths can help to produce four TZs for the proposed balanced filter. In fact, although these four coupling paths can form two individual cross-couplings, they perform for different sides of the differential signal. Specifically speaking, as S-R1-R2-L and S-L form cross-coupling to produce two TZs for the transmission of the positive (or negative) side of the differential signal, S'-R1-R2-L' and S'-L' will operate in a similar way for the transmission of the negative (or positive) side of differential signal. In addition, T-shaped spurlines at the source and load nodes of the proposed balanced filter can help to further enhance its stopband performance. In practical operation, T-shaped spurlines at the S and L nodes can produce an extra TZ for the transmission of the positive (or negative) side of differential signal, while T-shaped spurlines at the S' and L' nodes are beneficial for the negative (or positive) side. Comparing Figure 5 with Figure 8, it can be seen that the proposed balanced filter can be regarded as a combination of the single-ended filter and its mirror duplicate.

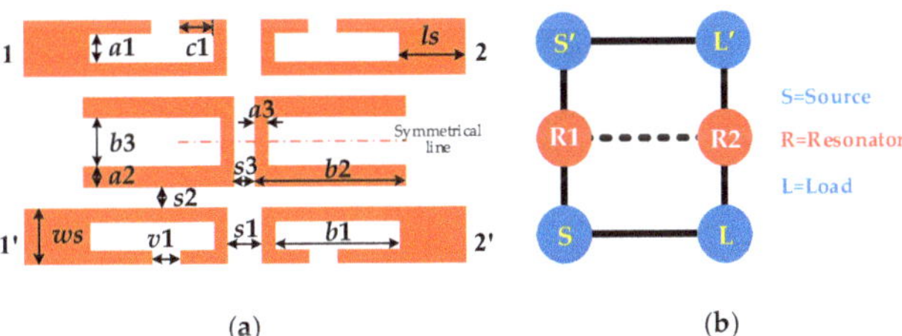

Figure 8. Configuration and coupling scheme of the proposed balanced filter: (**a**) Detailed configuration; (**b**) coupling scheme.

Before the balanced filter design, it is essential to investigate the influence of its constituent elements on its transmission properties. Influence from the USIR has been discussed in the realization of the single-ended filter; this can be applied to the balanced filter as well. This is mainly because the single-ended and balanced filters are with the same basic resonator and fundamental coupling scheme. Here, some numerical simulations have been carried out to study the influence of the T-shaped spurline on the transmission response of the balanced filter. In these simulations, the same commercial electromagnetic simulator as mentioned above is used, and the substrate is set with a thickness of 0.508 mm, a relative permittivity of 2.2, a loss tangent of 0.001, and a copper cover of 0.035 mm. Some geometrical parameters are fixed as well: $ws = 1.52$, $ls = 3$, $s1 = 0.6$, $s2 = 0.5$, $s3 = 0.8$, $a2 = 0.4$, $a3 = 0.2$, $b2 = 4$, $b3 = 2.2$ (units: mm). Figure 9 shows simulated $|S21|$ under differential mode (DM) operation with the variations of various geometrical parameters of the T-shaped spurline. For Figure 9a, $a1 = 0.7$ mm, $b1 = 3.7$ mm, $c1 = 1$ mm. In this case, as $v1$ becomes larger, the passband, TZ1, and TZ2 almost stay the same, while TZ3 moves to higher frequency. For Figure 9b, $a1 = 0.7$ mm, $b1 = 3.7$ mm, $v1 = 0.3$ mm. In this case, with a larger $c1$, TZ3 also shifts higher, whereas the passband, TZ1, and TZ2 remain nearly unchanged. Moreover, the variation of $c1$ will also have influence on the stopband performance. For Figure 9c, $a1 = 0.7$ mm, $c1 = 1$ mm, $v1 = 0.3$ mm. In this case, the passband, TZ1, and TZ2 just have very slight variety with the increasing of $b1$, while TZ3 shifts notably lower. Finally, for Figure 6d, $b1 = 3.7$ mm, $c1 = 1$ mm, $v1 = 0.3$ mm. In this case, the variation of $a1$ only has a slight influence on the passband, TZ1, TZ2, and TZ3. It can be derived from Figure 9 that TZ3 is contributed to by the spurline, while TZ1 and TZ2 are produced by the cross-coupling attributed to the source–load coupling scheme.

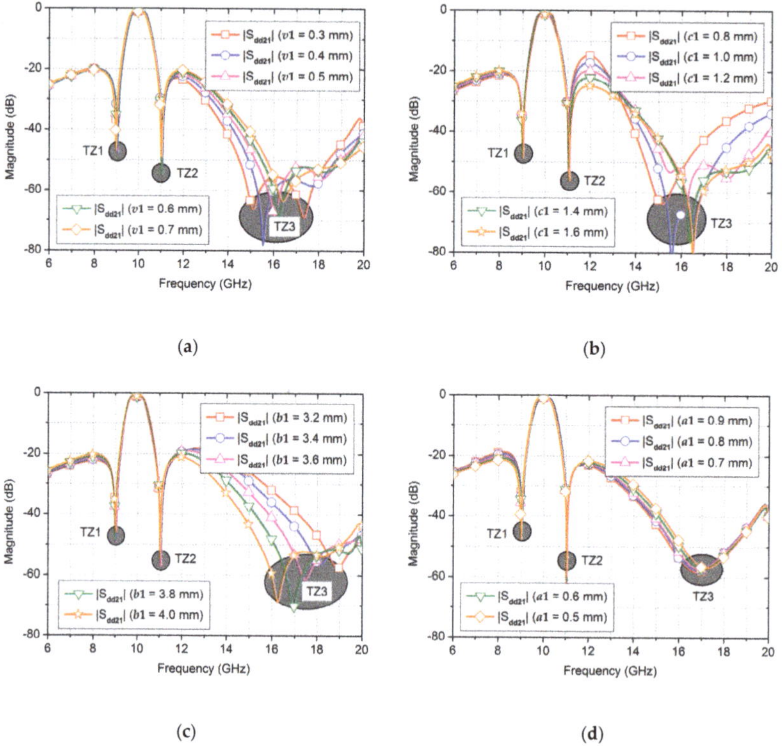

Figure 9. Simulated results of the proposed balanced filter with the variations of various geometrical parameters of the T-shaped spurlines: (**a**) $v1$; (**b**) $c1$; (**c**) $b1$; (**d**) $a1$.

Afterwards, the proposed balanced filter in Figure 8a is designed with the same specification parameters as the single-ended one (i.e., a central frequency of 10 GHz and a FBW of 7%). By using the same simulators mentioned above, its geometrical parameters are optimized to be: $ws = 1.52$, $ls = 1.0$, $a1 = 0.72$, $b1 = 3.8$, $c1 = 1.35$, $v1 = 0.37$, $s1 = 0.65$, $s2 = 0.51$, $s3 = 0.62$, $a2 = 0.4$, $b2 = 4.3$, $a3 = 0.2$, $b3 = 2.1$ (units: mm). The optimized balanced filter is fabricated by using the standard PCB process on a commercial Rogers RT/Duroid 5880 substrate with a relative permittivity of 2.2±0.02, a relative permeability of 1, a loss tangent of 0.0009, and a thickness of 0.508 mm. The measurement is carried out by using a Keysight N5245A vector network analyzer and a self-designed test fixture. Measured results of the fabricated BPF is illustrated in Figure 10, in which it can be seen easily that the measured results agree with the simulated ones well. According to Figure 10a, the prototyped balanced filter achieves a 1 dB fractional bandwidth about 6.1%, a minimum in-band insertion loss of 0.9 dB, and three TZs located at 9 GHz, 11 GHz, and 15.2 GHz, all under the DM operation. From the wideband response of |S21| under the DM operation given in Figure 10b, its upper stopband is far from up to 40 GHz (i.e., the fourth harmonic) with the suppression better than 20 dB. The measured in-band absolute group delay of S21 under the DM operation is less than 1.3 ns, with a small variation less than 0.3 ns, as shown in Figure 10c. It can be captured easily from Figure 10c that the measured in-band absolute group delay of S21 is about 0.25 ns larger than the simulated one, whereas the measured variation is nearly as same as the simulation. Such group delay performance illustrates that the fabricated balanced filter is suitable for digital data transmission applications in communication systems. Moreover, according to the results under the CM operation in Figure 10d, it can be seen that the measured insertion loss under the CM operation is over 17 dB, which means the measured CMRR is better than 16.1 dB. Additionally, it can be obtained easily from Figure 10e that the functional part of the fabricated balanced filter has a physical size of 9.4 mm × 5.7 mm, corresponding to an electrical size of $0.129 \ \lambda_g^2$.

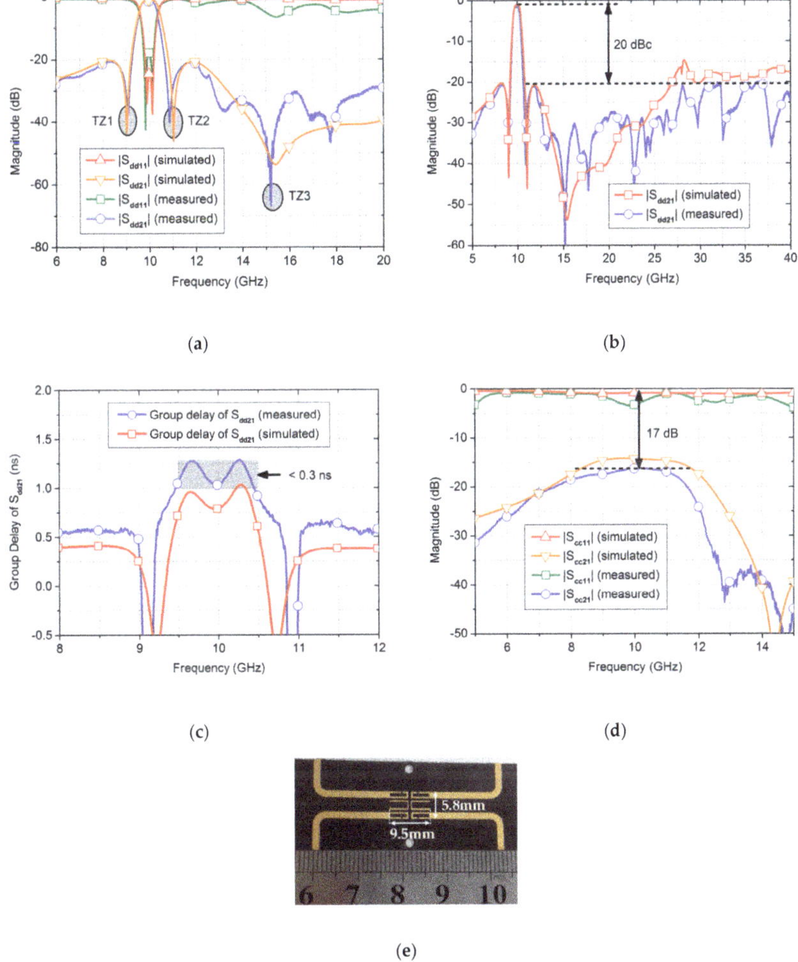

Figure 10. Measured results and photograph of the fabricated balanced filter: (**a**) |S21| and |S11| under differential mode (DM) operation; (**b**) wideband response of |S21| under DM operation; (**c**) |S21| and |S11| under CM operation; (**d**) measured group delay of S21 under DM operation; (**e**) photograph.

5. Discussion

Table 1 summarizes the comparison between the proposed balanced filter and some similar reported works. As compared with the works reported in [11,13,15], the proposed one has similar insertion loss and the largest number of TZs, which means the proposed one exhibits better selectivity performance. Moreover, the proposed balanced filter has wider a stopband, with about 20 dB out-of-band rejection as well. Even though the previously reported balanced filters listed in Table 1 have higher CMRR than the proposed one, they also suffer the issue of larger electrical sizes. For instance, compared with the balanced filter reported in [15], the proposed one achieves an electrical size reduction of approximately half. In addition, the differential filter reported in [13] is very unique since it utilizes lumped resistors in the filter to absorb CM noise. Hence, in the passband of this differential filter, both the transmission coefficient and reflection coefficient under the CM operation (i.e., S_{cc21} and S_{cc11}) have low values. More specifically, as reported in [13], the measured $|S_{cc11}|$ in the passband is about -25.38 dB, while the measured $|S_{cc21}|$ in the passband is slightly lower than the $|S_{cc11}|$. Meanwhile, the measured insertion loss under the DM operation is as small as 0.29 dB. Therefore, it can be obtained that the differential filter in [13] achieved a CMRR around 25 dB. It can be seen from Table 1 that the proposed balanced filter has the best selectivity and out-of-band rejection and the smallest electrical size.

Table 1. Comparison between the proposed balanced filter and some similar reported works.

Ref.	f_0, FBW [1] (GHz, %)	Insertion Loss (dB)	Number of TZ [2]	Stopband (dBc, $n \cdot f_0$)	CMRR [3] (dB)	Electrical Size ($\lambda_0^2 / \varepsilon_r$)
[11]	4, 16	1.5	0	20, 2	30	0.510
[13]	2, 55.5	0.29	1	N.A.	25	0.329
[15]	1.9, 47.4	1.5	2	22, 3.4	28	0.251
Ours	10, 6.1	0.9	3	20, 4	16.1	0.128

[1] FBW: fractional bandwidth. [2] TZ: transmission zero. [3] CMRR: common-mode rejection ratio.

6. Conclusions

USIR-based single-ended and balanced bandpass filters with compact transverse sizes, high selectivity, and wide stopbands were presented. Particularly, the balanced filter has exhibited good DM and CM performance, which is jointly attributed to the source–load-coupled spurlines and USIRs. Meanwhile, the out-of-band rejection over 20 dB under DM operation of the balanced filter was extended up to the fourth-order harmonic of the central frequency, with its CMRR being better than 16.1 dB through the passband as well. These results indicate that the proposed balanced filter is a promising alternative for microwave and millimeter circuit and system applications.

Author Contributions: Conceptualization, F.Y. and K.W.; investigation, Y.M.H. and T.H.; writing—original draft preparation, F.Y. and Y.M.H.; writing—review and editing, S.D. and M.B.

Funding: This research was funded by National Natural Science Foundation of China, grant no. 61601087; Sichuan Science and Technology Program, grant no. 2018GZ0518.

References

1. Rebeiz, G.M.; Entesari, K.; Reines, I.C.; Park, S.J.; Eltanani, M.A.; Grichener, A.; Brown, A.P. Tuning in to RF MEMS. *IEEE Microw. Mag.* **2009**, *10*, 55–72. [CrossRef]

2. Pozar, D.M. *Microwave Engineering*, 4th ed.; Wiley: New York, NY, USA, 2012; pp. 48–89. ISBN 978-0-470-63155-3.

3. Makimoto, M.; Yamashita, S. Bandpass filter using parallel coupled stripline stepped impedance resonators. *IEEE Trans. Microw. Theory Tech.* **1980**, *28*, 1413–1417. [CrossRef]

4. Pinel, S.; Bairavasubramanian, R.; Laskar, J.; Papapolymerou, J. Compact planar and vialess composite low-pass filters using folded stepped-impedance resonator on liquid-crystal-polymer substrate. *IEEE Trans. Microw. Theory Tech.* **2005**, *53*, 1707–1712. [CrossRef]

5. Roychoudhury, S.; Parui, S.K.; Das, S. Improvement of stopband using a spiral defected microstrip and defected ground structures. In Proceedings of the IEEE Applied Electromagnetics Conference, Kolkata, India, 18–22 December 2011.

6. Liu, H.; Knoechel, R.H.; Schuenemann, K.F. Miniaturized bandstop filter using meander spurline and capacitively loaded stubds. *ETRI J.* **2007**, *29*, 614–618. [CrossRef]

7. Liao, C.K.; Chang, C.Y. Design of microstrip quadruplet filters with source-load coupling. *IEEE Trans. Microw. Theory Tech.* **2005**, *53*, 2302–2308. [CrossRef]

8. Liu, H.; Liu, F.; Guan, X.; Ren, B.; Liu, T.; Wang, Y.; Xu, H. Wide-stopband superconducting bandpass filter using slitted stepped-impedance resonator and composite spurline structure. *IEEE Trans. Appl. Supercond.* **2018**, *28*, 1501508. [CrossRef]

9. Ponchak, G.E. Coplanar stripline spurline stub resonators with even-mode suppression for bandpass and bandstop filters. *IEEE Microw. Wirel. Compon. Lett.* **2018**, *28*, 1098–1100. [CrossRef]

10. Shi, J.; Xue, Q. Balanced bandpass filters using center-loaded half-wavelength resonator. *IEEE Trans. Microw. Theory Tech.* **2010**, *58*, 970–977.

11. Fernandez-Prieto, A.; Qian, S.L.; Hong, J.; Martel, J.F.; Medina, F.; Mesa, F.; Naqui, J.; Martin, F. Common-mode suppression for balanced bandpass filters in multilayer liquid crystal polymer technology. *IET Microw. Antennas Propag.* **2015**, *9*, 1249–1253. [CrossRef]

12. Sans, M.; Selga, J.; Velez, P.; Rodriguez, A.; Bonanche, J.; Borria, V.E.; Martin, F. Automated design of common-mode suppressed balanced wideband bandpass filters by means of aggressive space mapping. *IEEE Trans. Microw. Theory Tech.* **2015**, *63*, 3896–3908. [CrossRef]

13. Zhang, W.; Wu, Y.; Liu, Y.; Yu, C.; Hasan, A.; Ghannouchi, F.M. Planar wideband differential-mode bandpass filter with common-mode noise absorption. *IEEE Microw. Wirel. Compon. Lett.* **2017**, *27*, 458–460. [CrossRef]

14. Ebrahimi, A.; Baum, T.; Ghorbani, K. Differential bandpass filters based on dumbbell-shaped defected ground resonators. *IEEE Microw. Wirel. Compon. Lett.* **2018**, *28*, 129–131. [CrossRef]

15. Sans, M.; Selga, J.; Velez, P.; Bonanche, J.; Rodriguez, A.; Borria, V.E.; Martin, F. Compact wideband balanced bandpass filters with very broad common-mode and differential-mode stopbands. *IEEE Trans. Microw. Theory Tech.* **2018**, *66*, 737–750. [CrossRef]

16. Fernandez-Prieto, A.; Lujiambio, A.; Martel, J.; Medina, F.; Mesa, F.; Boix, R.R. Balanced-to-balanced microstrip diplexer based on magnetically coupled resonators. *IEEE Access* **2018**, *6*, 18536–18547. [CrossRef]

17. Gomez-Garcia, R.; Munoz-Ferreras, J.M.; Feng, W.; Psychogiou, D. Balanced symmetrical quasi-reflectionless single- and dual-band bandpass filters. *IEEE Microw. Wirel. Compon. Lett.* **2018**, *28*, 798–800. [CrossRef]

18. Feng, W.; Zhao, Y.; Che, W.; Gomez-Garcia, R.; Xue, Q. Single-ended-to-balanced filtering power dividers with wideband common-mode suppression. *IEEE Trans. Microw. Theory Tech.* **2018**, *66*, 5531–5542. [CrossRef]

19. Sagawa, M.; Makimoto, M.; Yamashita, S. Geometrical structure and fundamental characteristics of microwave stepped-impedance resonators. *IEEE Trans. Microw. Theory Tech.* **1997**, *45*, 1078–1085. [CrossRef]

20. Ting, H.-L.; Hsu, S.-K.; Wu, T.-L. Broadband eight-port forward-wave directional couplers and four-way differential phase shifter. *IEEE Trans. Microw. Theory Tech.* **2018**, *66*, 2161–2169. [CrossRef]

Review

Carbon-Based Composite Microwave Antennas

Nikolai A. Dugin [1], Tatiana M. Zaboronkova [2], Catherine Krafft [3,*] and Grigorii R. Belyaev [4]

[1]   Department of Propagation of Radiowaves and Radioastronomy, N.I. Lobachevsky University of Nizhny Novgorod, 23 Gagarin Ave., Nizhny Novgorod 603950, Russia; ndugin@yandex.ru
[2]   Department of Nuclear Physics, R.E. Alekseev State Technical University of Nizhny Novgorod, 24 Minin St., Nizhny Novgorod 603950, Russia; t.zaboronkova@rambler.ru
[3]   Laboratoire de Physique des Plasmas, Ecole Polytechnique, 91128 Palaiseau CEDEX, France
[4]   Department of Computer Science and Telecommunications, Volga State University of Water Transport, 5a Nesterova St., Nizhny Novgorod 603950, Russia; welles16@yandex.ru
*   Correspondence: catherine.krafft@lpp.polytechnique.fr

Received: 6 March 2020; Accepted: 25 March 2020; Published: 31 March 2020

Abstract: Applications of metamaterials to microwave antennas are reviewed over the past decade. The manufacturing of microwave antennas using graphene-containing carbon composite materials was developed and prototypes of dipole and horn antennas made from such materials were created. The radiation properties of the designed antennas and their metal analogs were measured and compared. The standing wave ratios, the radiation patterns and the amplitude-frequency characteristics were analyzed for horn antennas at frequencies 1.6 GHz and 5 GHz and for dipole antennas in the frequency range 0.2–0.6 GHz. The polarization characteristics of the horn antennas were studied. The effects of different carbon composite materials' structures (fiber or fabric) on the antennas' parameters were estimated. It is shown that antennas made from graphene-containing composite materials are able to operate efficiently and exhibit almost the same radiation properties as conventional metal antennas of the same geometry and size. However, the carbon-based antennas have much smaller weights and enhanced stability in a wide range of temperatures. In the future, such antennas should replace the conventional ones for many applications, especially for the excitation and reception of electromagnetic waves in space plasmas.

Keywords: microwave waveguide; dipole and horn antennas; carbon-composite material; graphene; antenna mesurements; electromagnetic waves

---

1. Introduction

Due to their wide domain of applications, composite materials attracted the attention of many researchers. Therefore a lot of efforts were made, including by the antenna community, to create various composite materials for industry [1–5]. Many works reported on the applications of composite metamaterials possessing unique properties in the microwave range 1–100 GHz [6–9]. The feasibility of left-handed materials, for which optical laws are significantly modified, was demonstrated by some authors [10]. Since pioneering works that raised much interest 50 years ago [10], different types of artificial materials were created for various applications, such as photonic crystals [11], electromagnetic band-gap [12], left-handed materials [13], graphene-containing materials [14], etc. The idea to use metamaterials in the framework of antenna technology was motivated by their ability to enhance the gain and the directivity of antennas while minimizing their size, to reduce the visibility of objects, etc. Today it is well-known that the technical designing of resonators, filters, high-gain horn antennas, transmission lines, and antenna covers is based on composite metamaterials [15–20].

As is well-known, carbon-based materials are widely used in civil industries such as, for example, in the fields of aerospace or shipbuilding. Moreover, the use of graphene-containing carbon composite materials (GCMs) in the design of antennas seems also very promising. The carbon composite materials have a long lifetime, an enhanced immunity to corrosion, the record-breaking durability-to-weight ratios, a high stability within a wide domain of temperatures [21,22]. They possess such important characteristics that they were used to create microwave antennas and antenna-feeder devices. For example, carbon materials are required for the manufacture of space, aircraft and helicopter antennas, in order to reduce their weight and to enhance their thermal expansion and their durability. The creation of composite graphene radio coatings might allow modernizing expensive radio telescopes. In this paper, an outline of recent advances concerning the carbon-based composite microwave antennas is given [23–26]. Moreover, the article includes significant new research related to dipole antennas with reflectors made from GCMs. Methods of manufacturing microwave antennas using GCMs have been developed [23] and samples of waveguides as well as microwave horn and dipole antennas were created. The radiation properties of such antennas were measured and compared with their metal analogs generally used for radio astronomical observations [24–26].

Therefore metamaterials can be used in various antenna systems for improving the directivity of electromagnetic waves, increasing the antennas' gain, obtaining excellent transmitting characteristics in wide frequency ranges, suppressing interactions between the antenna elements, innovating in antenna design, and reducing the weight and the size of current antennas. Furthermore, we shall discuss in more detail the applications of GCMs in antenna and radio engineering.

2. Carbon Composite Materials in Antenna Technology

Carbon composite materials are generally used for manufacturing the parts of high-loaded structures of space antennas. For example, it was proposed to use such materials to build the supporting structures of the umbrella-type spacecraft antenna [27] for which, however, a metallized coating was required to ensure the electrical conductivity of its reflecting surface. This coating consists of metal particles or plates, which make the antenna's radiation properties similar to those of the conventional metal prototypes [27]. A method of forming an antenna including the molding of a non-planar antenna surface with graphene coating was suggested by some authors [28]; after the graphene is coated on the antenna body, metallization, sputtering, or chemical plating is not required to ensure conductivity. It was also suggested to manufacture metamaterials consisting of concentric rings containing graphene, which have a higher conductivity than metallic components [29]. Therefore, this type of metamaterial cover suppresses the propagation loss of electromagnetic waves in antenna-fider devices.

The use of dielectric metamaterials containing graphene as elements of waveguides' or antennas' systems (such as inserts, corrugated structures, coatings, etc.) can improve their reflection and scattering properties [14]. Microwave transmission and reflection in multilayer composite materials consisting of cardboard, gypsum, and natural flaked graphite were studied [14,30]. The main specific characteristics of radio-absorbing materials are their thickness and the working range of the radiation wavelengths. Composite absorbing multilayer materials for the microwave range are filled with powdered metals, graphene, ferrites, or their mixtures. These absorbers may be manufactured in different forms as layers, columns, pyramids, and other complicated structures [31]. Cardboard was used as an outer layer (with the effective dielectric permittivity $\varepsilon_{eff} = 1.5$) that ensured the matching to free space. A minimum coefficient of reflection and transmission was observed for the four-layer samples with partly oriented graphite flakes, aimed at increasing the absorption coefficients of materials. The results obtained by some authors [14] showed the possibility to create effective radio-absorbing materials and coatings based on carbon composite materials widely used in the civil industry.

A model of gap-mode waveguide, which is a type of conductive tubular electromagnetic waveguide operating within the frequency range 1 GHz–10 THz, was proposed [32], for which the absorber provides a low-loss and low-dispersion propagation signal and improves the mechanical characteristics. It was stressed that anisotropic conductor walls (including organic conductors, graphene, carbon nanotubes) may have a specific conductivity in the direction of wave propagation [32].

Considering the polarization properties of the GCM in the microwave range [33], it was found that, when the GCM thin plate was rotated by $90°$, the ratio of the polarization coefficient to the field amplitude reached about 0.7–0.8, depending on the concentration of the binding substance in the composite material. Please note that the anisotropic properties of the artificial materials may influence on the radiation characteristics of the electromagnetic waves, including such important features as the input impedance, the radiation pattern, the antenna gain, etc. The radiation of thin strip metal antennas located on both cylindrical and plane surfaces of a uniaxial metamaterial was considered by some authors [34–36], and the influence of the material anisotropy on the radiation properties of these antennas was analyzed. It was shown that strip antennas located on anisotropic metamaterial slabs may have different characteristics depending on the properties of the artificial media. Therefore the anisotropy of composite materials may be used to obtain desired characteristics for composite microwave radio-technical devices.

The properties of GCMs enable their efficient use in the creation of ground-based and space-borne antenna systems and seem promising for the future antenna technologies. Despite a wide use of metamaterials in antenna techniques, little is known about the characteristics of microwave antennas made from carbon-based composite materials. As mentioned above, when using such materials for the reflectors of space antennas, their surface is covered by a material containing either metal or graphene particles (or both together) for providing electric conduction. Our recent studies have shown the possibility to use composite materials not only for manufacturing elements and coating microwave devices, but also for creating microwave waveguides [37] and antenna systems [23,24,26]. We have developed a method for creating microwave dipole and horn antennas made from a GCM [23], and have shown that such antennas exhibit almost the same characteristics as their metal analogs.

The L- and C-range horn antennas are operating at the frequencies 1.6 GHz and 5 GHz, respectively, whereas the dipole antennas operate in the lower microwave frequency range 0.2–1 GHz. Radiation properties of carbon-based antenna prototypes as the standing-wave ratio (SWR), the amplitude-frequency characteristics and the radiation pattern were measured and compared with those of their metal analogs [24–26]. The polarization characteristics of a GCM microwave horn antenna were studied; comparisons with a similar antenna made from metal alloys [38,39] showed that the polarization coefficient of the composite antenna was 1.5–2 times less than that of the metal analog. Polarization characteristics only exhibit differences between the GCM horn antennas and their metal analogs. The first antenna prototypes were made from a quasi-isotropic composite material. We used the carbon fiber from the brand *Zoltek Panex 35 (50K)* for manufacturing one prototype of composite-material-based antenna and carbon fabric for producing the second one. The carbon fabric is a plain-weave of carbon thread, with a unit cell size equal to 2 cm $\times$ 2 cm, a tensile strength of 4137 MPa and an electrical resistivity of $1.55 \cdot 10^{-5}$ Ωm (which ranges within 10^{-7}–10^{-3} Ωm for metals) [40]. To reach structural rigidity, a graphene-containing epoxy binding substance was used, i.e., an epoxy resin modified by graphene in an amount providing the necessary electrical conductivity.

Beside horn antennas, several prototypes of composite dipole antennas at 200 MHz, 600 MHz and 540 MHz were created. For the two first ones, the radiating elements (whiskers) are made from GCMs; for the third one, the reflector and both whiskers are made from GCM. For convenience, Table 1 presents the main parameters of various types of GCM antennas and of their metal analogs. More details can be found in Section 3 (dipole antennas) and Section 4 (horn antennas). Table 2 contains the mechanical properties of metal and GCM.

Table 1. Main antenna properties and parameters in the working frequency interval, for the composite prototype (with curricular winding of the fiber) and its metal analog (see Sections 3 and 4 for more details). DA: dipole antenna; HA: horn antenna; PU: polarization unit; $E_{max}^{(GCM)}$ and $E_{max}^{(Met)}$ are the maximum values of the field amplitude of the emitted signal for the GCM prototype and its metal analog, respectively; the sign "*" denotes the weight of the reflector of the dipole antenna at 540 MHz; the reflector of dipole antennas at 200 MHz and 600 MHz are made from metal; the weight is indicated for GCM antennas manufactured in laboratory conditions (in industrial conditions it may be two times smaller).

Antenna Properties	Material Type		DA 200 MHz	DA 600 MHz	DA 540 MHz	HA 1.6 GHz without PU	HA 1.6 GHz with PU	HA 5 GHz without PU
SWR_{min}	Metal		1.1	1.15	1.2	1.05	1.05	1.3
	GCM		1.07	1.13	1.2	1.1	1.15	1.25
Main-lobe width (degree)	Metal			86	80	58.6	58.6	60
of Radiation Pattern	GCM			90	75	61.4	61.4	60
$\left\| \dfrac{E_{max}^{(GCM)}}{E_{max}^{(Met)}} \right\|^2$	Metal		1	1	1	1	1	1
	GCM		1.08	0.93	1.03	0.9	0.88	1.1
Polarization	Metal						0.55–0.8	
	GCM	thread					0.3–0.4	
Weight (kg)	Metal				1.91 $(Al)^*$	8	8.2	0.51
	GCM				1.07 $(GCM)^*$	0.5	0.7	0.14
Size (m)			whiskers 0.75	whiskers 0.25	whiskers 0.28	opening 0.152	opening 0.152	opening 0.05

Table 2. Main properties of graphene-containing composite material antennas and corresponding metal analogs.

PARAMETERS	Zoltek Px 35	Brass	Al (Aluminium)
Electrical resistivity (Ωm)	$15.5 \cdot 10^{-6}$	$6 \cdot 10^{-8}$	$3 \cdot 10^{-8}$
Tensile strength (MPa)	4137	450	100
Tensile modulus (GPa)	242	100	75
Density (kg·m^{-3})	1810	8500	2700
Coefficient of thermal expansion ($^\circ$K^{-1})	$8 \cdot 10^{-8}$	$19.1 \cdot 10^{-6}$	$23.8 \cdot 10^{-6}$
Temperature of melting (decomposition) of metal (GCM) ($^\circ$C)	> 650	900	650

Thus, the carbon composite antennas have light weights, are of robust construction and low cost. The GCMs can be used as transparent, conductive and radio-absorbing coating reducing the visibility of objects by remote sensing [14]. For ground-based and space-born antenna systems, they ensure mechanical characteristics as enhanced immunity to corrosion, high strength in combination with low relative weight, high stability within wide domains of temperatures. Moreover, the GCM antennas are able to be used as receivers and transmitters of electromagnetic signals. Below we review the main properties of the dipole and horn microwave antennas made from GCMs.

3. Dipole Antennas

Two prototypes of composite dipole antennas have been realized and studied: (i) specialized antennas operating at two frequencies, 200 MHz and 600 MHz, with broad frequency bands, and (ii) standard dipole antennas operating at 540 MHz.

3.1. Specialized Dipole Antennas at 200 MHz and 600 MHz

As analogs of composite dipole antennas, we choose special metal dipoles operating at 200 MHz and 600 MHz within expanded bandwidths. A specialized composite dipole antenna operating at 600 MHz is shown in Figure 1. Due to the design features of the dipole antennas, their radiating elements (so-called whiskers) made from composite material have a metal thread for attaching to the feeder. The length of the radiating elements is half a wavelength; they are 75 cm and 25 cm long at 200 MHz and 600 MHz, respectively. The metal reflectors of the dipole prototypes and their metal analogs were the same. The insert of the whiskers of the prototypes was made from either duralumin or fluoroplastic. During its manufacturing, we used *Zoltek Panex* carbon fiber as well as epoxy resin modified by graphene powder as binding substance. Radiating elements were realized using either longitudinal laying or circular winding on the dielectric insert (see Figure 2).

Figure 1. View of the dipole antenna operating at 600 MHz with its radiation elements made from carbon fiber.

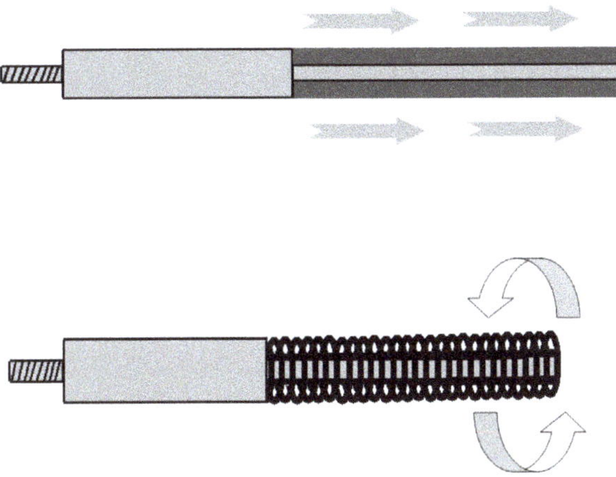

Figure 2. Radiation elements of the dipole antenna, with longitudinal laying on the dielectric insert (top panel) and circular winding on the dielectric insert (bottom panel).

The SWRs, the emitted power frequency dependence and the radiation patterns of the different antenna prototypes, and their metal analogs were measured. The influence of the types of laying the carbon fiber and of material of the dielectric insert on the antenna characteristics was estimated. As an example, Figure 3 shows the SWRs and the frequency variation of the power emitted by the dipole antennas operating within the 200 MHz frequency band, for three prototypes with different inserts, i.e., with (i) metal radiating elements, (ii) a longitudinal laying of the carbon fiber on the fluoroplastic insert, and (iii) a curricular winding of the fiber on the duralumin insert. Please note that the SWR and the emitted power frequency dependence of the dipole antenna corresponding to (ii) are slightly worse than for the other designed prototypes.

Figure 4 shows the radiation patterns and the emitted power frequency dependence of the dipole antenna operating within the 600 MHz frequency band, for four prototypes with different inserts, i.e., with (i) metal radiating elements, (ii) a longitudinal laying of the carbon fiber on the duralumin insert, (iii) a curricular winding of the fiber on the fluoroplastic insert, and (iv) a curricular winding of the fiber on the duralumin insert. One can observe that the frequency dependence of the emitted power and the radiation pattern of the various composite prototypes almost coincide.

The radiation properties of antennas were measured under laboratory conditions without using an anechoic chamber. Please note that the interpolation of the radiation patterns' measurements by Gaussian functions was not required. For the antennas operating at 200 MHz and 600 MHz, the main-lobe widths of the radiation patterns and the emitted power frequency dependence of their various prototypes coincide within the measurement uncertainties (around 10%). Thus, as it is seen in Figures 3 and 4, the electric properties weaklydepend on the type of dielectric material used for manufacturing the insert of the whiskers.

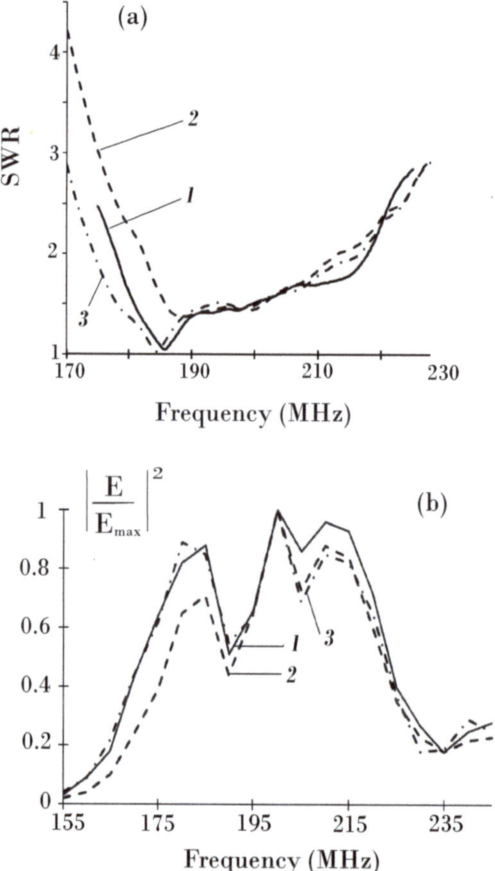

Figure 3. Dipole antenna operating at 200 MHz. **(a)** SWRs and **(b)** Square of the normalized field amplitude of the emitted signal as a function of frequency for 3 different prototypes with (i) metal radiating elements (curve 1, solid line), (ii) a longitudinal laying of the fiber on the fluoroplastic insert (curve 2, dashed line), and (iii) curricular winding of the fiber on the duralumin insert (curve 3, dashed-dotted line). E_{max} is the maximum value of the field amplitude.

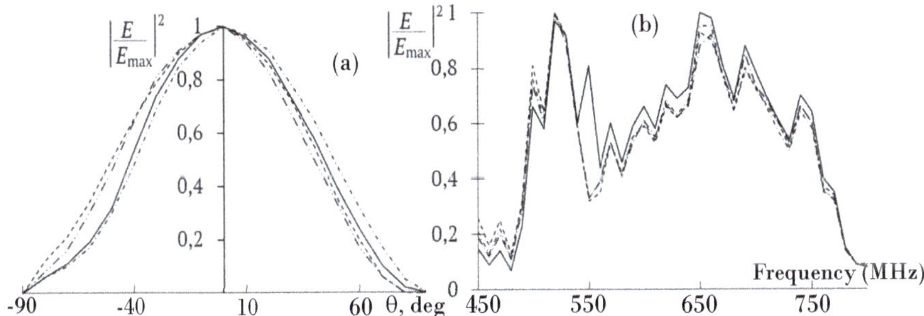

Figure 4. Dipole antenna operating at 600 MHz. (**a**) Radiation pattern and (**b**) Square of the normalized field amplitude of the emitted signal as a function of frequency for four different prototypes with (i) metal radiating elements (solid line), (ii) a longitudinal laying of the carbon fiber on the duralumin base (dashed line), (iii) a curricular winding of the fiber on the fluoroplastic insert (short dashed-dotted line), and (iv) a curricular winding of the fiber on the duralumin insert (long dashed-dotted line). E_{max} is the maximum value of the field amplitude.

3.2. Standard Dipole Antennas at 540 MHz

Several prototypes of a dipole antenna operating at 540 MHz and completely made from GCM were created. Their metal analogs were dipole mirror parabolic antennas operating at 540 MHz and 1040 MHz, with one central rod for attaching to the main mirror, as antennas generally used for radio astronomical measurements [41].

The whiskers of the standard dipole composite antennas were made from either carbon fiber or carbon fabric, using the same technology as for specialized dipole antennas. The reflectors, of circular shape, consisted of *Zoltek Panex* carbon fabric. The length of the whiskers is 28 cm, and the radius of the reflector is 30 cm. The weight of the composite material reflector is 1.5–2 times less than that of its metal analog. The main radiation properties of the prototypes were measured and compared with those of their metal analogs. The influence of the types of reflector and fiber laying on the antenna parameters were estimated. Figure 5 shows the SWRs and the emitted power frequency dependence of six different protopypes with the following characteristics: (i) metal radiating elements and metal reflector, (ii) metal radiating elements and carbon fabric reflector, (iii) curricular winding of the carbon fiber on the dielectric insert and metal reflector, (iv) longitudinal laying of the carbon fiber and metal reflector, (v) curricular winding of the fiber and carbon fabric reflector, and (vi) longitudinal laying of the fiber and carbon fabric reflector.

A good agreement between the measured characteristics of the five composite antenna prototypes and of their metal analog was observed, as shown in Figure 5, which is quite satisfactory for the laboratory tests.

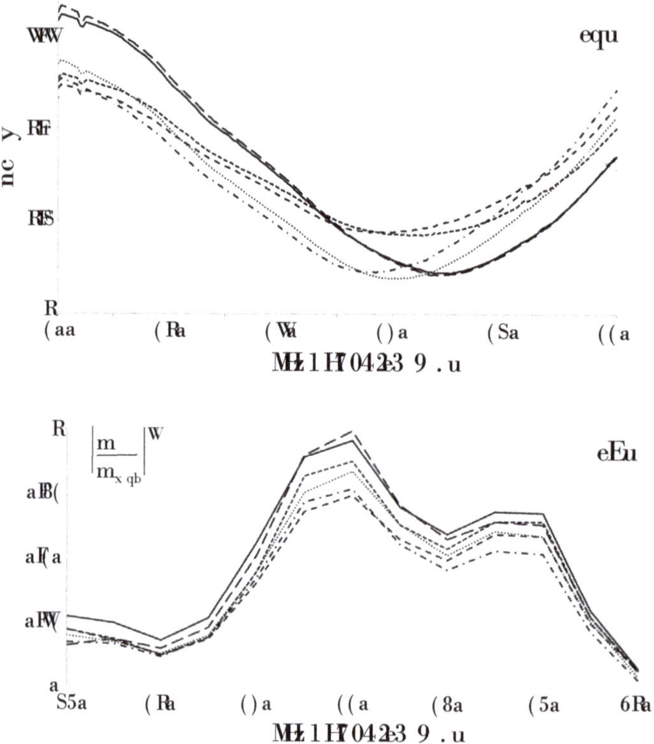

Figure 5. Dipole antenna at 540 MHz. (**a**) SWRs and (**b**) Square of the normalized field amplitude of the emitted signal as a function of frequency for six different protopypes with the following charcateristics: (i) metal radiating elements and metal reflector (solid line), (ii) metal radiating elements and carbon fabric reflector (long-dashed line), (iii) curricular winding of the carbon fiber and metal reflector (dashed line), (iv) longitudinal laying of the carbon fiber and metal reflector (dashed-dotted line), (v) curricular winding of the fiber and carbon fabric reflector (dotted line), and (vi) longitudinal laying of the fiber and carbon fabric reflector (short-dashed line). E_{max} is the maximum value of the field amplitude.

4. Horn Antennas

GCM microwave horn antennas were realized for operating frequencies belonging to two ranges: (i) the L–range with a central frequency of 1.6 GHz, and (ii) the C-range with a central frequency of 5 GHz. The antenna prototypes consist of a horn connected to a segment of circular waveguide. Their characteristic parameters were determined owing to a theoretical modeling considering a waveguide with circular cross section and finite wall thickness, immersed in vacuum. The conductivity of the waveguide's walls was described by a uniaxial permittivity tensor with nonzero diagonal elements. A dispersion equation was obtained taking into account the boundary conditions for the tangential components of the fields at the boundaries of the walls [38]. This equation allowed us to estimate the geometrical parameters of the waveguide segment of the horn antenna enabling an efficient radiation of the transverse fundamental mode H_{11} of the circular waveguide (or of the mode H_{01} of a rectangular waveguide) [42]. For our prototypes, the scale of the skin layer was much less than the thickness of the antenna walls.

4.1. Manufacturing Horn Antennas

During the manufacturing of the GCM horn antennas, we used either *Zoltek Panex* carbon fiber or fabric, as well as resin modified by graphene powder. Please note that the conductivity of the composite used was quasi-isotropic. To create the antenna, a blank matrix of duralumin was made, with an external size coinciding with the calculated geometric parameters of the device. Then the required number of layers of graphene-containing carbon fiber (or fabric) were applied to the external part of the blank using transverse winding (Figure 6). To ensure the required robustness of the construction, the thickness of the waveguide walls was chosen of the order of 3–5 mm. The blank matrix was separated from the specimen after the solidification of the antenna walls. The removal of the surface roughness of internal walls was reached owing to vacuum shaping. The manufacturing procedures of the *L*-range and the *C*-range antennas were the same [23,25]. Two orthogonal dipoles were used as excitation devices for all the carbon-based antennas and their metal analogs. Figure 7 shows the *C*-range antennas made from metal, fabric, and fiber. They consist of a horn connected to the segment of a circular waveguide of diameter 4.2 cm (with opening diameter of 5 cm); the length of the aperture is 4.5 cm. The weight of the metal C-range antenna (composite antenna) is 510 g (140 g).

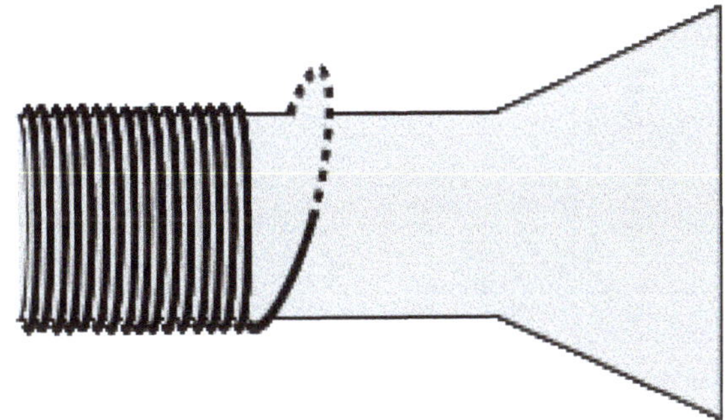

Figure 6. Transverse winding of the carbon fiber on the blank matrix.

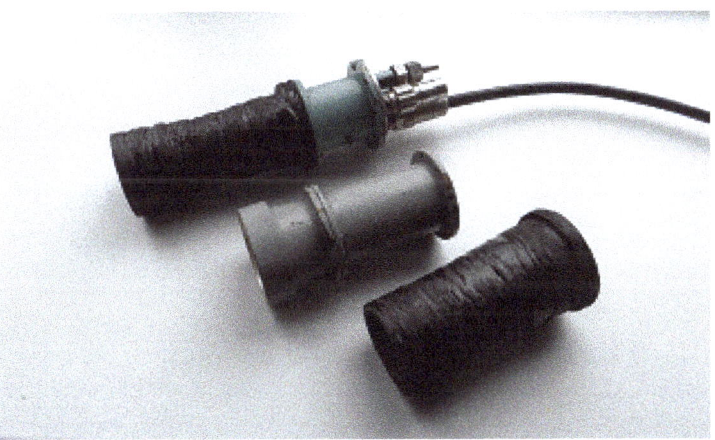

Figure 7. View of C-range horn antennas made from fabric (right), metal (center) and fiber (equipped with elements for antenna feeding, left).

The polarization characteristics of the composite antennas were studied under laboratory conditions using the rotation of a linearly polarized emitting antenna around its longitudinal axis. The composite antennas were used as receivers and equipped with a polarization unit based on a circular waveguide. Figure 8 shows *L*-range horn antennas with polarization units made of composite fiber and metal. The diameter of the circular waveguide and of the opening are is 12.5 cm and 15.2 cm, respectively; the length of the horn is 14.5 cm. The weight of the composite L-range antenna is 0.5 kg, whereas that of the metal one amounts about 8 kg.

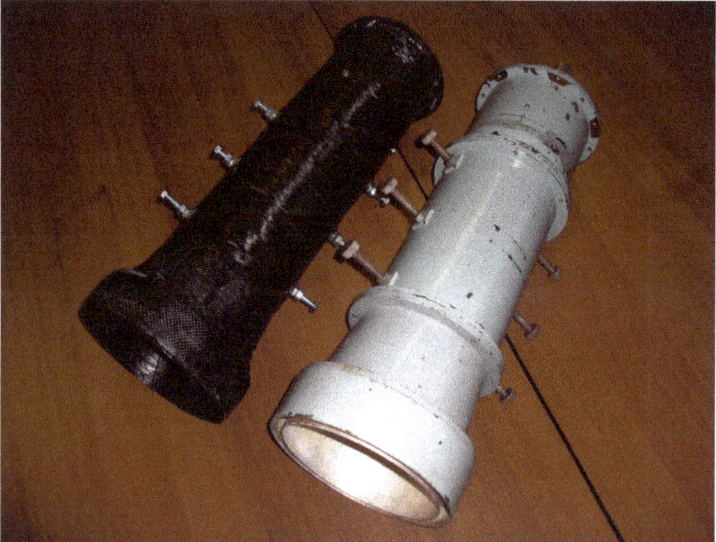

Figure 8. View of *L*-range horn antennas with polarization units made from fiber (left) and metal (right).

The radiation patterns, the SWRs, the emitted power frequency dependences and the polarization properties of the composite antennas were analyzed. Please note that these results can be used to estimate the antennas' gains.

4.2. Radiation Properties of the Horn Antennas

The measurements of the radiation properties of the horn antennas were performed under laboratory conditions without using an anechoic chamber. Receiving and emitting antennas were located far from each other. The characteristics of the GCM antenna, as the SWR, the emitted power frequency dependence and the radiation pattern, were compared with those of a metal antenna with the same geometry and size. Figures 9 and 10 present respectively the SWRs and the emitted power frequency dependence of *C*-range antennas made from fiber- and fabric-composites as well as of their metal analog. One can observe that the SWR's curves are smoother for the composite antennas compared to their metal analog, and that their minimum values are roughly the same. The behavior of the SWRs and the emitted power frequency dependence does not depend on the carbon materials.

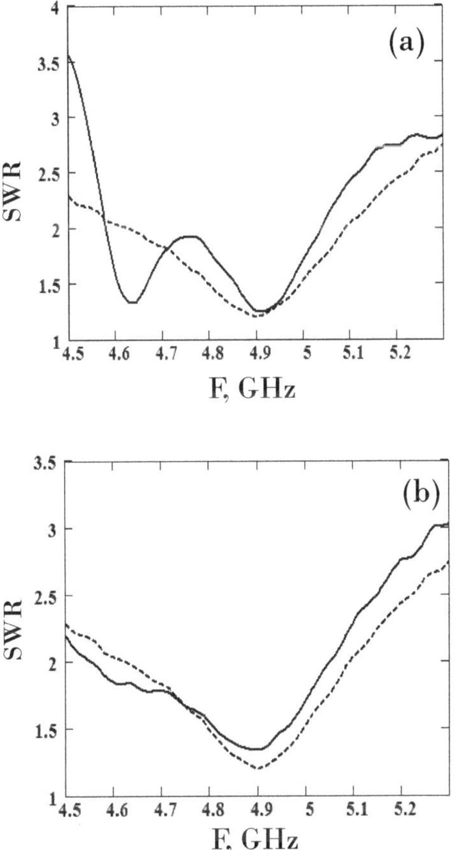

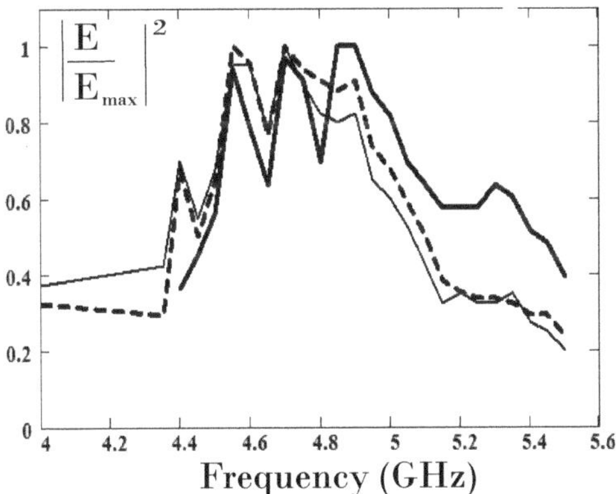

Figure 9. SWRs of *C*-range horn antennas made from (**a**) Fiber (dashed line) and metal (solid line), and (**b**) Fiber (dashed line) and fabric (solid line).

Figure 10. Square of the normalized field amplitude of the emitted signal as a function of frequency of *C*-range antennas made from fiber (dashed line), fabric (thin solid line), and metal (thick solid line). E_{max} is the maximum value of the field amplitude.

Please note that the main lobes of the radiation patterns of composite antennas and their metal analogs, measured in the frequency range 4.7–5 GHz, coincide very well within the limits of measurement uncertainties for both vertical and horizontal polarizations of the emitted signals [25].

Similar results were obtained for the *L*-range antennas. The SWRs of *L*-range antennas made from carbon-fiber and metal were measured without and with their polarization units, as can be observed in Figure 11 which shows that the presence of a polarization unit only slightly changes the width of the antennas' working frequency intervals where the SWR value is less than 1.4.

Figure 12 represents the main lobes of the radiation patterns of the metal and the composite *L*-range antennas. The measured values do not depend on the presence of a polarization unit. The difference about 7% between the half-widths of the radiation patterns of both antennas is not greater, under laboratory conditions, than the measurement uncertainties, which are due to the reflections of the signals on the surrounding equipment.

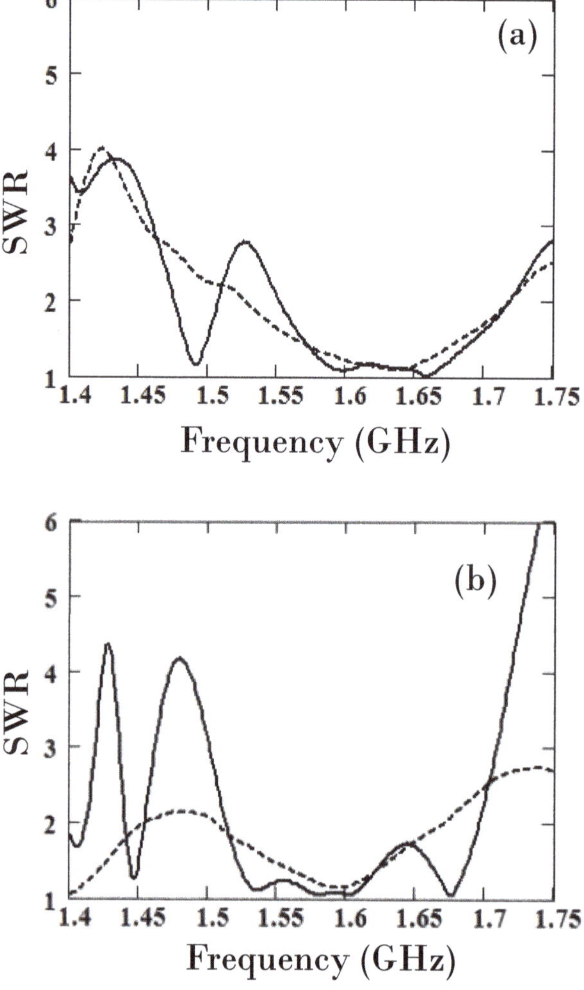

Figure 11. SWRs of *L*-range antennas made from metal (solid line) and fiber (dashed line), (a) without and (b) with a polarization unit.

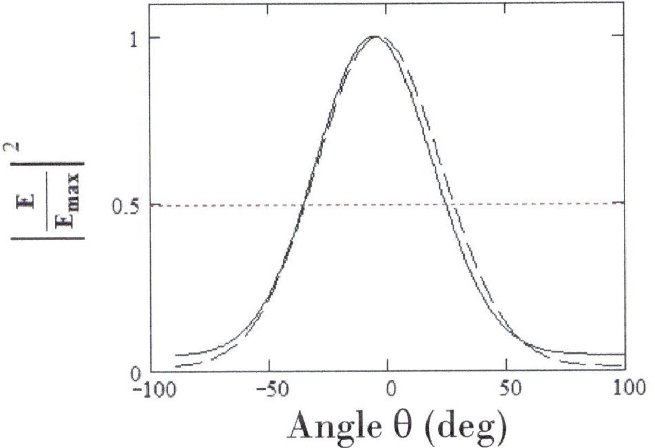

Figure 12. Radiation patterns of *L*-range antennas made from metal (solid line) and fiber (dashed line). E_{max} is the maximum value of the field amplitude.

4.3. Polarization Characteristics of Horn Antennas

The measurements of the polarization characteristics of the carbon fiber horn antenna and its metal analog were performed under laboratory conditions, using a polarization device built on the basis of a circular waveguide (see Figure 13). The antennas were used as receivers, whereas the emitter was a standard antenna that could be rotated around its axis.

As shown by the left panel of Figure 13, the horn antenna consists of two receiving dipoles (1), matching plexiglas plates (2), and six polarization rods (3). The plexiglas plates are used for a preliminary tuning of the antenna. The six rods inside the waveguide provide the phase shifts necessary for the generation of elliptically polarized waves. The cross section of the horn antenna exhibits the two orthogonal dipoles d_1 and d_2 (Figure 13, right-hand side panel). The angle between their axes and the rods is equal to $45°$; ϕ is the angle between their axes and the straight line OO′ which is the projection on the cross section of the plane of polarization of the input signal's electric field. Please note that the rotation of the emitting antenna provides rotated linearly polarized signals at the opening of the receiving horn antenna. The polarization characteristics of the composite and the metal antennas were measured by the two orthogonal dipoles. As is seen in Figure 11, the working frequency interval is equal to $1.55-1.7$ GHz (without a polarization unit) and $1.55-1.65$ GHz (in the presence of a polarization unit), for the composite antenna and its metal analog.

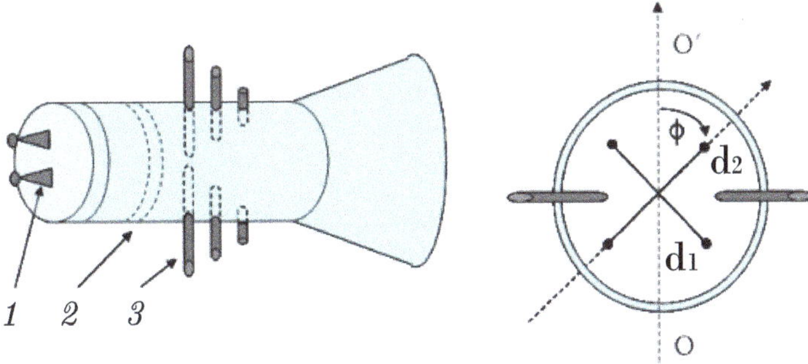

Figure 13. Geometry of the horn antenna with a polarization unit (left panel) and its cross-section showing the two orthogonal dipoles d_1 and d_2 used for the measure of the polarization (right panel).

Figure 14 shows the square $|\mathbf{E}/\mathbf{E}_{max}|^2$ of the normalized electric field amplitude of the composite antenna, for two orthogonal (to each other) polarizations of the signal corresponding to two orthogonal receiving dipoles (curves 1 and 2), at frequencies of 1.55 GHz and 1.575 GHz, as a function of the angle ϕ, respectively. Figure 15 shows a similar dependence for the metal antenna at 1.55 GHz.

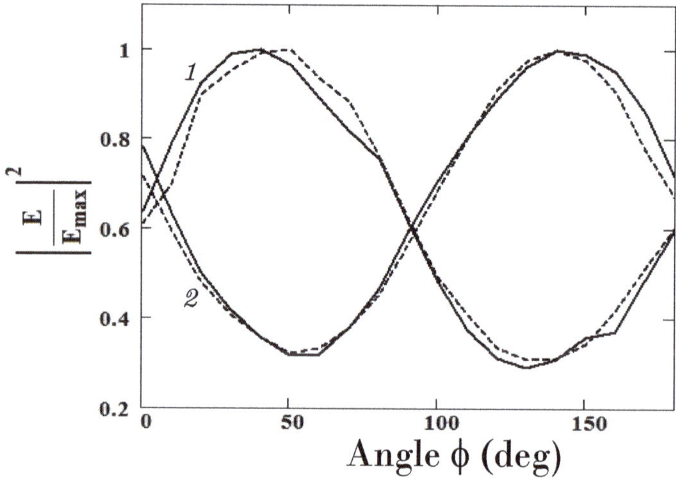

Figure 14. Square of the normalized amplitude of the signal received by the composite horn antenna as a function of the rotation angle ϕ at 1.55 GHz (solid line) and 1.575 GHz (dashed line).

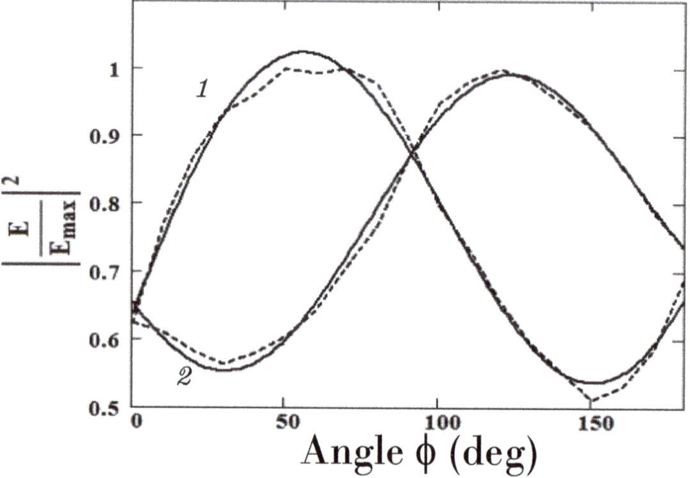

Figure 15. Square of the normalized amplitude of the signal received by the metal horn antenna as a function of the rotation angle ϕ at 1.55 GHz: measured signal (dashed line) and interpolation by a sinusoidal function of ϕ (solid line).

One can see in Figures 14 and 15 that the ellipticity coefficient $|\mathbf{E}_{min}/\mathbf{E}_{max}|^2$ is close to 0.3 (0.55) for the composite (metal) antenna. Five frequencies lying in the range 1.55–1.65 GHz were chosen for measuring the polarization characteristics. It should be noted that the ellipticity of the composite antenna weakly depends on the signal's frequency. For the metal antenna, the ellipticity varies within the interval 0.55−0.8 and reaches a maximum near 1.6 GHz. Variations of the antenna ellipticity can be due to the different damping factors of the left- and right-hand circularly polarized waves. The difference between composite and metal horn antennas is determined by their polarization coefficients, which are 1.5–2 times smaller for composite antennas than for metal ones. They may be explained by ohmic losses and wall conductivity anisotropy. Please note that the results obtained for the composite antenna made from carbon fabric have shown that the composite horn antenna and its metal analogue possess almost the same polarization characteristics.

5. Conclusions

It is shown that microwave dipole and horn antennas made from carbone composite materials modified by graphene structures are capable of operating efficiently and that their main radiation properties are identical to those of their metal analogs. Composite antennas' prototypes designed under laboratory conditions exhibit temperature stability, durability, and reduced weight compared to metal antennas. For example, the weights of the *L*-range metal and composite antennas are 8 kg and 0.5 kg, respectively. Moreover, the composite dipole and horn antennas have almost the same amplitude frequency response functions and radiation patterns as their metal analogs. Composite and metal horn antennas are different only as to their polarization characteristics. It should be noted that the anisotropy of the conductivity of GCMs can be used to obtain the desired characteristics of composite radio-technical devices. In the future, with the improvement of antenna technology, the GCM antenna devices should have substantial advantages over the metal analogs and should be able to replace them.

Over the last decade, one of the main goals of the antenna industry was to create microwave antennas with desirable radiation patterns whereas reducing their weight and cost. This can be achieved by designing new types of antennas made from GCMs.

6. Patents

Dugin, N.A.; Zaboronkova, T.M.; Chugurin, V.V.; Myasnikov, E.N. Antenna-feeder the microwave device from graphene-containing carbon composite material and its manufactoring, Russian Federation Patent N–RU2016/2577918 (C1), 2016-02-18.

Author Contributions: Conceptualization, N.A.D. and T.M.Z.; methodology, N.A.D. and T.M.Z.; software, G.R.B.; validation, N.A.D.; formal analysis, T.M.Z.; investigation, N.A.D., T.M.Z. and G.R.B.; resources, N.A.D. and T.M.Z.; data curation, N.A.D.; writing–original draft preparation, N.A.D. and T.M.Z.; writing–review and editing, T.M.Z. and C.K.; visualization, G.R.B. and C.K.; supervision, N.A.D.; project administration, N.A.D. and T.M.Z.; funding acquisition, N.A.D. All authors have read and agreed to the published version of the manuscript.

Funding: This research was funded by the Russian Science Foundation (grant number 14-12-00510) and by the Ministry of Education and Science of the Russian Federation (project N-3.2722.2017/4.6).

Conflicts of Interest: The authors declare no conflict of interest.

Abbreviations

The following abbreviations are used in this manuscript:

SWR   Standing wave ratio
GCM   Graphene-containing carbon composite material

References

1. Novoselov, K.S.; Jiang, D.; Schedin, F.; Booth, T.J.; Khotkevich, V.V.; Morozov, S.V.; Giem, A.K. Two-dimensional atomic crystals. *Proc. Nat. Acad. Sci. USA* **2005**, *102*, 10451–10453. [CrossRef] [PubMed]
2. Novoselov, K.S.; Geim, A.K.; Morozov, S.V.; Jiang, D.; Zhang, Y.; Dubbonos, S.V.; Grigorieva, I.V.; Firsov, A.A. Electric field effect in atomically thin carbon films. *Science* **2004**, *306*, 666–669. [CrossRef] [PubMed]
3. Berger, C.; Song, Z.; Li, T.; Li, X.; Ogbazghi, A.Y.; Feng, R.; Dai, Z.; Marchenkov, A.N.; Conrad, E.H.; First, P.N.; et al. Ultrathin epitaxial graphite: 2D electron gas properties and a route toward graphene-based nanoelectronics. *J. Phys. Chem. B* **2004**, *108*, 19912–19915. [CrossRef]
4. Hernandez, Y.; Nicolosi, V.; Lotya, M.; Blighe, F.M.; Sun, Z.; De, S.; McGovern, I.T.; Holland, B.; Byrne, M.; Gun'Ko, Y.K.; et al. High-yield production of graphene by liquid-phase exfoliation of graphite. *Nature Nanotech.* **2008**, *3*, 563–568. [CrossRef]
5. Engheta, N.; Ziolkowski, R.W. (Eds.) *Metamaterials: Physics and Engineering Explorations*; Wiley-IEEE: Hoboken, NJ, USA, 2006.
6. Slyusar, V.I. Metamaterials in antenna technology: history and basic principles. *Electronika NTV* **2009**, *7*, 70–79. (In Russian)
7. Lier, E. Review of soft and hard horn antennas, including metamaterial-based hybrid-mode horns. *IEEE Antennas. Propag. Mag.* **2010**, *52*, 31–39. [CrossRef]
8. Tang, M.C.; Xiao, S.; Wang, B.; Guan, J.; Deng, T. Improved performance of a microstrip phased array using broadband and ultra-low-loss metamaterial slabs. *IEEE Antennas. Propag. Mag.* **2011**, *53*, 31–41. [CrossRef]
9. Vendik, I.B.; Vendik, O.G. Metamaterials and their application in microwaves: A Review. *Tech. Phys.* **2013**, *58*, 1–24. [CrossRef]
10. Veselago, V.G. The electrodynamics of substances with simultaneous negative values of ε and μ. *Sov. Phys. Uspekhi* **1967**, *10*, 509–514. [CrossRef]
11. Temelkuara, B.; Bayindir, M.; Ozbay, E.; Biswas, R.; Sigalas, M.M.; Tuttle, G.; Ho, K.M. Photonic crystal-based resonant antenna with a very high directivity. *J. Appl. Phys.* **2000**, *87*, 603–605. [CrossRef]
12. Ge, Z.C.; Zhang, W.X.; Liu, Z.G.; Gu, Y.Y. Broadband and high-gain printed antennas constructed from Fabry-Perot resonator structure using EBG or FSS cover. *Microw. Opt. Technol. Lett.* **2006**, *48*, 1272–1274. [CrossRef]
13. Marques, R.; Martin, F.; Sorolla, M. *Metamaterials with Negative Parameters: Theory, Design and Microwave Applications*; Wiley Series in Microwave and Optical Engineering, Wiley-Blackwell: Hoboken, NJ, USA, 2008.
14. Bychkov, I.V.; Zotov, I.S.; Fedii, A.A. Studing microwave transmission and reflection in multilayer $CaSO_4·2H_2O$—Graphite composites. *Tech. Phys. Lett.* **2011**, *37*, 689–691. [CrossRef]
15. Weily, A.R.; Horvath, L.; Esselle, K.P.; Sanders, B.C.; Bird, T.S. A planar resonator antenna based on a woodpile EBG material. *IEEE Trans. Antennas Propag.* **2005**, *53*, 216–223. [CrossRef]
16. Caiazzo, M.; Maci, S.; Engheta, N. A metamaterial surface for compact cavity resonators. *IEEE Antennas Wirel. Propag. Lett.* **2004**, *3*, 261–264. [CrossRef]
17. Zhigang, X.; Xu, H. Low refractive metamaterials for gain enhancement of horn antenna. *J. Infrared Millim. Terahertz Waves* **2009**, *30*, 225–232.
18. Lashab, M.; Hraga, H.I.; Read, A.A.; Zebiri, C.; Benabdelaziz, F.; Jones, S.M.R. Horn antennas loaded with metamaterial for UWB applications. *PIERS Online* **2011**, *7*, 161–165.
19. Lai, A.; Caloz, C.; Itoh, T. Composite right/left-handed transmission line metamaterials. *IEEE Microw. Mag.* **2004**, *5*, 34–50. [CrossRef]
20. Hu, J.; Yan, C.S.; Lin, Q.C. A new patch antenna with metamaterial cover. *J. Zhejiang Univ. Sci. A* **2006**, *7*, 89–94. [CrossRef]
21. Rybin, V.V.; Kuznetsov, P.A.; Ulin, I.V.; Farmakovskii, B.V.; Bakhareva, V.E. Nanomaterials structural and functional class. *Vopr. Mater.* **2006**, *1*, 169–178. (In Russian)
22. Maher, F.E.; Strong, V.; Dubin, S.; Kaner, R.B. Laser scribing of high-performance and flexible graphene-based electrochemical capacitors. *Science* **2012**, *335*, 1326–1330.
23. Dugin, N.A.; Zaboronkova, T.M.; Chugurin, V.V.; Myasnikov, E.N. Antenna-Feeder the Microwave Device From Graphene-Containing Carbon Composite Material and Its Manufactoring. RF Patent N–RU2016/2577918 (C1), 18 February 2016.

24. Dugin, N.A.; Zaboronkova, T.M.; Myasnikov, E.N.; Belyaev, G.R.; Lobastov, V.V. Electrodynamic characteristics of dipole antennas made of graphene-containing carbon composite materials. *J. Commun. Technol. Electron.* **2018**, *63*, 864–867. [CrossRef]
25. Dugin, N.A.; Zaboronkova, T.M.; Myasnikov, E.N. Using carbon-based composite materials for manufactoring C-range antenna devices. *Latv. J. Phys. Tech. Sci.* **2016**, *53*, 17–24.
26. Dugin, N.A.; Zaboronkova, T.M.; Myasnikov, E.N. Antenna-waveguide microwave devices of carbon composition materials. *Tech. Phys. Lett.* **2016**, *42*, 598–600. [CrossRef]
27. Ovechkin, G.I.; Golovanova, V.V.; Dvirnyi, G.V. Transformable Antenna of the Umbrella-Type Spacecraft. RF Patent N-RU2011/2427949 (C1), 20 April 2011.
28. Chung-Yen Yang Method of Forming Antenna by Utilizing Graphene. U.S. Patent N-US2013/0004658 (A1), 3 January 2013.
29. Gi, C.C.; Soon, O.S. Metamaterial And Manufacturing Method at the Same. U.S. Patent N-US2012/013989 (A1), 25 January 2012.
30. Bychkov, I.V.; Zotov, I.S.; Fedii, A.A. Reflection properties of electromagnetic crystal based composite material. In Proceedings of the 3rd International Congress on Advanced Electromagnetic Materials in Microwaves and Optics, London, UK, 31 August–4 September 2009; p. 602.
31. Kazantseva, N.E.; Ryvkina, N.G.; Chmutin, I.A. Promising materials for microwave absorbers. *J. Commun. Technol. Electron.* **2003**, *48*, 173–184.
32. Giboney, K.S. Gap-Mode Waveguide. U.S. Patent N-US2012/0243823 (A1), 27 September 2012.
33. Dugin, N.A.; Zaboronkova, T.M.; Chugurin, V.V.; Myasnikov, E.N. Study of electromagnetic properties of multylayer graphene fractal structures by remote method. In Proceedings of the 21th annual International Conference on Advanced Laser Technologies, Budva, Montenegro, 16–20 Spetember 2013; p. 192.
34. Kudrin, A.V.; Zaitseva, A.S.; Zaboronkova, T.M.; Krafft, C. Analysis of a strip antenna located on the interface between a uniaxial plasma and an isotropic medium. In Proceedings of the 11th European Conference on Antennas and Propagation, Paris, France, 19–24 March 2017; pp. 3574–3578.
35. Kudrin, A.V.; Zaitseva, A.S.; Zaboronkova, T.M.; Krafft, C. Current distribution and input impedance of a circular loop antenna located on the surface of a gyromagnetic cylinder. In Proceedings of the International Conference Days on Diffraction 2018, St-Petersburg, Russia, 4–8 June 2018; pp. 192–197.
36. Kudrin, A.V.; Zaboronkova, T.M.; Zaitseva, A.S.; Spagnolo, B. Theory of strip antenna located at a plane interface of a uniaxial metamaterial and an isotropic magnetodielectric. *IEEE Trans. Ant. Prop.* **2020**, *68*, 195–206. [CrossRef]
37. Zaboronkova, T.M.; Dugin, N.A.; Myasnikov, E.N. Microwave horn antenna made of a graphene-containing carbon composite material. In Proceedings of the 9th European Conference on Antennas and Propagation, Lisbon, Portugal, 13–17 April 2015; pp. 7228220-1–7228220-2.
38. Dugin, N.A.; Belyaev, G.R.; Zaboronkova, T.M.; Lobastov, V.G. Polarization characteristics of graphene-containing composite horn antenna. In Proceedings of the International Conference Days on Diffraction 2018, St-Petersburg, Russia, 4–8 June 2018; pp. 88–92.
39. Dugin, N.A.; Zaboronkova, T.M.; Myasnikov, E.N.; Belyaev, G.R. Electrodynamic characteristics of horn microwave antennas made of graphene-containing carbon–composite materials. *Tech. Phys.* **2018**, *63*, 268–273. [CrossRef]
40. Available online: https://zoltek.com/products/px35/ (accessed on 20 January 2020).
41. Tseitlin, N.M. *Methods of Measurement of Antenna Characteristics*; Radio i Svyaz': Moscow, Russia, 1985. (In Russian)
42. Felsen, L.B.; Marcuvitz, N. *Radiation and Scattering of Waves*; Prentice-Hall: Englewood Cliffs, NJ, USA, 1973.

 *electronics*

Article

Material Identification Using a Microwave Sensor Array and Machine Learning

Luke Harrsion, Maryam Ravan, Dhara Tandel, Kunyi Zhang, Tanvi Patel and Reza K. Amineh *

Department of Electrical and Computer Engineering, New York Institute of Technology, New York, NY 10023, USA; lharriso@nyit.edu (L.H.); mravan@nyit.edu (M.R.); dtandel@nyit.edu (D.T.); kzhang11@nyit.edu (K.Z.); tpatel60@nyit.edu (T.P.)
* Correspondence: rkhalaja@nyit.edu; Tel.: +1-646-273-6204

Received: 16 January 2020; Accepted: 5 February 2020; Published: 8 February 2020

Abstract: In this paper, a novel methodology is proposed for material identification. It is based on the use of a microwave sensor array with the elements of the array resonating at various frequencies within a wide range and applying machine learning algorithms on the collected data. Unlike the previous microwave sensing systems which are mainly based on a single resonating sensor, the proposed methodology allows for material characterization over a wide frequency range which, in turn, improves the accuracy of the material identification procedure. The performance of the proposed methodology is tested via the use of easily available materials such as woods, cardboards, and plastics. However, the proposed methodology can be extended to other applications such as industrial liquid identification and composite material identification, among others.

Keywords: machine learning; material identification; microwave sensor array

---

1. Introduction

Microwave sensing has been utilized in a broad range of applications from noise measurement systems [1] and spatial displacement measurement [2] to single-cell viability detection [3] and material characterization [4,5].

One of the applications of microwave sensing is in material identification, which is a systematic approach to identify the particular grade of materials for reverse engineering, alteration, or repair of the existing assets or products and the use of substitute materials. In general, in microwave sensing and spectroscopy systems, the aim is to characterize the spectral behavior of the materials in a frequency range below 100 GHz. This can result in a low-cost integrated system for material identification and characterization. Conventionally, the sensing part of a microwave material characterization system is constructed using either a single resonator [6,7] or a transmission line (TL) [8].

In single-resonator sensors, the high quality factor (Q) of these devices renders the resonance frequency change due to the exposure to the material under test (MUT) measurable. However, this high sensitivity is only over a limited frequency band. If the tested MUTs have similar properties over the resonance band, they cannot be distinguished well. Here, we review some of the recent works in material identification using single-resonant-based sensors. In [9], a slot-loaded microstrip patch sensor antenna was proposed to enhance sensitivity in measuring the permittivity of planar materials. There, the antenna had two resonance frequencies, and the sensitivities of the $|S_{11}|$ responses to changes in the permittivities of the MUTs were studied. In [10], a microwave ring-resonator sensor was presented to evaluate the moisture content, or, more precisely, the water holding capacity, of broiler meat over a four-week period. There, the developed sensor showed significant changes in its resonance frequency and return loss due to the reduction in water holding capacity in the studied duration. In [11], a metamaterial-based microwave sensor with a complementary split ring resonator (CSRR) was implemented for dielectric characterization of liquids. Liquid samples placed inside glass capillary tubes modify the resonant frequency and Q-factor of the CSRR sensor. Thereby, a relation between the sensor resonant frequency, Q-factor, and complex permittivity of the liquid samples was established. In [12], a method was presented to estimate the complex permittivity of liquids based on an embedded microstrip line with a CSRR etched in the ground plane, beneath the conductor strip. A liquid container surrounding the CSRR was added to the structure in order to load the sensing element, i.e., CSRR, with the liquid under test. The complex permittivity of the liquid was inferred from the frequency response of the structure, particularly from the change in the response at resonant frequency.

On the other hand, transmission-line-based sensors are suitable for broadband sensing with reduced sensitivity. According to the above-mentioned discussions, tackling the tradeoff between the sensitivity and bandwidth to achieve broadband material identification is challenging. Thus, in this paper, a material identification system is proposed based on a microwave sensor array with the elements of the array resonating at different frequencies, covering a wide band. This allows us to benefit from the excellent sensitivity of the high-Q resonators while, similar to the TL approach, collecting information over a wide bandwidth. A machine learning algorithm is then developed to identify different materials based on the data collected by the microwave sensor array. We use an automatic framework that consists of a multiclass support vector machine (SVM) classification based on a decision tree approach [13].

The performance of the proposed material identification system is validated via testing of three groups of materials made of cardboard, wood, and plastic. The results show satisfactory performance of the system in distinguishing these three groups of materials. The advantages of the proposed material identification system include its compactness, low cost, and low error rate, and its contactless, reusable, easy-to-fabricate, and easy-to-work operation.

2. Microwave Sensor Array

In this section, we review the design of a microwave sensor array first proposed in [14] for water quality testing. The array is composed of five resonating elements, as shown in Figure 1. These resonating elements are based on a microstrip transmission line loaded with CSRRs designed at different frequencies within the range of 1 GHz to 10 GHz. The use of planar CSRR sensors offers several advantages such as low cost, portability, noninvasiveness, and flexibility of sample preparation. This is why they have been widely used for sensing applications (see, e.g., [15–18]).

The sensor array in Figure 1 allows for measuring changes in dielectric properties over a wide frequency range from 1 GHz to 10 GHz, specifically at five frequencies around 1 GHz, 3 GHz, 5 GHz, 7 GHz, and 9 GHz. Changes in the dielectric properties can then be monitored through measuring the resonance frequency shifts for the resonator array.

The design of the sensor array was implemented in Altair FEKO software [19]. It is high-frequency electromagnetic simulation software based on the method of moments (MoM). The sensor array was designed on a Rogers RO4350 substrate with dielectric properties of $\varepsilon_r = 3.66$ and $\tan \delta = 0.0031$. The width W_s, length L_s, and thickness of the substrate are 20 mm, 56 mm, and 0.75 mm, respectively. The microstrip line was designed to have a characteristic impedance of 50 Ω. This corresponds to the width of strip line being $W_m = 1.68$ mm. This strip line was placed at the center of the front surface of the substrate. On the back side of the substrate, the five CSRRs were etched out of the ground plane. Figure 1 shows the front and back sides of the designed sensor array. The CSRRs, as shown in Figure 1b, were named Sensor 1 to Sensor 5 from left to right and correspond to resonance frequencies of 1.36 GHz, 3.09 GHz, 5 GHz, 6.82 GHz, and 8.91 GHz. Figure 2 shows the main design parameters for each CSRR structure, including the length of the outer ring L, width of the rings W, the track width between adjacent rings b, and the width of narrow lines a. As a rule of thumb, a larger value of L corresponds to a lower resonance frequency, and as W increases, the resonance frequency decreases. Tables 1 and 2 show the optimal design parameters and center-to-center distances for the sensor array, respectively.

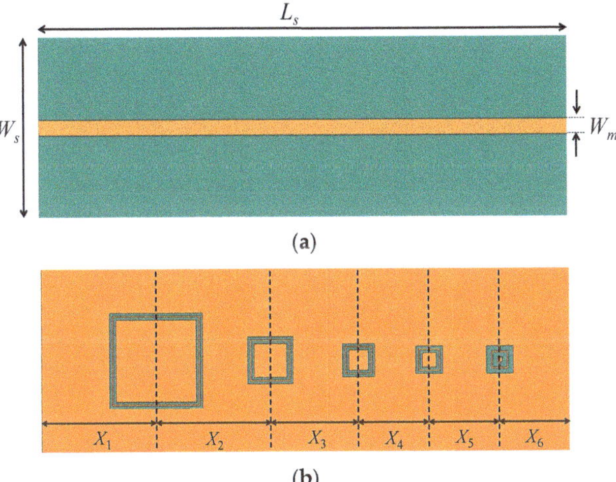

(a)

(b)

Figure 1. Microwave sensor array (**a**) front surface and (**b**) back surface with Sensor 1 to Sensor 5, shown from left to right, corresponding to resonance frequencies of 1.36 GHz, 3.09 GHz, 5 GHz, 6.82 GHz, and 8.91 GHz, respectively.

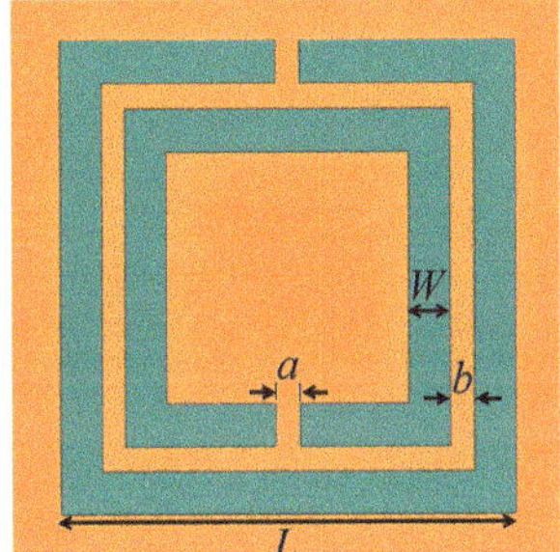

Figure 2. Parametric model of each complementary split ring resonator (CSRR).

Table 1. Dimensions of the designed CSRRs.

	f_r (GHz)	W (mm)	L (mm)	a (mm)	b (mm)
Sensor 1	1.36	0.27	10.5	0.16	0.16
Sensor 2	3.09	0.27	5.15	0.16	0.16
Sensor 3	5	0.27	3.65	0.16	0.16
Sensor 4	6.82	0.27	3	0.16	0.16
Sensor 5	8.91	0.54	3	0.16	0.16

Table 2. Distance between CSRRs as shown in Figure 1b.

X_1 (mm)	X_2 (mm)	X_3 (mm)	X_4 (mm)	X_5 (mm)	X_6 (mm)
11	13	10	8	8	6

The sensor array was fabricated using printed circuit board technology (PCB) as shown in Figure 3. Sub-Miniature A (SMA) connectors were soldered on both ends of the microstrip transmission line, and the whole device was installed on a wooden stand piece to ensure a mechanically stable platform for conducting the measurements.

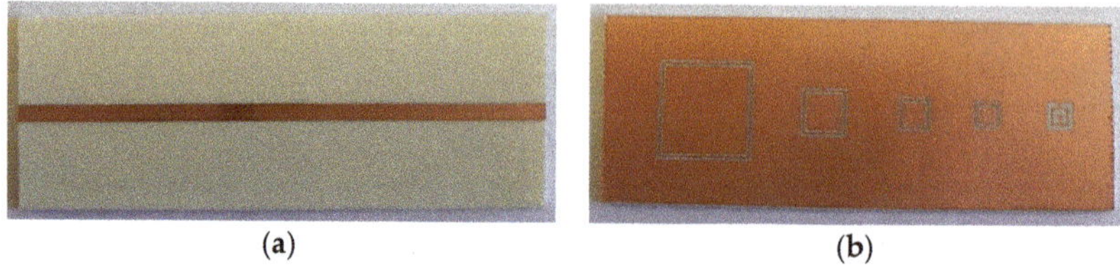

(a)   (b)

Figure 3. Fabricated sensor array (**a**) front surface and (**b**) back surface.

In order to record the response of the sensor array to the materials, the variation of the transmission S-parameter $|S_{21}|$ versus frequency was measured without any MUT using an E5063A vector network analyzer (VNA) from Keysight Technologies. Figure 4 shows a comparison of the simulated and measured $|S_{21}|$ for the sensor array. A reasonable match can be observed between the simulated and measured results. The mismatches may be due to the fabrication tolerances and soldering. Please note that the difference between the measured $|S_{21}|$ in [14] and that here is due to the fact that the measured $|S_{21}|$ in [14] is for a sensor array with a plexiglass container placed on the back side of it for measuring water samples, while the measured $|S_{21}|$ reported here in Figure 4 is for a bare sensor array. Furthermore, in Figure 4, an additional resonance is observed for the measured $|S_{21}|$ at around 9.6 GHz. It is worth noting that this extra resonance is not utilized in our material characterization, described later, since it is very close to the other resonance frequency at 9.1 GHz for which the data will be used in our processing. In fact, the data from this additional resonance at 9.6 GHz would be very similar to those from 9.1 GHz due to approximately similar material properties at these two close frequencies. This justifies the use of data only at 9.1 GHz.

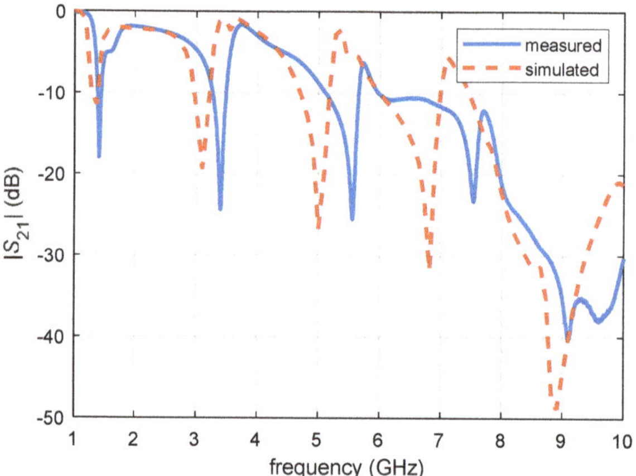

Figure 4. Simulated and measured $|S_{21}|$ for the sensor array.

3. Machine Learning Algorithm

The use of an array of resonator sensors allow us to process the collected data using proper machine learning algorithms. As shown later, such a system provides significantly more accurate results compared to the use of a single resonator sensor. In the following, we describe the three MUT categories and the machine learning approach utilized to distinguish them.

3.1. Three Groups of MUTs

To validate the performance of the proposed material identification system, three groups of MUTs were measured with the microwave sensor array. These three MUT groups were (1) 50 cardboard samples, (2) 50 wood samples, and (3) 50 plastic samples. Figure 5 shows pictures of a few samples from each group. Figure 6 shows the measurement setup in which the MUTs were placed on the back side of the sensor array (the side containing the CSRR elements) and $|S_{21}|$ versus frequency was measured using an E5063A ENA from Keysight Technologies.

(a)

(b)

(c)

Figure 5. Pictures of a few samples measured for each category: (**a**) cardboard samples, (**b**) wood samples, and (**c**) plastic samples.

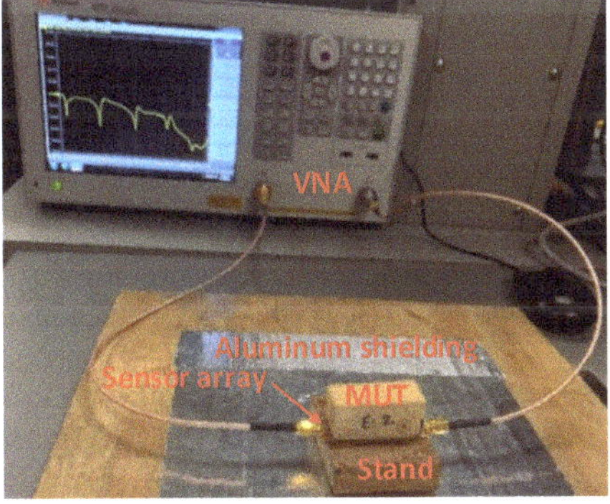

Figure 6. Measurement of $|S_{21}|$ for the sensor array using a vector network analyzer (an E5063A ENA from Keysight Technologies).

3.2. Feature Selection

To evaluate the performance of the microwave sensor array in distinguishing the three groups of MUTs described above, we used the shifts in the five resonant frequencies in the $|S_{21}|$ response due to exposure to the MUT as our features. We then applied different combinations of the five features to the classifier, from selecting only one feature to using all five features. Therefore, the total number of combinations that were used for classification was 31 (see Table 3).

Table 3. Relationships between combination indices and combinations of sensors.

Combination Index	Combinations of Sensors	Combination Index	Combinations of Sensors
1	1	17	1, 2, 4
2	2	18	1, 2, 5
3	3	19	1, 3, 4
4	4	20	1, 3, 5
5	5	21	1, 4, 5
6	1, 2	22	2, 3, 4
7	1, 3	23	2, 3, 5
8	1, 4	24	2, 4, 5
9	1, 5	25	3, 4, 5
10	2, 3	26	1, 2, 3, 4
11	2, 4	27	1, 2, 3, 5
12	2, 5	28	1, 2, 4, 5
13	3, 4	29	1, 3, 4, 5
14	3, 5	30	2, 3, 4, 5
15	4, 5	31	1, 2, 3, 4, 5
16	1, 2, 3		

3.3. Outlier Detection and Removal

In order to detect the data samples that were away from other samples in each class, i.e., outliers, we used the k-Nearest Neighbor (KNN) outlier detection algorithm available in the PyOD toolbox (Python Toolbox for Scalable Outlier Detection) [20]. It uses the math behind the KNN classification algorithm. Indeed, for any data point, the distance to its k-many nearest neighbors could be viewed as the outlying score. We used the median distance to these k neighbors to detect outliers, where $k = 5$ was considered in this study.

3.4. Decision-Tree-Based Support Vector Machine

In this study, we use a decision-tree-based support vector machine (DSVM) approach for classifying the three categories of cardboard, wood, and plastic samples. The rationale behind this is that combining decision tree architecture with binary SVMs allows us to benefit from the advantages of the efficient computation of decision trees and the high classification accuracy of SVMs. The chosen kernel function was a Gaussian radial basis function using Scikit-learn, the machine learning library in Python [21].

4. Results

In this section, we evaluate the performance of the proposed approach in distinguishing the three groups of MUTs discussed in the previous section. First, in order to detect possible outliers, KNN outlier detection was applied to data samples in each group. We found four outliers in cardboard data samples, three outliers in wood data samples, and three outliers in plastic data samples that were then removed from the dataset. Therefore, the total number of data samples was reduced to 140, including 46 cardboard samples, 47 wood samples, and 47 plastic samples.

The median ± median absolute deviation (MAD) of each resonant frequency shift for the three MUTs, along with the corresponding p values using the Wilcoxon rank sum test over each pair of MUTs, is shown in Table 4. From Table 4, the median resonant frequency shift is significantly different in cardboard samples in comparison to wood samples for Sensors 1 to 4 and plastic samples for Sensors 2 to 5. Furthermore, the median resonant frequency shift is significantly higher in wood samples in comparison to plastic samples for all five sensors, which suggests the possibility to distinguish these three MUTs with a machine learning algorithm.

Table 4. Median (± median absolute deviation (MAD)) values of resonant frequency shift across all MUT samples, along with Wilcoxon rank sum test p values comparing each resonant frequency shift for each pair of MUTs.

Resonant Frequency Shift (MHz)	Cardboard	Wood	Plastic	p Values (Cardboard, Wood)	p Values (Cardboard, Plastic)	p Values (Wood, Plastic)
Δf_1	81.00 ± 11.70	94.50 ± 18.00	73.80 ± 16.20	0.0012	0.18	3.28×10^{-4}
Δf_2	150.75 ± 31.95	122.40 ± 31.50	84.60 ± 41.40	0.0141	5.13×10^{-4}	0.032
Δf_3	225.00 ± 57.15	162.00 ± 63.90	92.70 ± 62.10	0.0026	7.10×10^{-5}	0.026
Δf_4	241.20 ± 75.15	147.60 ± 86.40	52.20 ± 58.50	6×10^{-4}	1.34×10^{-4}	0.027
Δf_5	331.65 ± 163.80	308.70 ± 206.10	96.30 ± 145.80	0.78	9.14×10^{-4}	0.004

The hierarchical cluster analysis step for the DSVM machine learning approach is shown in Figure 7. Since the cardboard samples are filled with air gaps, their electrical properties are highly different from those of wood and plastic samples. Therefore, at the top of the tree (i.e., the root node), the first binary classifier (SVM1) is trained to classify the cardboard class (a terminal node) as a negative class and the remaining merged two classes (wood + plastic) as a positive class. Then, the second binary classifier in the tree (SVM2) is trained to classify the elements of plastic as a negative class and the elements of wood as a positive class.

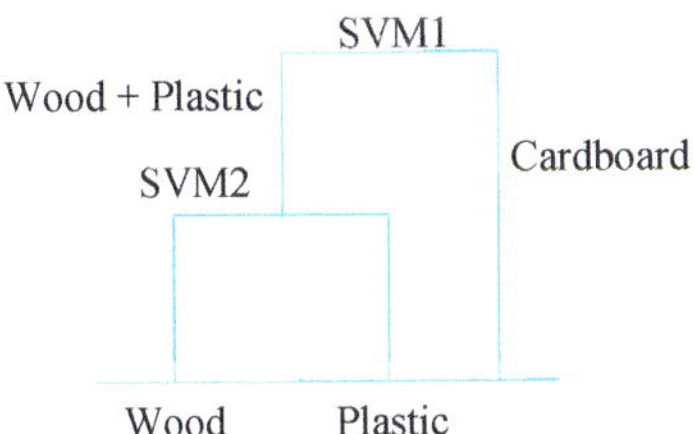

Figure 7. A decision-tree-based support vector machine (SVM) generated for the three classes of cardboard, wood, and plastic.

Leave-one-out (LOO) and stratified k-fold class validation (SKF) procedures [22] were used to evaluate the classification performance. The LOO procedure is an iterative process where, in each iteration, all the features associated with one particular data sample are taken as a test dataset and are omitted from the training set. The iterations repeat until all data samples have been taken as a test dataset once. The SKF procedure is also an iterative process. In this procedure, first the data samples are split into k-many groups (folds) where in each fold the same proportion of data samples is considered for each class. Then, at each iteration, one fold is taken as a test dataset and is omitted from the training set. The iteration is repeated until all folds have been taken as a test dataset once. In this study, the value of k was 5. The performance of the DSVM approach was then compared with that of some widely used machine learning approaches: decision tree (DT), random forest (RF), k-nearest neighbors (KNN), Gaussian naive Bayes (GNB), and multilayer Perceptron (MLP), where 10 trees was considered for the RF approach, 5 neighbors was considered for the KNN approach, and 1 hidden layer with 5 neurons was considered for the MLP approach to get the best corresponding classification performance.

To demonstrate the classification accuracy for each of the 31 combinations of the 5 features, Figures 8–11 depict the total classification performance versus feature combination index when classifying cardboard versus the rest and wood versus plastic using LOO and SKF cross-validation approaches. The relationship between the combination index and the combination of sensors is shown in Table 3, where the first index belongs to the case where the resonance frequency shift in the first sensor is used as the feature, the second index belongs to the case where the resonance frequency shift in the second sensor is used as the feature, and so on. From these figures, the classification performance was less than 70% when using the resonant frequency shift of each single sensor for all approaches using both LOO and SKF cross-validation. However, using LOO cross-validation for classifying cardboard versus the rest, the performance was higher than 80% for 3 to 14 out of the 31 combinations that involve at least two sensors, where the lowest number belonged to the GNB approach and the highest number belonged to the DSVM, MLP, and KNN approaches (Figure 8). Furthermore, using SKF cross-validation, the performance was higher than 80% for 1 to 14 out of the 31 combinations, where the lowest number belonged to the GNB and DT approaches and the highest number belonged to the DSVM and KNN approaches (Figure 10). Using LOO cross-validation, the classification performance for classifying wood and plastic was higher than 75% for 0 to 10 out of the 31 combinations that involve at least two sensors, where the lowest number belonged to the GNB, KNN, and RF approaches and the highest belonged to the DSVM approach (Figure 9). Furthermore, using SKF cross-validation, the performance was higher than 75% for 0 to 7 out of the 31 combinations, where the lowest number belonged to the GNB, DT, and RF approaches and the highest number belonged to the DSVM and MLP approaches (Figure 11).

The highest classification performance for classifying cardboard versus the rest using the LOO cross-validation approach belonged to KNN (87%) for feature combination index 27, where the selected features were the resonance frequency shifts in Sensors 1, 2, 3, and 5 (see Table 3), followed by the DSVM approach (85%) for feature combination indices 16, 18, 20, 21, 27, and 28 and by the MLP (85%) approach for feature combination indices 18, 27, and 29. However, the best classification performance for classifying cardboard versus the rest using SKF and also for classifying wood from plastic using both LOO and SKF cross-validation belonged to the DSVM approach at multiple feature combinations with common feature combination index 18 (the resonance frequency shifts in Sensors 1, 2, and 5). This confirms that the DSVM approach has the highest and most consistent overall performance in terms of the selected features.

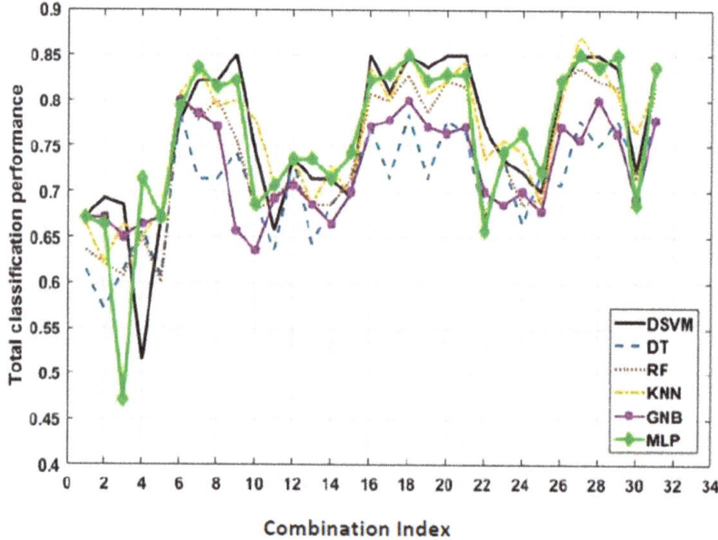

Figure 8. Total classification accuracy versus combination index for classifying cardboard from the rest for different classification approaches using LOO cross-validation.

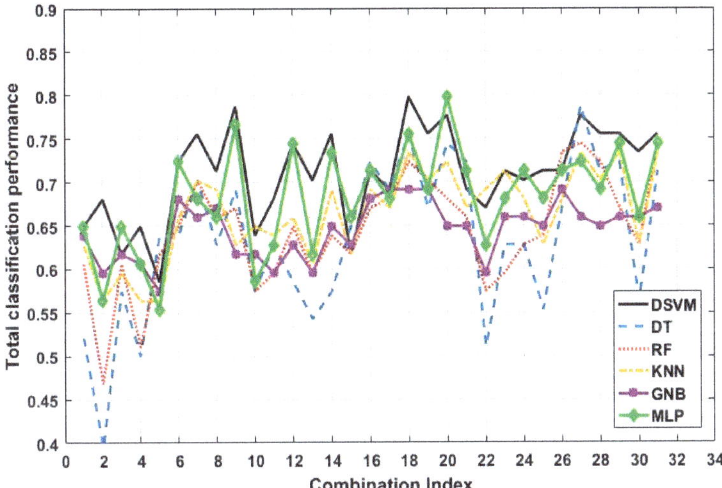

Figure 9. Total classification accuracy versus combination index for classifying wood and plastic for different classification approaches using LOO cross-validation.

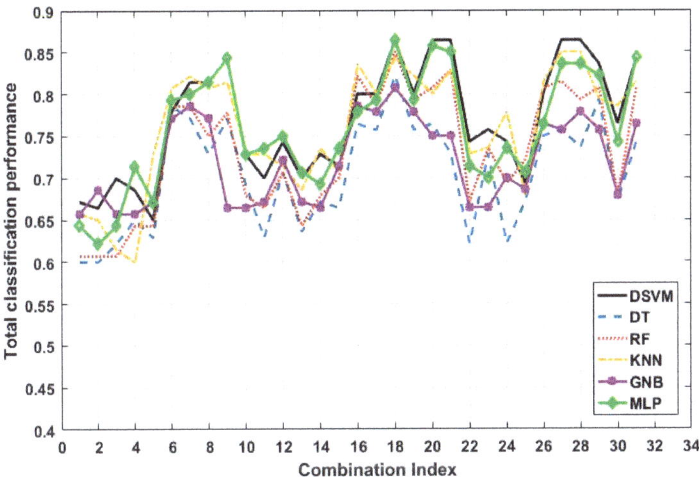

Figure 10. Total classification accuracy versus combination index for classifying cardboard from the rest for different classification approaches using SKF cross-validation.

Tables 5 and 6 show the corresponding confusion matrix along with the sensitivity, specificity, and total accuracy of the DSVM approach for feature combination index 18 using the LOO cross-validation approach. From Table 5, the highest classification performance was 85% for classifying cardboard versus the rest. Furthermore, from Table 6, wood and plastic were classified with a highest classification performance of 79.8%. The corresponding classification performance using SKF cross-validation is shown in Tables 7 and 8, where the highest classification performance was 86.4% for classifying cardboard versus the rest and 81.9% for classifying wood and plastic. These results demonstrate that the proposed procedure can classify both cardboard from the rest and wood from plastic with high accuracy.

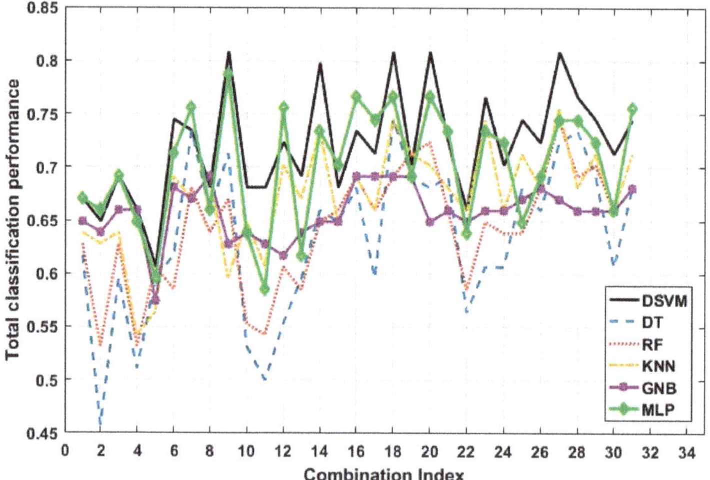

Figure 11. Total classification accuracy versus combination index for classifying wood and plastic for different classification approaches using SKF cross-validation.

Table 5. DSVM classification performance for classifying cardboard vs. the rest using LOO cross-validation.

Class	Wood + Plastic	Cardboard	Sensitivity	Specificity	Total Accuracy
Wood + Plastic	84	10			
Cardboard	11	35	89.4%	76.1%	85%

Table 6. DSVM classification performance for classifying wood vs. plastic using LOO cross-validation.

Class	Wood	Plastic	Sensitivity	Specificity	Total Accuracy
Wood	39	8			
Plastic	11	36	83.0%	76.6%	79.8%

Table 7. DSVM classification performance for classifying cardboard vs. the rest using SKF cross-validation.

Class	Wood + Plastic	Cardboard	Sensitivity	Specificity	Total Accuracy
Wood + Plastic	86	8			
Cardboard	11	35	91.5%	76.1%	86.4%

Table 8. DSVM classification performance for classifying wood vs. plastic using SKF cross-validation.

Class	Wood	Plastic	Sensitivity	Specificity	Total Accuracy
Wood	40	7			
Plastic	10	37	85.1%	78.7%	81.9%

5. Conclusions

This study for the first time proposed a material identification methodology using a wideband microwave sensor array and machine learning. The sensor array is composed of five planar resonating elements (sensors) designed at different frequencies within the range of 1 GHz to 10 GHz and has several advantages such as low cost, portability, noninvasiveness, reusability, and flexibility of sample preparation. A machine learning approach was then applied to the data collected from these five sensors to classify three MUTs: cardboard, wood, and plastic. The utilized features were the resonant frequency shifts in the five sensors due to exposure to MUTs. In this study, we utilized a DSVM classifier to first classify cardboard from the rest (wood + plastic) and then classify wood and plastic. We examined all 31 combinations of the 5 measured frequency shifts. We then compared the performance of the DSVM with that of widely used machine learning approaches: DT, RF, KNN, GNB, and MLP. The results showed that the DSVM approach had the highest and most consistent classification performance with both LOO and SKF cross-validation when using the resonant frequency shifts in Sensors 1, 2, and 5 as selected features. This demonstrated that the proposed approach could classify the three MUTs with high accuracy. By collecting more samples, the proposed approach can be used as an accurate automatic technique for identifying materials that are used in buildings, bridges, and structures such as swimming pool panels, racing car bodies, bathtubs, storage tanks, etc.

Author Contributions: Conceptualization, R.K.A.; methodology, R.K.A. and M.R.; software, L.H., M.R., and K.Z.; validation, L.H., M.R., K.Z., D.T., and T.P.; formal analysis, L.H. and M.R.; investigation, all the authors; resources, R.K.A. and M.R.; data curation, L.H. and M.R.; writing—original draft preparation, R.K.A., M.R., and L.H.; writing—review and editing, R.K.A. and M.R.; visualization, all of the authors; supervision, R.K.A. and M.R.; project administration, R.K.A. and M.R.; funding acquisition, R.K.A. All authors have read and agreed to the published version of the manuscript.

Funding: This research was funded by a New York Institute of Technology's Institutional Support for Research and Creativity (ISRC) Grant.

Conflicts of Interest: The authors declare no conflict of interest.

References

1. Ivanov, E.N.; Tobar, M.E.; Woode, R.A. Microwave interferometry: Application to precision measurements and noise reduction techniques. *IEEE Trans. Ultrason. Ferroelect. Freq. Control* **1998**, *45*, 1526–1536. [CrossRef] [PubMed]

2. Kim, S.; Nguyen, C. A displacement measurement technique using millimeter-wave interferometry. *IEEE Trans. Microw. Theory Techn.* **2003**, *51*, 1724–1728.

3. Yang, Y.; Zhang, H.; Zhu, J.; Wang, G.; Tzeng, T.; Xuan, X.; Huang, K.; Wang, P. Distinguishing the viability of a single yeast cell with an ultra-sensitive radio frequency sensor. *Lab Chip* **2010**, *10*, 553–555. [CrossRef] [PubMed]

4. Schueler, M.; Mandel, C.; Puentes, M.; Jakoby, R. Metamaterial inspired microwave sensors. *IEEE Microw. Mag.* **2012**, *13*, 57–68. [CrossRef]

5. Jilani, M.T.; Rehman, M.Z.; Khan, A.M.; Khan, M.T.; Ali, S.M. A brief review of measuring techniques for characterization of dielectric materials. *Int. J. Inf. Elect. Eng.* **2012**, *1*, 1–5.

6. Cherpak, N.T.; Barannik, A.A.; Prokopenko, Y.V.; Smirnova, T.A.; Filipov, Y.F. *A New Technique of Dielectric Characterization of Liquids*; Springer: Amsterdam, The Netherlands, 2005.

7. Bernard, P.A.; Gautray, J.M. Measurement of dielectric constant using a microstrip ring resonator. *IEEE Trans. Microw. Theory Techn.* **1991**, *39*, 592–595. [CrossRef]

8. Cataldo, A.; Benedetto, E.D.; Cannazza, G. *Quantitative and Qualitative Characterization of Liquid Materials*; Springer: Berlin, Germany, 2011; pp. 51–83.

9. Yeo, J.; Lee, J. Slot-loaded microstrip patch sensor antenna for high-sensitivity permittivity characterization. *Electronics* **2019**, *8*, 502. [CrossRef]

10. Jilnai, M.; Wen, W.; Cheong, L.; Rehman, M.Z. A microwave ring-resonator sensor for non-invasive assessment of meat aging. *Sensors* **2016**, *16*, 52. [CrossRef] [PubMed]

11. Chuma, E.L.; Iano, Y.; Fontgalland, G.; Roger, L.L.B. Microwave sensor for liquid dielectric characterization based on metamaterial complementary split ring resonator. *IEEE Sens. J.* **2018**, *18*, 9978–9983. [CrossRef]

12. Su, L.; Mata-Contreras, J.; Vélez, P.; Martín, F. Estimation of the complex permittivity of liquids by means of complementary split ring resonator (CSRR) loaded transmission lines. In Proceedings of the IEEE MTT-S International Microwave Workshop Series on Advanced Materials and Processes for RF and THz Applications (IMWS-AMP), Pavia, Italy, 20–22 September 2017.

13. Lajnef, T.; Chaibi, S.; Ruby, P.; ˊAguera, P.E.; Eichenlaub, J.B.; Samet, M.; Kachouri, A.; Jerbi, K. Learning machines and sleeping brains: Automatic sleep stage classification using decision-tree multi-class support vector machines. *J. Neurosci. Methods* **2015**, *250*, 94–105. [CrossRef] [PubMed]

14. Zhang, K.; Amineh, R.K.; Dong, Z.; Nadler, D. Microwave sensing of water quality. *IEEE Access* **2019**, *7*, 69481–69493. [CrossRef]

15. Boybay, M.S.; Ramahi, O.M. Material characterization using complementary splitring resonators. *IEEE Trans. Instrum. Meas.* **2012**, *61*, 3039–3046. [CrossRef]

16. Lee, C.-S.; Yang, C.-L. Thickness and permittivity measurement in multi-layered dielectric structures using complementary split-ring resonators. *IEEE Sens. J.* **2014**, *14*, 695–700. [CrossRef]

17. Lee, C.-S.; Yang, C.-L. Complementary split-ring resonators for measuring dielectric constants and loss tangents. *IEEE Microw. Wirel. Compon. Lett.* **2014**, *24*, 563–565. [CrossRef]

18. Ansari, M.A.H.; Jha, A.K.; Akhtar, M.J. Design and application of the CSRR-based planar sensor for noninvasive measurement of complex permittivity. *IEEE Sens. J.* **2015**, *15*, 7181–7189. [CrossRef]

19. Altair FEKO Software. Available online: https://altairhyperworks.com/product/FEKO (accessed on 8 February 2020).

20. Zhao, Y.; Nasrullah, Z.; Li, Z. PyOD: A Python Toolbox for Scalable Outlier Detection. *J. Mach. Learn. Res.* **2019**, *20*, 1–7.

21. Pedregosa, F.; Varoquaux, G.; Gramfort, A.; Michel, V.; Thirion, B.; Grisel, O.; Blondel, M.; Prettenhofer, P.; Weiss, R.; Dubourg, V.; et al. Scikit-learn: Machine Learning in Python. *J. Mach. Learn. Res.* **2011**, *12*, 2825–2830.

22. Hastie, T.; Tibshirani, R.; Friedman, J. *The Elements of Statistical Learning: Data Mining, Inference, and Prediction*, 2nd ed.; Springer: Berlin, Germany, 2009.

 *electronics* 

Article

Analysis of Propagation Characteristics along an Array of Silver Nanorods Using Dielectric Constants from Experimental Data and the Drude-Lorentz Model

Chujing Zong and Dan Zhang *

College of Information Science and Technology, Nanjing Forestry University, Nanjing 210037, China; zcj9066@163.com
* Correspondence: zhangdan@njfu.edu.cn; Tel.: +86-25-8542-7877

Received: 5 October 2019; Accepted: 31 October 2019; Published: 3 November 2019

Abstract: In this study, the Fourier series expansion method (FSEM) was employed to calculate the complex propagation constants of plasma structures consisting of infinitely long, silver nanorod arrays in the range of 180–1900 nm, and the characteristics of the complex propagation constant were analyzed in depth. According to the results of FSEM using dielectric constants from Johnson experimental data, a multi-mode frequency band appears in the propagation stage, which can be adopted to achieve a multi-mode communication, multi-mode transceiver, integrated filter with single multi-mode combination. In the meantime, the comparison between the three sets of results with only single mode transmission of the generalized multipole technique (GMT) using dielectric constants from Johnson experimental data, FSEM using dielectric constants from Palik experimental data, and FSEM using dielectric function from Drude–Lorentz model suggested that the results of the four sets of complex propagation constants were well consistent with each other. Furthermore, a finite array of only 40 silver nanorods was studied, and the ability of guided waves when a finite array is excited by a plane wave at a specific wavelength was explored. According to different guiding abilities—propagation, attenuation, and cut off, it can be applied to waveguides, sensor, filters, etc.

Keywords: Fourier series expansion; nanorod; multimode; propagation characteristics; guided wave

---

1. Introduction

In the past few decades, numerous studies have focused on the research and the development of novel structures based on multi-layer arrays of the periodic distributions of photonic crystals [1,2]. After that discovery, the field of photonic crystals garnered great interest among researchers, and various analytical methods and numerical methods have been proposed to analyze photonic crystal waveguides [3–9]. The periodic distribution of scattering elements (a simple single-layer structure) that may effectively guide electromagnetic waves along the chain aroused attention from scientists, so has been extensively studied [10–12]. The systematic study on the bandgap structure of two-dimensional periodic dielectric materials is conducted using different methods [13–18]. Simple design and manufacturing can be considered a major advantage of this structure. However, to verify its good guiding attributes, detailed theoretical and numerical studies are necessary.

In recent years, in the direction of reducing the size of photonic devices below the diffraction limit, it was not enough to take dielectric material as the object of research. Therefore, it led to the development of metal plasma-resonant waveguides. From the structural perspective, the waveguide consists of several coupled nanoparticles arranged in the array. Near-field interactions between these closely spaced nanoparticles causes light to propagate along the array's axis. The structures that guide these collective oscillations have been analyzed in numerous papers [19–23]. However, since metallic losses are rather large in these waveguides, the calculation of the attenuation constant needs to be considered. Numerous studies (e.g., the analysis of the optical interactions of nanostructures formed on metal films) focus on the application of metals to photonic crystals. The absorption characteristics of one-dimensional quasi-periodic photonic crystals with thin metal layers and dielectric Fibonacci layers were investigated. The full-vector modal solver, using the finite element method, was employed to numerically study the polarization characteristics of wire-selectively filled photonic crystal fibers into the cladding pores. The possibility of selective polarization in photonic crystal structure was analyzed using a single molecular layer of metal nanoparticles. Using the transfer matrix method and some studies of photonic crystals in three dimensions, the transmission properties of a one-dimensional photonic crystal made of alternate layers of an isotropic ordinary dielectric and a graphene-based hyperbolic metamaterial were studied theoretically [24–29].

In the previous study, a Fourier series expansion method was proposed as a feasible method to study the propagation mode and leaky mode of an infinitely long circular chain structure. Additionally, to study the propagation characteristics and electric field distribution characteristics in three Bragg propagation periods, the model of an infinitely long square single-chain and multi-chain periodic structure was used.

In this paper, the complex propagation constants of an infinitely long silver chain structure with a wavelength in the range of 180–1900 nm are studied in detail. The correctness of the method (FSEM) can be verified by the four results that are very consistent. A special phenomenon of single frequency multimode was found in the results of FSEM using dielectric constants from Johnson experimental data (FSEM(J)), and the complex propagation constants of the three methods of generalized multipole technique using dielectric constants from Johnson experimental data (GMT(J)), FSEM using dielectric constants from Palik experimental data (FSEM(P)), and FSEM using dielectric function from Drude–Lorentz model (FSEM(D-L)) with only the single mode transmission were compared. It was found that the results of the four sets of complex propagation constants agree well. Furthermore, the guided wave capability of different nanowavelengths under the finite-length chain structure of 40 circular nano-columns was studied. By studying the waveguide capabilities of a particular wavelength, different stages can be applied to different devices. Propagation phase can be used for waveguide, sensor, etc. The cut-off phase can be used for filters, nanomirrors, nanoprobes, etc. In particular, the multi-mode stage that occurs in FSEM (J) can be used for multi-mode, communication integrated filters of a single and multi-mode combination, etc.

2. Formulation of the Problem

A two-dimensional infinite periodic chain composed of circular rods is periodic along the x-axis and with the lattice constant h (Figure 1). The scatters are infinitely long in the z-direction, and they are parallel to each other. The circular rods, having radius r, are assumed to be pure dielectrics, with a relative dielectric permittivity ε. An array with this structure repeats the same configuration with a fictitious period Λ along the y-direction, and then the original structure is approximated by the array's unit cell located in $0 \leq y \leq \Lambda$. Assume that we induce the propagation of E (E_z, H_x, H_y) and H (H_z, E_x, E_y) waves. The details of the formulation are omitted in this communication due to the space limitations; see our previous reports for details [30,31]. Surface plasmon is a large amount of collective free-electron vibration occurring at the metal–medium interface and excited only by H waves (magnetic field components perpendicular to the nanowire cross section) [32].

First, the major idea of the formulation should be described. In the case of the H wave, Maxwell equations are written as follows:

$$v(y)\frac{\partial H'_z}{\partial y} = ik_0\varepsilon(y)E_x, \tag{1}$$

$$-\frac{\partial H'_z}{\partial x} = ik_0\varepsilon(y)E_y = ik_0\frac{D_y}{\varepsilon_0}, \tag{2}$$

$$\frac{\partial E_y}{\partial x} - v(y)\frac{\partial E_x}{\partial y} = -ik_0 H'_z, \tag{3}$$

$$D_y = \varepsilon_0 \varepsilon(y) E_y, \tag{4}$$

where $H'_z = (\mu_0/\varepsilon_0)^{1/2} H_z$, $v(y) = [1 + i\sigma(y)]^{-1}$ denotes the stretched coordinate variable [24] characterizing the assumed Perfectly Matched Layers (PMLs), where $\sigma(y) = \sigma_{\max}(1 - y/w)^d$ is the conductivity function. Under the fictitious periodicity of the system, the electric and magnetic fields are approximated by a truncated Fourier series:

$$H'_z(x,y) = \sum_{m=-M}^{M} h_{z,m}(x)e^{i\varphi_m y}, \tag{5}$$

$$E_x(x,y) = \sum_{m=-M}^{M} e_{x,m}(x)e^{i\varphi_m y}, \tag{6}$$

$$E_y(x,y) = \sum_{m=-M}^{M} e_{y,m}(x)e^{i\varphi_m y}, \tag{7}$$

$$D_y(x,y) = \sum_{m=-M}^{M} d_{y,m}(x)e^{i\varphi_m y}, \tag{8}$$

where D_y is the electric displacement vector, $\varphi_m = 2\pi m/\Lambda$. Equation (2) is substituted into (1), and the orthogonality of the Fourier bases is exploited. For H modes with field components (H_z, E_x, E_y), H_z acts as the leading field. To overcome the discontinuity of the electric field normal to the core-cladding boundaries, Li's factorization rule [33] should be followed by the Fourier series expansion of E_y field. A set of linear equations to determine the Fourier coefficients, $\{h_{z,m}(x)\}$ and $\{e_{y,m}(x)\}$ are derived as

$$\frac{\partial^2}{\partial x^2}h_z(x) = -k_0^2 C \cdot h_z(x), \tag{9}$$

$$e_y(x) = -i\frac{1}{k_0}\hat{N} \cdot \frac{\partial}{\partial x}h_z(x), \tag{10}$$

$$h_z(x) = \left[h_{z,-M} \ldots h_{z,0} \ldots h_{z,M}\right]^T, \tag{11}$$

$$e_y(x) = \left[e_{y,-M} \ldots e_{y,0} \ldots e_{y,M}\right]^T, \tag{12}$$

$$C = \hat{N}^{-1}(I - VAN^{-1}VA), \tag{13}$$

$$[N]_{mm'} = \frac{1}{\Lambda}\int_0^\Lambda \varepsilon(y)e^{-i(\varphi_m - \varphi_{m'})y}dy, \tag{14}$$

$$[\hat{N}]_{mm'} = \frac{1}{\Lambda}\int_0^\Lambda \frac{1}{\varepsilon(y)}e^{-i(\varphi_m - \varphi_{m'})y}dy, \tag{15}$$

$$[V]_{mm'} = \frac{1}{\Lambda}\int_0^\Lambda v(y)e^{-i(\varphi_m - \varphi_{m'})y}dy, \tag{16}$$

$$[A]_{mm'} = \frac{\phi_m}{k_0}\delta_{mm'}, \tag{17}$$

where $\delta_{mm'}$ denotes Kronecker's delta; the superscript T denotes the transpose of the indicated vector; k_0 is the wavenumber in a free space; and $\varepsilon(y)$ is the dielectric permittivity along the y-axis within the period $0 \leq y \leq \Lambda$. The eigenvalue $k_n = \xi_n^2$ ($n = 1, 2, 3, \cdots, 2M + 1$) of matrix C and the eigenvectors P_n

determine the propagation constant ξ_n and the field distributions for the guided and radiation modes in the assumed waveguide, respectively. The solutions to Equations (8) and (9) are written as

$$\begin{bmatrix} h_z(x) \\ e_y(x) \end{bmatrix} = F_H U(x - x\prime) \cdot a(x\prime), \tag{18}$$

$$F_H = \begin{bmatrix} P & P \\ \hat{N}PB & -\hat{N}PB \end{bmatrix}, \tag{19}$$

$$U(x) = \begin{bmatrix} U^{(+)}(x) & 0 \\ 0 & U^{(-)}(x) \end{bmatrix}, \tag{20}$$

$$P = [p_1 \, p_2 \, \cdots \, p_{2M+1}], \tag{21}$$

$$U^{(\pm)}(x) = [\exp(\pm ik_0\xi_n x)\delta_{nn'}], \tag{22}$$

$$B = [\xi_n \delta_{mm'}], \tag{23}$$

$$a(x) = [a^{(+)}(x) \, a^{(-)}(x)], \tag{24}$$

$$a^{\pm}(x) = [a_1^{\pm}(x) \, a_2^{\pm}(x) \, \cdots \, a_{2M}^{\pm}(x) \, a_{2M+1}^{\pm}(x)], \tag{25}$$

where $a_n^{(\pm)}(x)$ denotes the amplitudes of the forward and backward propagating n-th modes. The elements of $U(x)$, P, and $a^{(\pm)}(x)$ are rearranged in the order of guided and radiation modes with decreasing $\mathrm{Re}[\xi_n]$.

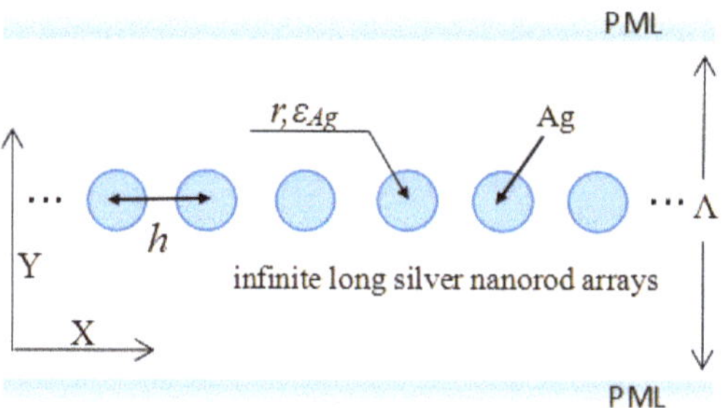

Figure 1. Infinite periodic chain of circular rods along the x-axis with lattice constant h. Radius and dielectric permittivity of rods are r and ε, respectively.

Next, each circular rod is divided into a sufficient number of thin, parallel rectangular rods, as shown in Figure 2, and the unit cell of the periodic chain in the x-direction replaced by a cascading connection of layered, parallel, planar waveguides [31]. In each waveguide section, the solutions to Equations (1)–(8) are given by Equations (18)–(25). The boundary conditions for H_z and E_y at each step-discontinuity are fulfilled by equating the Fourier coefficients on both sides of the section. This relation leads to the scattering matrix S_j defined at the interface $x = x_j$ as follows:

$$\begin{bmatrix} a_{j-1}^{(-)}(x_j - 0) \\ a_j^{(+)}(x_j + 0) \end{bmatrix} = S_j \begin{bmatrix} a_{j-1}^{(+)}(x_j - 0) \\ a_j^{(-)}(x_j + 0) \end{bmatrix}, \tag{26}$$

where

$$S_j = \begin{bmatrix} R_j^{(-)} & T_j \\ T_j & R_j^{(+)} \end{bmatrix}$$
$$= \begin{bmatrix} P_{j-1} & -P_j \\ P_{j-1}B_{j-1} & P_jB_j \end{bmatrix}^{-1} \begin{bmatrix} -P_{j-1} & P_j \\ P_{j-1}B_{j-1} & P_jB_j \end{bmatrix} \tag{27}$$

Using the scattering matrix S_{j+1} at the interface $x = x_{j+1}$ defined in the same way as Equation (25) and considering the modes propagation over the distance $x_{j+1} - x_j$, it yields:

$$S_{j+1} = \begin{bmatrix} R_{j+1}^{(-)} & T_{j+1} \\ T_{j+1} & R_{j+1}^{(+)} \end{bmatrix} \tag{28}$$

with

$$\hat{R}_{j+1}^{(-)} = \hat{R}_j^{(-)} + \hat{T}_j X_{j+1} U_j^{(+)} R_{j+1}^{(-)} U_j^{(+)} \hat{T}_j, \tag{29}$$

$$\hat{T}_{j+1} = T_j U_j^{(+)} Y_{j+1} \hat{T}_j, \tag{30}$$

$$\hat{R}_{j+1}^{(+)} = \hat{R}_{j+1}^{(-)} + \hat{T}_{j+1} U_j^{(+)} Y_{j+1} \hat{R}_j^{(+)} U_j^{(+)} \hat{T}_{j+1}, \tag{31}$$

$$X_{j+1} = \left(I - U_j^{(+)} R_{j+1}^{(-)} U_j^{(+)} \hat{R}_j^{(+)} \right)^{-1}, \tag{32}$$

$$Y_{j+1} = \left(I - \hat{R}_j^{(+)} U_j^{(+)} R_j^{(-)} U_j^{(+)} \right)^{-1}, \tag{33}$$

$$U_j^{(+)} = U^{(+)}(x_{j+1} - x_j), \tag{34}$$

where I denotes the unit matrix; $U^{(+)}(x)$ is defined by Equation (22). Equations (29)–(31) express the recursive relations to connect the scatterings matrices defined at each step of discontinuity of the waveguide.

When the waveguide changes continuously along the optical propagation, the transition section is approximated by a considerable number of step-discontinuities. If N step discontinuities at $x = x_j$ $(j = 1, 2, \ldots, N)$ exist along the waveguide, $(N - 1)$ times the recursion process will lead to the generalized scattering matrices S_N [31] for the whole system as follows:

$$\begin{bmatrix} a_1^{(-)}(x_1 - 0) \\ a_N^{(+)}(x_N + 0) \end{bmatrix} = S_N \begin{bmatrix} a_1^{(+)}(x_1 - 0) \\ a_N^{(-)}(x_N + 0) \end{bmatrix}. \tag{35}$$

The generalized scattering matrix over one period h in the x direction is calculated following the procedure, where N denotes the number of the thin rectangular rods within a unit cell. The result is adopted to obtain the transfer matrix K over the unit cell that satisfies the following relation [31]:

$$\begin{bmatrix} a^+(h) \\ a^-(h) \end{bmatrix} = K \begin{bmatrix} a^+(0) \\ a^-(0) \end{bmatrix}. \tag{36}$$

Thus, the propagation constant $\gamma_k = \beta_k + i\alpha_k$ of the k-th mode is determined by

$$\gamma_k = -i \log \chi_k / h, \tag{37}$$

where χ_k denotes the k-th eigenvalue of the transfer matrix K.

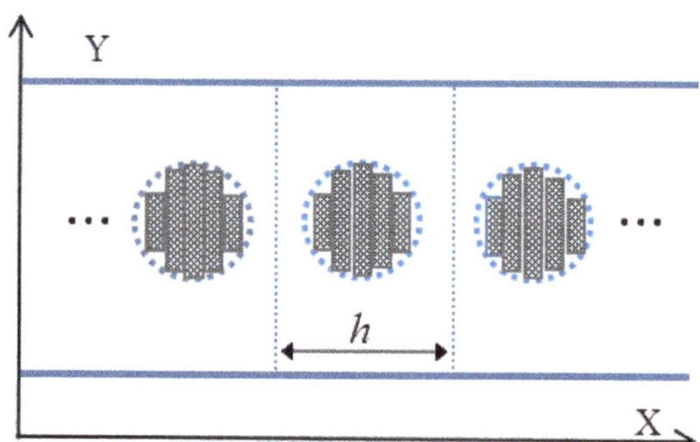

Figure 2. Discretization of circular rods by a sufficient number of thin, parallel rectangular rods.

3. Results and Discussion

This section compares the trend of complex propagation constants of the four methods and explore the differences and specificities of the four outcomes first. The four methods are: FSEM using dielectric constants from Johnson experimental data (FSEM(J)), generalized multipole technique using dielectric constants from Johnson experimental data (GMT(J)), FSEM using dielectric constants from Palik experimental data (FSEM(P)), and FSEM using dielectric function from Drude–Lorentz model (FSEM(D-L)).

The phase constants and attenuation constants of the infinitely long circular chain structure in H mode were analyzed. The radius of the circle was 25 nm, and the period length h was 55 nm. The fictitious period Λ was set to $60h$, the truncation number was M = 150, the number of thin parallel rectangular rods was 20, and the thickness, σ_{max}, and d of the PML were chosen to be h, 8.0, and 2.1, respectively.

Two different experimental datasets of metallic silver were used: the wavelength ranges of Johnson experimental data and the Palik experimental data were from 180–1900 nm, and from 180–1300 nm, respectively [34,35].

In this paper, the Drude–Lorentz model was used to characterize the optical dielectric function $\varepsilon(\omega)$ of the silver nanowire, and an optimized fit of silver dispersion was used [36–38]:

$$\varepsilon(\omega) = \varepsilon_\infty - \frac{\omega_p^2}{i\omega\gamma + \omega^2} - \frac{\Delta\Omega_L^2}{\left(\omega^2 - \Omega_L^2\right) + i\Gamma_L\omega}, \tag{38}$$

where Ω_L and Γ_L denote the frequency and the spectral width of the Lorentz oscillator, respectively; ω_p and γ are the usual constants of the Drude model; Δ is the strength of the Lorentz oscillator. These parameters were optimized to fit the experimental permittivity data of silver [34]. The results of this optimization are listed in Table 1.

Table 1. Values of the optimized parameters to fit the experimental data of silver.

ε_∞	$\omega_p/2\pi/\text{THz}$	$\gamma/2\pi/\text{THz}$	Δ	$\Gamma_L/2\pi/\text{THz}$	$\Omega_L/2\pi/\text{THz}$
2.4064	2214.6	4.8	1.6604	620.7	1330.1

In order to make the method of this paper more persuasive and correct, first, the same Johnson experimental data was used, and the results of the complex propagation constants of the FSEM(J) was compared with those of the complex propagation constant of GMT(J) in the comparative paper [39]. The parameters used were consistent with the paper [39]. Second, the results of complex propagation constants obtained from two different metal silver experimental datasets (Johnson and Palik) and D-L were compared by the method of FSEM (the radius of the circle was 25 nm, and the period length h was 55 nm), as shown in Figure 2. The permittivity values of the two sets of silver experimental data and Drude–Lorentz model were also obtained. The inset of Figure 3 shows the values for the dielectric function of silver, as used in these computations.

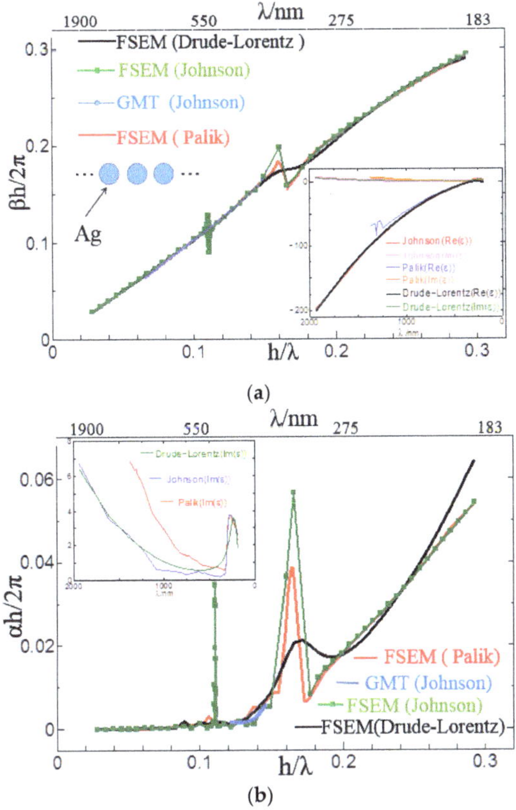

Figure 3. Phase constant and attenuation constant of infinite long silver nanorod arrays with $r = 25$ nm and $h = 55$ nm corresponding to the four results. The inset shows the dielectric function, as reported in [34–38]. (**a**) Phase constant; (**b**) attenuation constant.

It can be seen from the comparison results in Figure 3 that whether the method is different or the experimental data of the dielectric constant is different, the complex propagation constants obtained by the four methods are quite consistent, with only slight differences in the details, which can confirm the correctness of the method used in this paper. By contrast, we can find special physical phenomena such as multimode. According to the study on the FSEM (J) method, the phase constant and the attenuation constant have jumps in two places: the first jump's range is around 496 nm, which is caused by multimode causes; the second jump's range is around 340 nm, attributed to fluctuations of the permittivity and the effects of the structure. The jump variation of the dielectric constant can be seen in the illustration in Figure 3. A special phenomenon of single frequency multimode in the result of FSEM (J) is a special physical phenomenon. The results of GMT(J), FSEM(P), and FSEM(D-L) are only single mode. The reason why there are no multimodes in the results of the GMT(J) and FSEM(P) is that the frequency points measured by the experiment are different and the method is different.

The reason why FSEM using the Drude–Lorentz model especially, whose parameters are optimized to fit the Johnson experimental permittivity data of silver, does not have the same multimode as FSEM (J), is that there is a certain gap between values of the two dielectric constants. That can be seen from the inset-graph of Figure 3. In general, the emergence of multimode is due to different methods, different frequency points of measurement, and different values of dielectric constant.

In the meantime, through Figure 3a, when the FSEM method was used to study two different experimental data (Johnson and Palik), it was found that propagation constants were consistent.

The comparison of the above four ways suggest that only the FSEM (J) data of the four sets of data show the new physical phenomenon of multi-mode, and the distribution of the other three sets of data are in single-mode form. On the whole, the calculation results of the four sets of complex propagation constants can be of high consistency even using different research methods, different sets of experimental data, and calculated dielectric constant data.

Next, to study the guided wave ability of the finite array excited by plane waves at a specific wavelengths, 40 silver cylinders were selected as experimental objects. The results of the four methods show that $h/\Lambda < 0.145$ is the propagation stage, and it is in an attenuated or cut-off state when $h/\Lambda > 0.145$. The results obtained by FSEM(J) were analyzed in detail. To make the analysis more convincing, field pattern distribution and the propagation distribution of a few special points and the representative points at different stages are presented; $0.1101 < h/\Lambda < 0.1116$ (that is 500 nm $> \Lambda > 492$ nm) is a multi-mode stage. At that stage, the even mode distribution map for each frequency has 2–4 different modes. When $h/\Lambda = 0.1108$, i.e., the point 496 nm where the phase constant value is the largest in the first jump portion, its even mode distribution map displays four different modes; the order is 10^{-3} and 10^{-2}, though there are orders 10^{-2}. At this frequency, the light can be propagated because the formation of the propagation map may cover four modes, or the first mode (order 10^{-3}) may be selected. The field pattern distribution and the propagation distribution representing the wavelength of 496 nm are given below, as shown in Figures 4 and 5. It can be used for multi-mode communication, integrated filters of single multi-mode combination, etc., in that frequency range. The other propagation stages, only exist in the form of a single mode that can be used for a waveguide, sensor, etc. Below, we give an example: at $h/\Lambda = 0.133$; i.e., $\Lambda = 413$ nm; field pattern distribution and propagation distribution representing the wavelength of 413 nm (Figures 6 and 7). The second jump is caused by the jitter variation of the permittivity and the effect of the structure. In this jump range, the highest point is $h/\Lambda = 0.165$, the wavelength is 330 nm, and the order is 10^{-1}. It is suggested that when $h/\Lambda > 0.145$, the trend of the imaginary part increases, decreases, and then increases, so the corresponding propagation distribution is attenuation–cutoff–attenuation–cutoff. Figures 8 and 9 show the field pattern distribution and propagation distribution at a wavelength of 292 nm, and the propagation distribution of this frequency is attenuated. The field pattern distribution and propagation distribution for a wavelength of 191 nm are shown in Figures 10 and 11, suggesting that light does not propagate at this frequency. It can be used for filters, nanomirrors, nanoprobes, etc., when $h/\Lambda > 0.145$. For the other three sets of results, it was found that they just exist in the form of single mode, and that there is no multimode. When using the Palik experimental data, by observing the dielectric constant, it was found that there are two places to jump: one of them is around the wavelength of 340 nm, which can be clearly seen from the complex propagation constant of Figure 2; the other part of the jump is around 1200 nm. Those results are because the wavelength is relatively large, the frequency is very small, and the corresponding order is about 10^{-5}, so no obvious jump change can be seen from the complex propagation-constant graph. Figures 12 and 13 show the field pattern distribution and propagation distribution corresponding to the frequency of 496 nm using the Palik experimental data with only single mode, which can be compared with Figures 4 and 5.

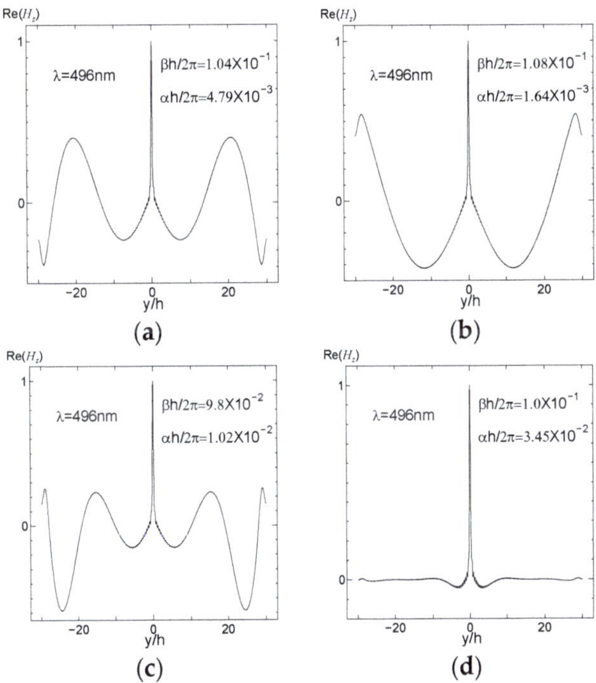

Figure 4. Distribution of the *Hz* field as a function of the y coordinate representing the wavelength of 496 nm. (**a**) Even mode 1; (**b**) even mode 2; (**c**) even mode 3; (**d**) even mode 4.

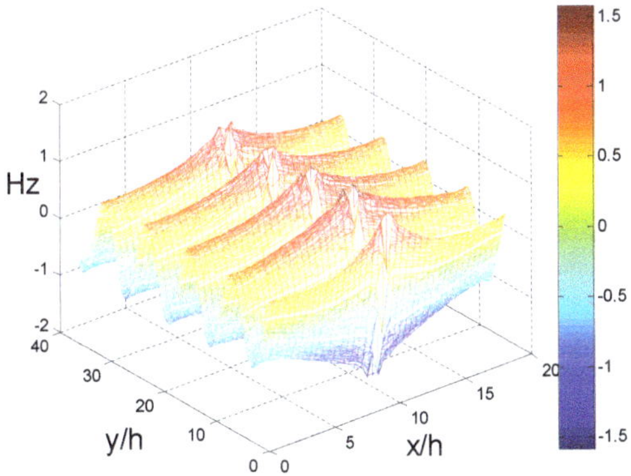

Figure 5. Near field propagation distribution representing the wavelength of 496 nm.

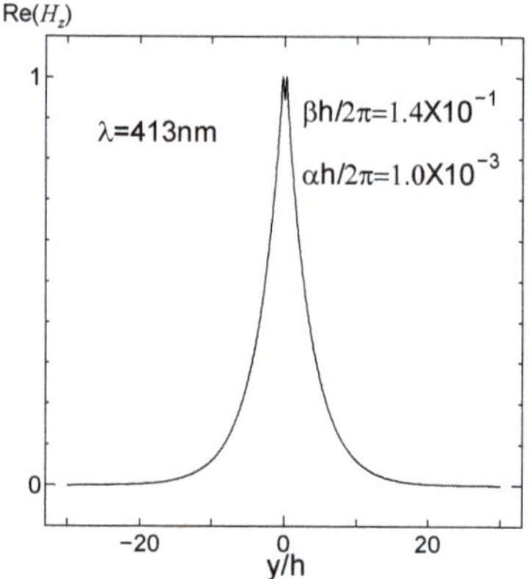

Figure 6. Distribution of the *Hz* field as a function of the y coordinate representing the wavelength of 413 nm.

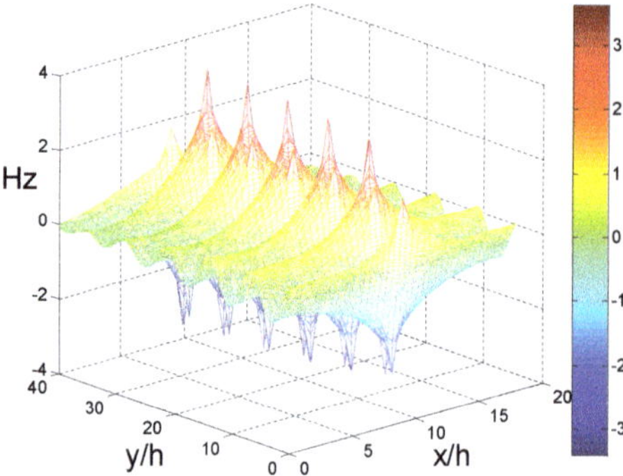

Figure 7. Near field propagation distribution representing the wavelength of 413 nm.

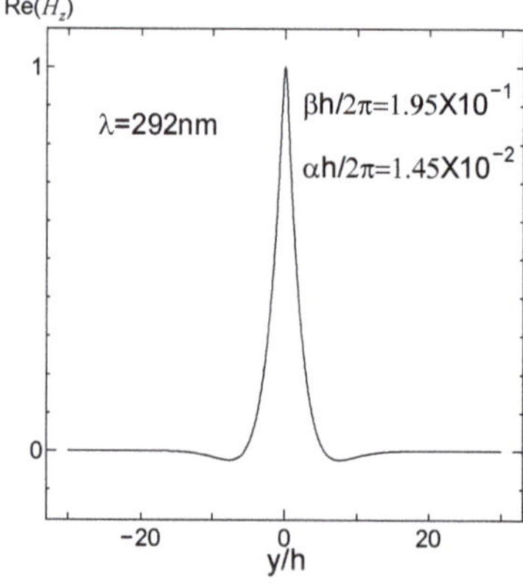

Figure 8. Distribution of the *Hz* field as a function of the y coordinate representing the wavelength of 292 nm.

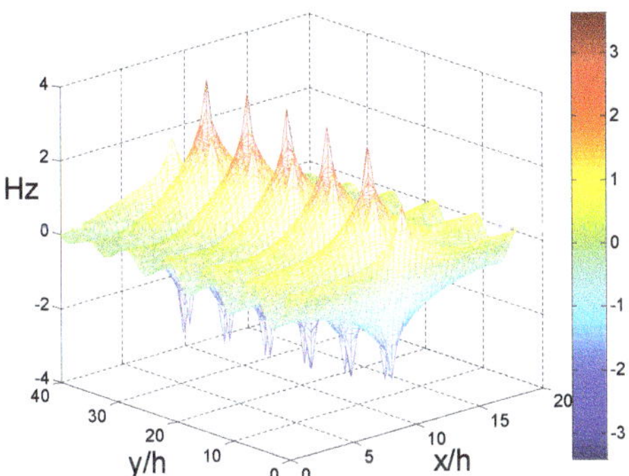

Figure 9. Near field propagation distribution representing the wavelength of 292 nm.

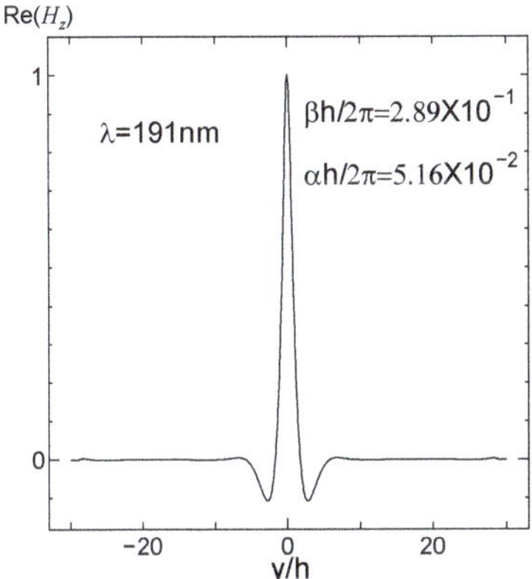

Figure 10. Distribution of the *Hz* field as a function of the y coordinate representing the wavelength of 191 nm.

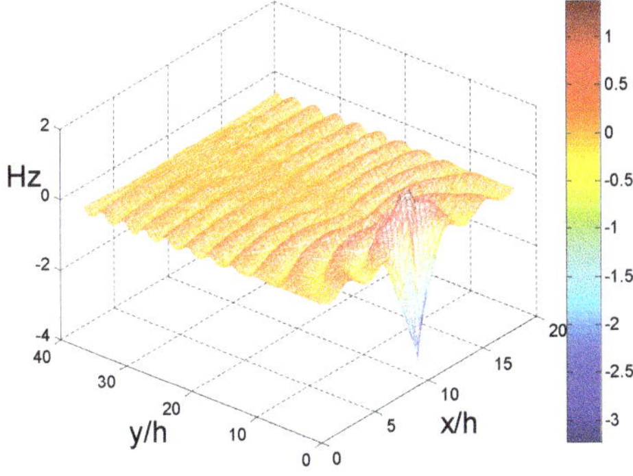

Figure 11. Near field propagation distribution representing the wavelength of 191 nm.

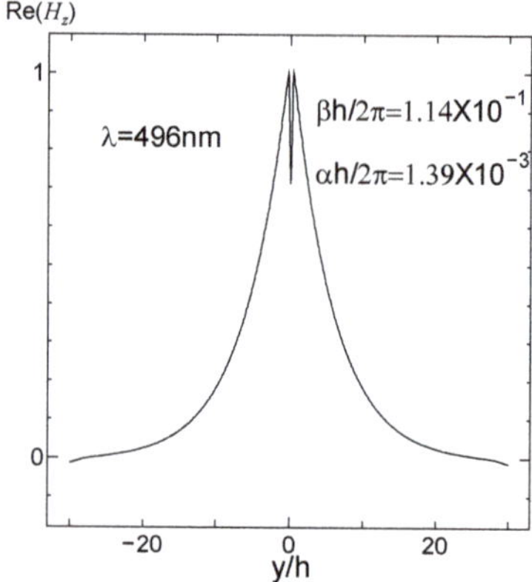

Figure 12. Distribution of the *Hz* field as a function of the y coordinate representing the wavelength of 496 nm (Palik experimental data).

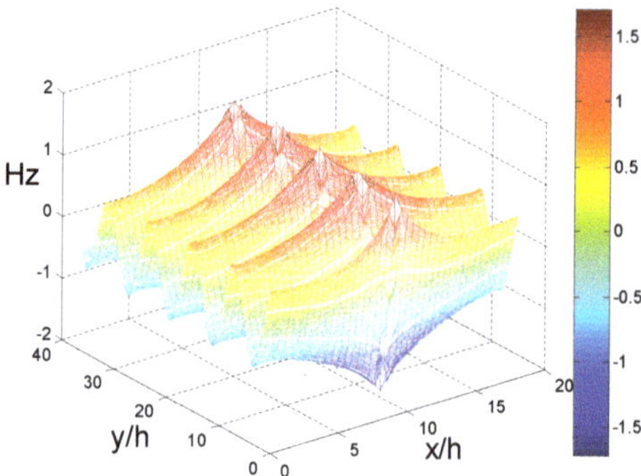

Figure 13. Near field propagation distribution representing the wavelength of 496 nm (Palik experimental data).

4. Conclusions

In this paper, the complex propagation constants of metallic silver circular chain structures were studied in depth. According to the complete result plot, the propagation region ($h/\Lambda < 0.145$) and the attenuation-cutoff region ($h/\Lambda > 0.145$) were obtained. In the case of FSEM(J), it was found that in the frequency range of 492 nm $< \Lambda <$ 500 nm, there is a special phenomenon of multi-mode, which can be applied to new multi-mode communication and multi-mode transceivers, single-mode combined integrated filters, and other devices. However, in the other three results of FSEM(P), GSM(J), and FSEM(D-L), only single mode exists. In the study on a limited array of only 40 silver nanorods, the ability of the finite array to be guided by plane waves at specific wavelengths was explored. According to different guided wave capabilities (e.g., propagation, attenuation, and cutoff), it can be used in waveguide, multi-mode communication, integrated filters of a single and multi-mode combination, sensor, filters, nanomirrors, and other devices.

Author Contributions: Data curation, C.Z.; Project administration, D.Z.

Funding: This research was funded by the National Natural Science Foundation of China, grant number 31170668, and the Postgraduate Research and Practice Innovation Program of Jiangsu Province.

Conflicts of Interest: The authors declare no conflicts of interest.

References

1. Yasumoto, K.; Miyamoto, T.; Momoda, M. *Full-Wave Analysis of Optical Waveguides Using Periodic Boundary Conditions*; SPIE: Bellingham, WA, USA, 1999; Volume 3666, pp. 170–176.
2. Miyamoto, T.; Momoda, M.; Yasumoto, K. Full-vectorial analysis of connection problem in optical fiber. *IEEJ Trans. Fundam. Mater.* **2002**, *122*, 39–46. [CrossRef]
3. Twersky, V. Multiple scattering of waves and optical phenomena. *J. Opt. Soc. Am.* **1962**, *52*, 145. [CrossRef] [PubMed]
4. Ohtaka, K.; Numata, H. Multiple scattering effects in photon diffraction for an array of cylindrical dielectric. *Phys. Lett. A* **1979**, *73*, 411–413. [CrossRef]
5. Cao, Q.; Lalanne, P. Negative Role of Surface Plasmons in the Transmission of Metallic Gratings with Very Narrow Slits. *Phys. Rev. Lett.* **2002**, *88*, 057403. [CrossRef] [PubMed]
6. Yasumoto, K. (Ed.) *Electromagnetic Theory and Applications for Photonic Crystals*; CRC Press: Boca Raton, FL, USA, 2005.
7. Yasumoto, K.; Jandieri, V.; Liu, Y. Coupled-mode formulation of two-parallel photonic-crystal waveguides. *J. Opt. Soc. Am. A* **2013**, *30*, 96–101. [CrossRef] [PubMed]
8. Takakura, Y. Optical Resonance in a Narrow Slit in a Thick Metallic Screen. *Phys. Rev. Lett.* **2001**, *86*, 5601–5603. [CrossRef] [PubMed]
9. Yang, F.; Sambles, J.R. Resonant transmission of microwaves through a narrow metallic slit. *Phys. Rev. Lett.* **2002**, *89*, 063901. [CrossRef] [PubMed]
10. Nemec, H.; Kuzel, P.; Coutaz, J.-L.; Ctyroky, J. Transmission properties and band structure of a segmented dielectric waveguide for the terahertz range. *Opt. Commun.* **2007**, *273*, 99–104. [CrossRef]
11. Zhang, D.; Mase, A. A Formula for Fourier Series Expansion Method with Complex Coordinate Stretching Layers. *J. Infrared Millim. Terahertz Waves* **2011**, *32*, 196–203.
12. Yasumoto, K.; Watanabe, K.; Ishihara, J. Numerical analysis of optical waveguides using Fourier series expansion: Application of perfectly matched layer. *Proc. Int. Symp. Recent Adv. Microw. Technol.* **1999**, *1999*, 589–592.
13. Zhang, D.; Jandieri, V.; Yasumoto, K. Modal analysis of wave guidance by a periodic chain of circular rods. In Proceedings of the 2016 Progress in Electromagnetics Research Symposium (PIERS), Shanghai, China, 8–11 August 2016.
14. Benisty, H. Modal analysis of optical guides with two-dimensional photonic band-gap boundaries. *J. Appl. Phys.* **1996**, *79*, 7483–7492. [CrossRef]
15. Yasumoto, K.; Jia, H.; Sun, K. Rigorous analysis of two-dimensional photonic crystal waveguides. *Radio Sci.* **2005**, *40*, 1–7. [CrossRef]
16. Xu, Y.; Lee, R.K.; Yariv, A. Adiabatic coupling between conventional dielectric wave-guides and waveguide with discrete translational symmetry. *Opt. Lett.* **2000**, *25*, 755–757. [CrossRef] [PubMed]
17. Happ, T.D.; Kamp, M.; Forchel, A. Photonic crystal tapers for ultracompact mode conversion. *Opt. Lett.* **2001**, *26*, 1102–1104. [CrossRef] [PubMed]
18. Talneau, A.; Lalanne, P.; Agio, M.; Soukoulis, M.C. Low-reflection photonic-crystal taper for efficient coupling between guide sections of arbitrary widths. *Opt. Lett.* **2002**, *27*, 1522–1524. [CrossRef] [PubMed]
19. Akahara, J.; Yamagishi, S.; Taki, H.; Morimoto, A.; Kobayashi, T. Guiding of a one-dimensional optical beam with nanometer diameter. *Opt. Lett.* **1997**, *22*, 475–477. [CrossRef] [PubMed]
20. Maier, S.A.; Kik, P.G.; Atwater, H.A.; Meltzer, S.; Harel, E.; Koel, B.E.; Requicha, A.A.G. Local detection of electromagnetic energy transport below the diffraction limit in metal nanoparticle plasmon waveguides. *Nat. Mater.* **2003**, *2*, 229–232. [CrossRef] [PubMed]
21. Quinten, M.; Leitner, A.; Krenn, J.R.; Aussenegg, F.R. Electromagnetic Energy Transport via Linear Chains of Silver Nanoparticles. *Opt. Lett.* **1998**, *23*, 1331–1333. [CrossRef] [PubMed]
22. Brongersma, M.L.; Hartman, J.W.; Atwater, H.A. Electromagnetic Energy Transfer and Switching in Na noparticle Chain Arrays below the Diffraction Limit. *Phys. Rev. B* **2000**, *62*, R16356–R16359. [CrossRef]
23. Krenn, J.R.; Lamprecht, B.; Ditlbacher, H.; Schider, G.; Salerno, M.; Leitner, A.; Aussenegg, F.R. Non-diffraction-limited light transport by gold nanowires. *Europhys. Lett.* **2002**, *60*, 663–669. [CrossRef]
24. Kim, J.W.K. Excitation and Propagation of Surface Plasmons in a Metallic Nanoslit Structure. *IEEE Trans. Nanotechnol.* **2008**, *7*, 229–236.

25. Gong, Y.K.; Liu, X.M.; Wang, L.R.; Lu, H.; Wang, G.X. Multiple responses of TPP-assisted near-perfect absorption in metal/Fibonacci quasiperiodic photonic crystal. *Opt. Express* **2011**, *19*, 9759–9769. [CrossRef] [PubMed]

26. Nagasaki, A.; Saitoh, K.; Koshiba, M. Polarization characteristics of photonic crystal fibers selectively filled with metal wires into cladding air holes. *Opt. Express* **2011**, *19*, 3799–3808. [CrossRef] [PubMed]

27. Madani, A.; Entezar, S.R. Optical properties of one-dimensional photonic crystals containing graphene-based hyperbolic metamaterials. *Photonics Nanostruct.-Fundam. Appl.* **2017**, *25*, 58–64. [CrossRef]

28. Zhang, H.F. Three-dimensional function photonic crystals. *Phys. B Phys. Condens. Matter* **2017**, *525*, 104–113. [CrossRef]

29. Miyamoto, T.; Momoda, M.; Yasumoto, K. Numerical Analysis for Three-Dimensional Optical Waveguides with Periodic Structure Using Fourier Series Expansion Method. *Electron. Commun. Jpn.* **2003**, *86*, 22–31. [CrossRef]

30. Jia, H.; Zhang, D.; Yasumoto, K. Fast analysis of optical waveguides using an improved Fourier series method with perfectly matched layer. *Microw. Opt. Technol. Lett.* **2005**, *46*, 263–268. [CrossRef]

31. Zhang, D.; Jia, H. Numerical analysis of leaky modes in two-dimensional photonic crystal waveguides using Fourier series expansion method with perfectly matched layer. *IEICE Trans. Electron.* **2007**, *90*, 613–622. [CrossRef]

32. William, L.B.; Alain, D.; Thomas, W.E. Surface plasmon subwavelength optics. *Nature* **2003**, *424*, 824–830.

33. Li, L. Use of Fourier series in the analysis of discontinuous periodic structures. *J. Opt. Soc. Am. A* **1996**, *13*, 1870–1876. [CrossRef]

34. Palik, E.D. *Handbook of Optical Constants of Solids*; Academic: San Francisco, CA, USA, 1998.

35. Johnson, P.; Christy, R. Optical Constants of the Noble Metals. *Phys. Rev. B* **1972**, *6*, 4370–4379. [CrossRef]

36. Laroche, T.; Girard, C. Near-field optical properties of single plasmonic nanowires. *Appl. Phys. Lett.* **2006**, *89*, 233119. [CrossRef]

37. Jiang, H.L.; Liu, Q.N. Study on characteristics of optical tamm state by drude-lorentz model. *Semicond. Optoelectron.* **2016**, *37*, 218–222.

38. Su, J.; Sun, C.; Wang, X.Q. A metallic dispersion model for numerical simulation. *J. Optoelectron. Laser* **2013**, *24*, 408–411.

39. Rahbarihagh, Y.; Kalhor, F.; Rashed-Mohassell, J.; Shahabadi, M. Modal Analysis for a Waveguide of Nanorods Using the Field Computation for a Chain of Finite Length. *Appl. Comput. Electromagn. Soc. J.* **2014**, *29*, 140–148.

Article

LED Arrays of Laser Printers as Valuable Sources of Electromagnetic Waves for Acquisition of Graphic Data

Ireneusz Kubiak [1,*] and Joe Loughry [2]

[1] Electromagnetic Compatibility Department and Laboratory, Military Communication Institute, 05-130 Zegrze Poludniowe, Poland
[2] Daniel Felix Ritchie School of Engineering and Computer Science, University of Denver, Denver, CO 80208, USA; joe.loughry@cs.du.edu
* Correspondence: i.kubiak@wil.waw.pl; Tel.: +48-261-885 537

Received: 15 August 2019; Accepted: 23 September 2019; Published: 23 September 2019

Abstract: Classified information may be derivable from unintended electromagnetic signals. This article presents a technical analysis of LED arrays used in monochrome computer printers and their contribution to unintentional electromagnetic emanations. Analyses were based on realistic type sizes and distribution of glyphs. Usable pictures were reconstructed from intercepted radio frequency (RF) emanations. We observed differences in the legibility of information receivable at a distance that we attribute to different ways used by printer designers to control the LED arrays, particularly the difference between relatively high voltage single-ended waveforms and lower-voltage differential signals. To decode the compromising emanations required knowledge of—or guessing—printer operating parameters including resolution, printing speed, and paper size. Measurements were carried out across differences in construction and control of the LED arrays in tested printers.

Keywords: pattern recognition; image processing; graphic information; LED array; laser printer; compromising emanations; electromagnetic infiltration; reconstruction; non-invasive data acquisition

1. Introduction

Printers translate the symbolic form of data processed by computers into a graphical form during their operation. As with every electronic device, printers are sources of electromagnetic emanations. Besides control signals, which carry no information (e.g., directing the operation of stepper motors or heaters), there are other signals (useful signals) that are correlated with the information being processed. Such emissions are called sensitive or valuable or compromising emanations from the point of view of electromagnetic protection of processed information (Figure 1).

Like other devices included in a computer system [1,2], the printer can be subject to electromagnetic infiltration, or eavesdropping [3,4]. Therefore, efforts to reduce the level of susceptibility to electromagnetic eavesdropping are initiated for such devices. Organizational and technical solutions are the most often-used methods for limiting infiltration sensitivity of devices [5]. Technical solutions are limited to changes in the design of devices that typically increase the cost of such devices and sometimes limit their functionality. Therefore, it is desirable to find solutions that avoid these drawbacks and at the same time allow "safe" processing of classified information [6,7].

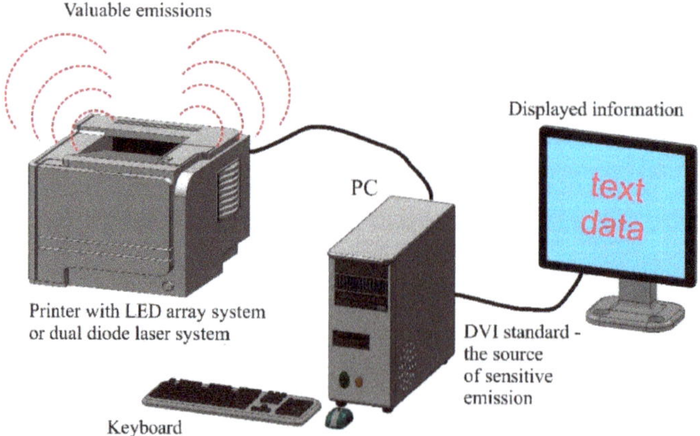

Figure 1. Laser printer as a source of valuable emissions.

One technical method that is commonly used in the field of electromagnetic compatibility—both to reduce the amount of electromagnetic interference emitted from the device and the susceptibility of the device to electromagnetic disturbance—is the use of differential-mode signals.

In this paper, analysis of useful signals [8] in the operation of LED arrays used inside printers (Figure 2) shows that such a design was used by printer *B* in its photoconductor exposure system. Is this sufficient, however, to foil non-invasive information gathering? The clear answer is that the solution adopted in the design of printer *B* (Figure 2a) significantly reduces the susceptibility of the device to infiltration, in comparison to printer *A* (Figure 2b). Moreover, the level of electromagnetic emission of printer *A* is higher than that of a typical single or dual diode laser printers [9].

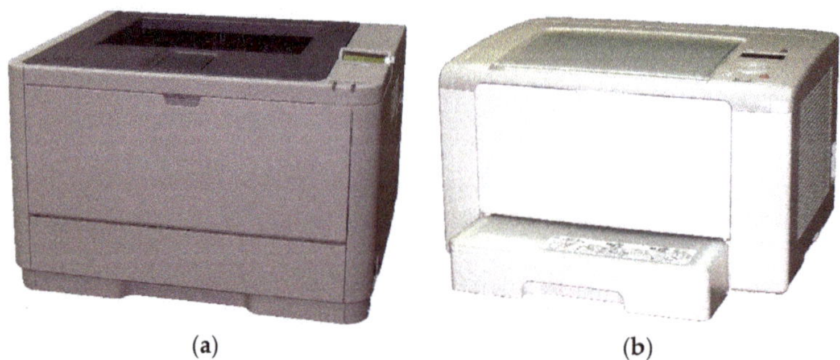

<table>
<tr><td>(a)</td><td>(b)</td></tr>
</table>

Figure 2. Two printers (**a**) printer A and (**b**) printer B were tested for sensitive emissions.

2. Materials and Methods

The analyses were carried out on two printers that use LED array technology. Different ways of controlling the LED array—chosen by the printer's designer—affect the number of useful signals (Figure 3) and the structure of those signals. In the case of printer, *A*.; we can distinguish four useful signals and six control signals in the cable (Figure 3a). The next ten wires are ground wires. Printer *B* has eight useful signals (four differential pairs) in the cable leading to its LED array (Figure 3b).

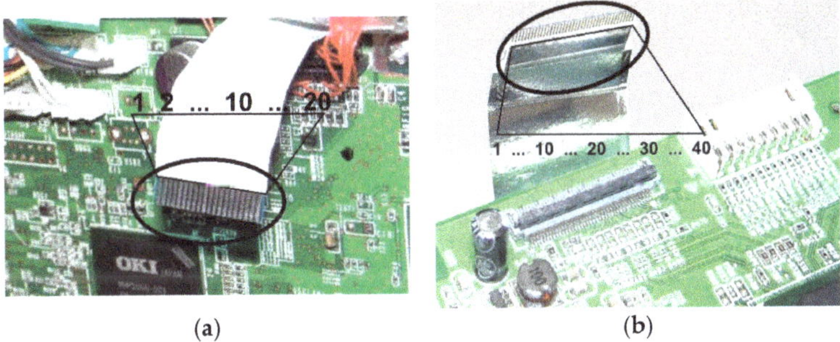

(a)   (b)

Figure 3. Ribbon cable supplying useful signals to the LED array: (**a**) Printer, *A.*; (**b**) Printer, *B.*

The other signals are control wires and ground wires (32 in all). By probing signal wires while exercising the printer, we were able to learn the structure of the control signals, how the LED array is controlled, and the way in which different print quality options are achieved depending on the operating mode and the toner-save option. Each of the tested printers uses different methods of controlling the LED array, which can affect the level of electromagnetic emanations. Examples of waveforms of useful and control signals for printer *A* are shown in the oscilloscope traces of Figures 4 and 5.

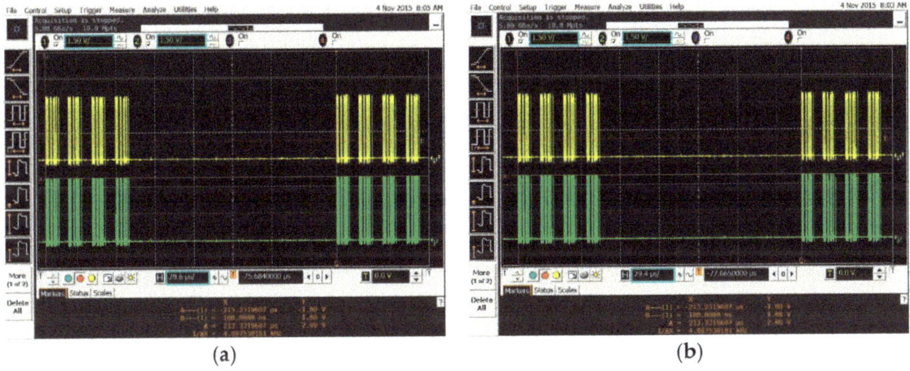

(a)   (b)

Figure 4. Waveforms of useful signals on pins 2 (lower trace) and 5 (upper trace) of printer *A* for: (**a**) the 300 dpi mode and the Best option, (**b**) the 300 dpi mode, and the Eco option.

The structure of the useful signals, based on the example of the signal on pin 2, does not change for the 300 dpi and 600 dpi operating modes of the printer. In the case of the 1200 dpi mode, the frequency of signal repetition increases by a factor of two. The amplitude is constant at approximately 3.5 V (Table 1).

Table 1. Parameters of useful signals of printer *A* in relation to printing parameters.

Operating Mode	Parameters of Useful Signal	
	Frequency [kHz]	Amplitude [V]
300 dpi, Eco	~4.7	3.5
300 dpi, Best	~4.7	3.5
600 dpi, Eco	~4.7	3.5
600 dpi, Best	~4.7	3.5
1200 dpi, Eco	~9.4	3.5
1200 dpi, Best	~9.4	3.5

The structure of the signal (waveform shape and duty cycle) does not change. This proves that the level of risk of electromagnetic emanations that are correlated with the processed information is not affected by print quality options (resolution and toner-save), in contrast to the situation found earlier with single and dual diode laser printers [4].

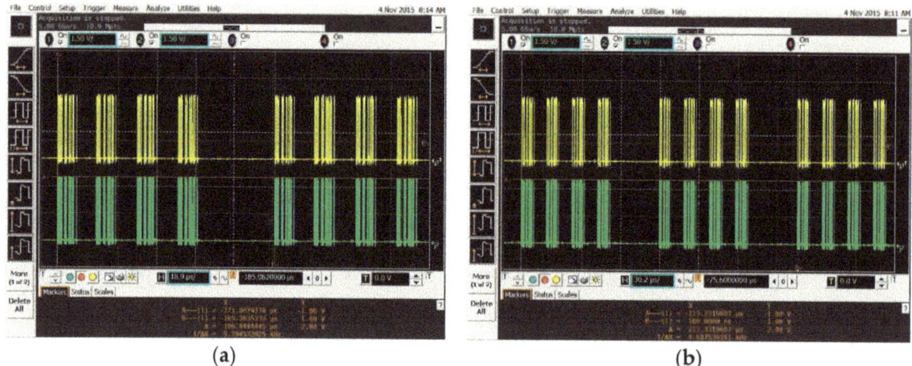

(a)   (b)

Figure 5. Waveforms of useful signals on pins 2 (lower trace) and 3 (upper trace) of printer *A* for: (**a**) the 1200 dpi mode and the Best option, (**b**) the 1200 dpi mode, and the Eco option.

For laser printers, changes of operating mode print quality options *do* have an effect on the structure of useful signals and thus the characteristics of sensitive RF emissions [10,11]. Information about the operating mode and print quality for printer *A* is encoded in the structure of the control signals (Figures 6–8). The amplitude of these signals is approximately 4 to 5 V. The pulse repetition frequency also changes depending on operating mode and use of the toner-save option. However, these signals carry no information at all about the information being printed [12,13].

Moreover, the amplitude of the control signals is higher than that of the useful signals. The control signals might be considered a serendipitous source of masking emissions which disturb the reception of sensitive emanations. This phenomenon is advantageous from an electromagnetic protection point of view [14–16].

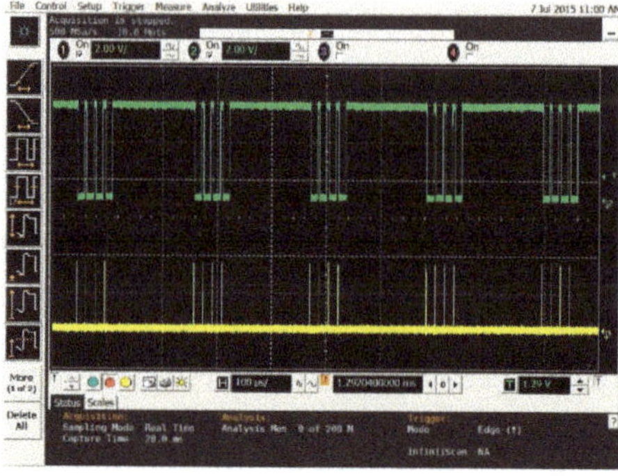

Figure 6. Waveforms of control signals on pins 6 (lower trace) and 9 (upper trace) of printer *A* for the 300 dpi mode and the Best option.

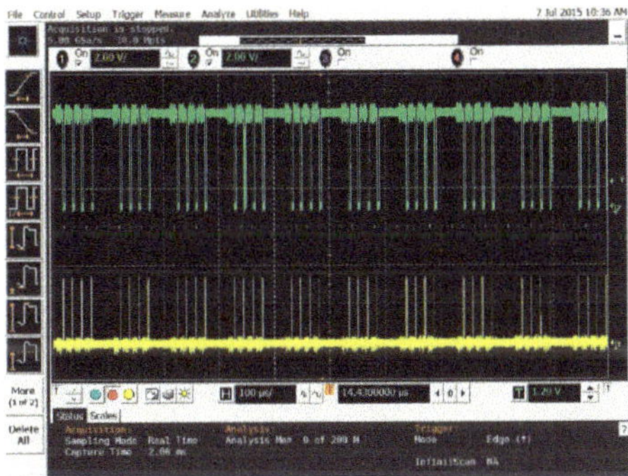

Figure 7. Waveforms of control signals on pins 6 (lower trace) and 9 (upper trace) of printer *A* in 1200 dpi mode with the Best option.

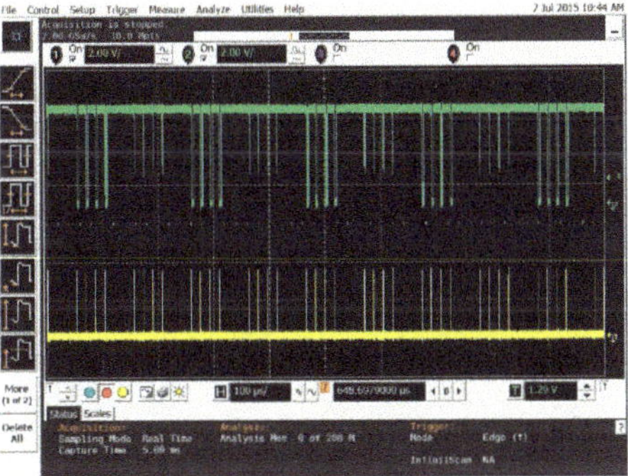

Figure 8. Waveforms of control signals on pins 6 (lower trace) and 9 (upper trace) of printer *A* for the 1200 dpi mode and the Eco option.

A completely different method of control of the LED array was implemented in printer *B* despite using the same xerographic technology in its photosensitive drum. Here, some information about modes of operation and the toner-save option is visible in the useful signal. The amplitude of the signal is approximately 250 mV. The amplitude is less than a tenth of similar signals in printer, *A*. Moreover, the signaling method is differential. Figures 9–11 show example waveforms of useful signals.

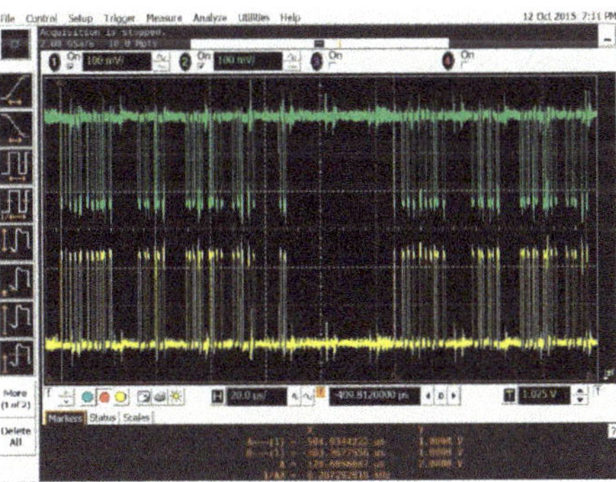

Figure 9. Waveforms of useful signals (one of the differential pairs) of printer *B* for the 1200 dpi mode and the Best option.

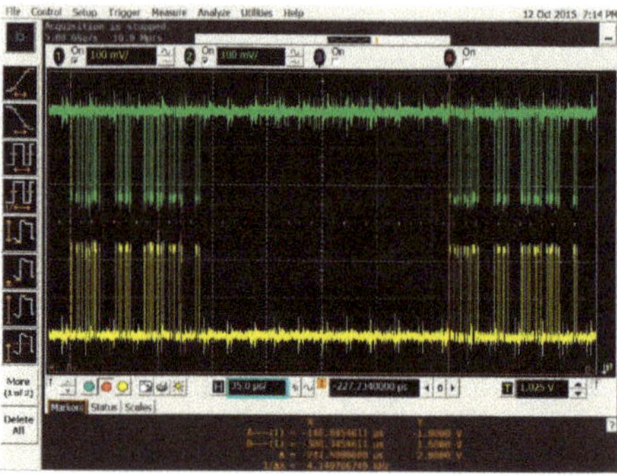

Figure 10. Waveforms of useful signals (one of the differential pairs) of printer *B* for the 1200 dpi mode and the Eco option.

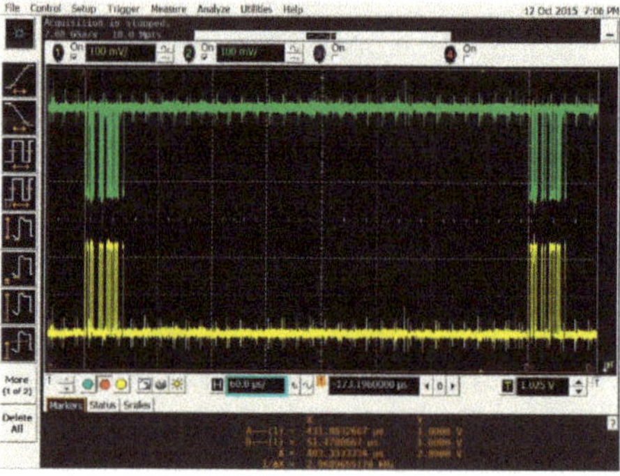

Figure 11. Waveforms of useful signals (one of the differential pairs) of printer *B* for the 600 dpi mode and the Eco option.

For these signals, the pulse repetition frequency changes when printing mode of operation and printout quality are changed (Table 2). The structure of these signals (duty cycle) does not change.

Table 2. Parameters of useful signals from printer *B* in relation to printing parameters.

| Operating Mode | Signal Parameters | | |
| | | Amplitude (mV) | |
	PRF of Differential Signals (kHz)	First Differential Pair (1, 3, 5, 7)	Second Differential Pair (2, 4, 6, 8)
600 dpi Eco	2.07	−250	+250
600 dpi Best	4.14	−250	+250
1200 dpi Eco	4.14	−250	+250
1200 dpi Best	8.28	−250	+250

By analysis of the parameters of useful signals, we can derive an important property crucial to reconstructing images from intercepted RF signals that contain printed data. In the case of printer, *B*.; a change of printing quality (from Eco to Best and vice versa), for a fixed printing mode, causes predictable changes of the pulse repetition rate of the useful signal. For printer, *A*.; changes to these parameters (printing mode and printing quality) are not reflected in the behavior of the useful signal.

3. Results

3.1. Reconstructed Images from Sensitive Emissions

Images of printed data were recreated from recorded RF useful signals transmitted in the wires supplying signals to the LED array. The test signal bandwidth was determined according to the equation:

$$B = \frac{W \cdot L \cdot (dpi)^2}{t} \tag{1}$$

where:

B—is the signal bandwidth for printing one pixel;
W—is the width of the printing area in inches;
L—is the length of the printing area in inches;
dpi—is the printing resolution in dots per inch, and
t—is the time to print one page.

We have to know the printing parameters in order to reconstruct the original information. These parameters are the length of printer video line (in pixels), and the number of video lines on a sheet of paper. As we can see, full reconstructed images can have very large dimensions; for example, for a:

- resolution of 1200 × 1200 dpi,
- printing speed of 30 pages per minute,
- paper size of A4 (about 8.27 by 11.69 inches, that is 9924 × 14 028 pixels), and
- three samples per pixel collected,

We obtain a data size of about 450 MB. Therefore, further analyses are based on fragments of images [17,18].

Fragments of these images are presented in Figures 12 and 13. The reconstructed glyphs contained in the image are constructed from horizontal lines at intervals equal to the width of the line [5] apart from the repetition frequency of the useful signal and use of options Best or Eco and for the default printing resolution of printer, B.

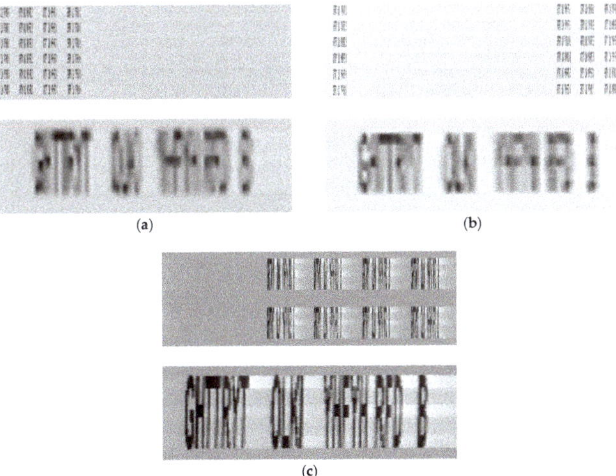

Figure 12. Examples of images reconstructed from useful signals (printer A) for (**a**) 300 dpi with toner-save, (**b**) 600 dpi without toner-save, and (**c**) 1200 dpi with toner-save.

Figure 13. Examples of images recreated from useful signals (printer *B*) for (**a**) 600 dpi without toner-save, (**b**) 600 dpi with toner-save, (**c**) 1200 dpi without toner-save, and (**d**) 1200 dpi with toner-save.

In two-diode laser printers, a phenomenon occurs that causes the reconstructed images from sensitive emissions to contain only single points, corresponding to the beginning and endpoint of each horizontal line comprising the printed glyphs (essentially run-length encoding the reconstructed images) [19].

However, these useful signals were not differential signals. In the case of printer, *B.*; despite the predictable structure of character glyphs, the differential signaling used tends to help protect printed data against electromagnetic infiltration [20].

3.2. Levels of Electromagnetic Emissions

Printer *B* uses differential transmission of useful signals. Its primary aim is probably to lower the levels of electromagnetic emission and increase resistance to external disturbances. Since the differential signal is responsible for the transfer of information from the printer's raster image processor (RIP) to the LED array, its characteristics correspond to characteristics of the processed information. Therefore, the solution adopted by the printer's designer (in the form of a differential signal) also reduces the levels of electromagnetic emission correlated with printed data [21]. Such a solution was not used in printer, *A.*; despite the fact that the amplitude of useful signals are over ten times higher than in the case of printer, *B.* This necessarily translates into a level of electromagnetic emission. The number of wires carrying useful signals is also half that of printer, *B.*

An anechoic chamber (Figure 14) was used to test these predictions. During the tests, sensitive emissions were measured with a bandwidth of 1 MHz in the frequency range from 2 MHz to 1 GHz. This frequency range was selected as a result of many years of experience in testing of laser printers and display screens. The aforementioned bandwidth value is the most effective for sensitive emissions from laser printers. During the tests, a TEMPEST DSI 1550A receiving system (20 Hz to 22 GHz), which can be seen in Figure 15, was used.

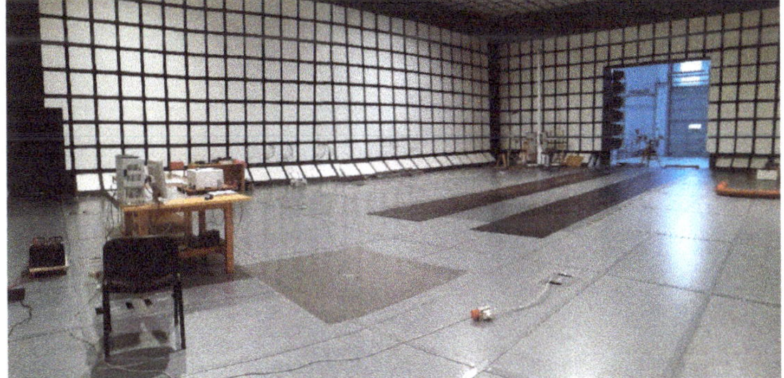

Figure 14. An anechoic chamber where the tests were carried out.

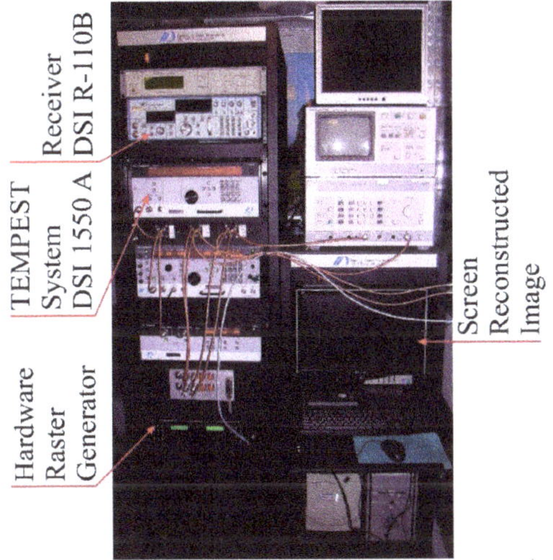

Figure 15. TEMPEST Test System model DSI 1550A.

The printers to be tested were connected to a TEMPEST computer, which is certified for electromagnetic safety. The reason for using a TEMPEST computer for this purpose is because typical computers have much higher levels of electromagnetic emissions. These emissions can "cover" the target emissions. When that happens, the electromagnetic infiltration process becomes impossible. TEMPEST printers exist and would be used along with TEMPEST computers in a TEMPEST environment. However, we aimed to measure the compromising emanations of normal, non-TEMPEST printers. (TEMPEST computers and printers are much more expensive, and hence are not used in most places.) The potential RF masking effect of a non-TEMPEST computer located near a non-TEMPEST printer is a valid objection, and we acknowledge it. However, our methodology, including testing in an anechoic chamber, was designed to lower the noise floor as much as possible in order to gather the cleanest possible data. Results of the TEMPEST measurements are shown in Figure 16.

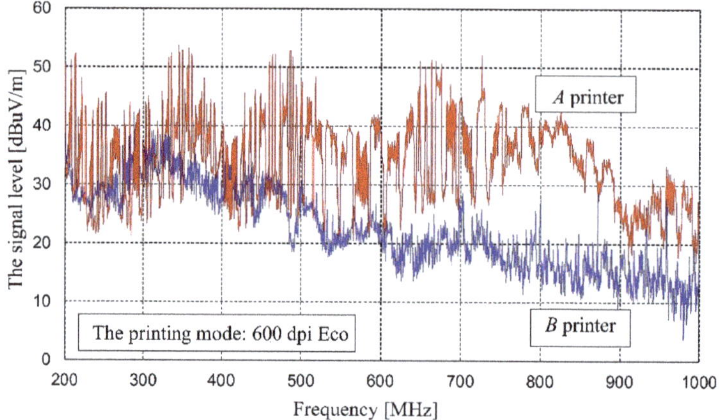

Figure 16. Radiated disturbances measured from the *A* and *B* printers, both operating in 600 dpi mode (with Eco option), *BW* = 1 MHz.

4. Discussions

4.1. Printer A

The useful signals are sent by four wires. The parameters of the signals are constant regardless of the printing mode and toner-save option. The only change relates to the frequency of repetition of the useful signals for the 1200 dpi mode, which is twice as high as for the two lower modes (300 dpi and 600 dpi). The reconstructed images, regardless of the operating mode of the printer, are visually similar, precluding identification of the operating mode of the printer (Figures 17 and 18). At the same time, the operating mode does not change the radiated characteristic of the emission source or the level of susceptibility to infiltration. In this printer, information about print quality is sent by additional control wires. In this case, different printing modes generate control signals having different timing structures.

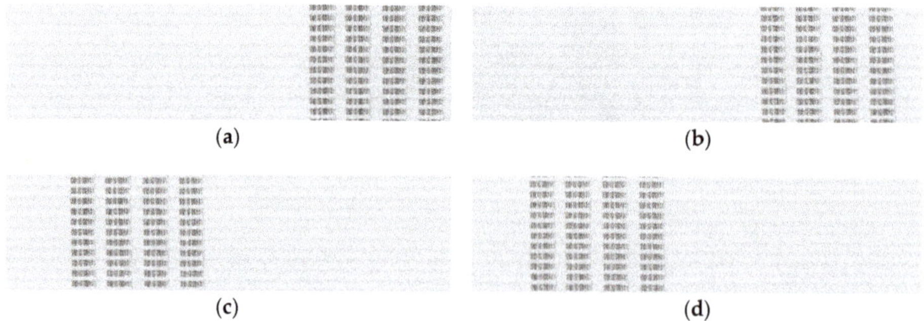

Figure 17. Fragments of reconstructed images from sensitive emissions of printer *A*: (**a**) 300 dpi mode without toner-save, (**b**) 300 dpi mode with toner-save, (**c**) 600 dpi mode without toner-save, and (**d**) 600 dpi mode with toner-save. Measured frequency of sensitive emission: f_0 = 525 MHz, *BW* = 5 MHz. Image is inverted.

As we can see, the xerographic printing technology of a photosensitive drum illuminated by LED arrays, which is used in these printers, has significant characteristics from the point of view of electromagnetic protection of the processed data. The reconstructed data do not directly contain characteristics that would facilitate their identification, as is the case with conventional laser printers for the same printing modes [22]. Even the use of digital image processing—such as an extension of pixel amplitude histogram, pixel amplitude thresholding, logical filtering, or edge detection filtering—failed to yield satisfactory results [5,23,24].

(a)   (b)

Figure 18. Fragments of reconstructed images from sensitive emissions of printer *A*: (**a**) 1200 dpi mode without toner-save, (**b**) 1200 dpi mode with toner-save, measured frequency of sensitive emission: $f_0 = 525$ MHz, $BW = 5$ MHz. Image is inverted.

4.2. Printer B

Printer *B* uses a similar xerographic exposure process comprising a photosensitive drum as printer, *A*. However, in the structure of the original image, we can distinguish horizontal gaps spaced at the same interval as one line—this is the Eco option in action—which has a beneficial effect, from the point of view of the defender, against the effectiveness of a side-channel attack (SCA). This printer uses differential signaling, which gives the attacker a much lower amplitude signal that is also missing some signal features that would facilitate electromagnetic eavesdropping.

In order to prove the conclusions above, sensitive emission signals were recorded, and then images reconstructed from the recordings. Undoubtedly the image in Figure 19 contains some glyphs [25]. However, due to the elimination of a number of distinctive features caused by differential transmission and reduction of the repetition frequency of the signal (the Eco option), these elements prevent reading any information related to the printed data.

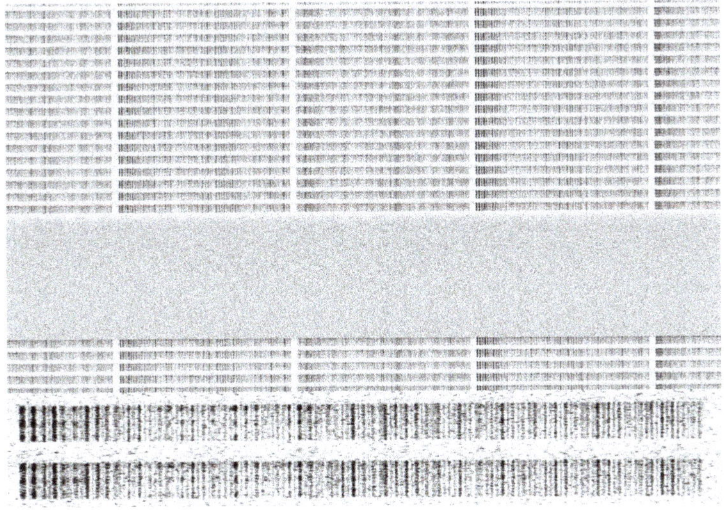

Figure 19. Printer *B* with LED array, 600 × 600 dpi with toner-save (images is inverted). Measured frequency of sensitive emission $f_0 = 384$ MHz, $BW = 2$ MHz.

4.3. Generality

There are four commonly used technologies underlying almost all modern printers: impact, thermal, inkjet or dye sublimation, and xerographic (toner-based) raster engines—what we think of as laser printers (pen plotters, wet electrolytic processes, and older technologies are not used as often anymore.) The majority of printers used in offices are the xerographic type examined here, and universally they are built on a selection from only a few photosensitive drum engines made by a few manufacturers. This inevitably constrains the design performance envelope of active optical components used to write images on the photosensitive drums. Only two photonics technologies are used: serial (a scanning laser beam) or parallel (LED arrays). Both have technical and cost advantages and disadvantages. Both are about equally used. They have noticeably different characteristics from an RF emanations point of view, and we have fully described the behavior of LED array-based printers here. RF emanations from scanning laser-based printers have been fully described previously [14,23,26].

4.4. Further Exploration

One of the anonymous reviewers suggested an extremely interesting line of inquiry. Would it be possible to reverse the effect, and somehow inject a signal from a distance into the printer to affect the single-ended or differential signals applied to the LED arrays, either control or data or both? To do so, in this case, was outside the administrative limits imposed by the agreement for use of the anechoic chamber for the duration of our experiments but we would like to try it.

Glyphs in some of the recovered images (Figure 19) seem almost legible and it is exactly this kind of high volume, noisy data that are sometimes amenable to analysis by neural networks and machine learning (ML). It was a different anonymous reviewer who put us onto the ML line of attack [27].

5. Conclusions

We offer the results of tests conducted on useful and control signals to the LED array for two different printers representative of different internal design decisions. The tests were carried out from an electromagnetic-protection-of-information point of view. The dependency of the structure of signals on the printing mode and toner-save option was shown. In general, the use of LED array technology in printers increases the level of electromagnetic protection of information (as compared to laser printers). The level of protection from RF electromagnetic eavesdropping is greater than for printers employing a dual diode laser system [27] and it does not require changes of construction in the printers. Printers using the dual diode laser system use serial signal transmission, compared with the parallel mode of LED printers. That design decision is advantageous to the electromagnetic eavesdropper (Figure 20).

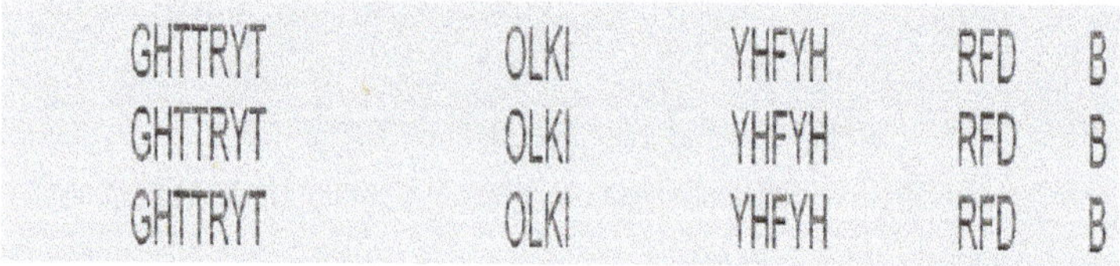

GHTTRYT   OLKI   YHFYH   RFD   B
GHTTRYT   OLKI   YHFYH   RFD   B
GHTTRYT   OLKI   YHFYH   RFD   B

Figure 20. Example of reconstructed image from sensitive emission for a two-diode laser printer, 660 dpi mode with the "Eco" option turned on. Measured frequency of sensitive emission: $f_0 = 444$ MHz, $BW = 5$ MHz. Image is inverted.

The LED array system requires parallel signal transmission. Overlapping signals in time cause successful reception and decoding of sensitive emissions to be very difficult. The reconstructed images from valuable emissions obtained from LED-array based printers can be seen to contain glyphs, but they are not legible.

Printer B goes further by using differential signaling. This method, if adopted, significantly reduces the level of useful electromagnetic emission (from the perspective of an eavesdropper) and thus reduces the effectiveness of the attack. The reconstructed images cannot be read by humans. Therefore the resistance level of printer *B* to electromagnetic eavesdropping is much higher than printer *A*—and that of typical laser printers (with the dual diode laser system). On the basis of recorded signals and reconstructed images, we may draw the conclusion that the method works. However, the low quality of the decoded data stands in the way of easy and simple interpretation.

The collected information related to printout quality and its impact on the forms of recreated data are presented in Table 3. The results obtained by analysis of the *A* and *B* printers were compared with results of analogous analyses of printers using a dual diode laser system. In summary, the best approach to increase resistance to electromagnetic infiltration is the LED array system.

Table 3. Comparison of the quality of reconstructed data—depending on resolution (DPI) and the use of "Best" or "Eco" options—for laser printers that use a dual diode laser system or an LED array from an electromagnetic protection point of view.

Type of Printer	600 dpi		1200 dpi	
	Best	Eco	Best	Eco
Dual diode printer *	W1	K1	W1	K1
Dual diode printer **	W1	W1	K1	W1
Dual diode printer ***	W1	W1	W1	W1
LED array printer *A*	K2	K2	K2	K2
LED array printer *B*	W2	W2	W2	W2

* Producer 1, ** Producer 2, *** Producer 3.

Legend:
K1—only the edges of glyphs appear in the reconstructed image, but the information is legible;
K2—visible filled glyphs appear in the reconstructed image, but the information is not legible;
W1—visible filled glyphs appear in the reconstructed image, and the information is legible;
W2—glyphs in the reconstructed image are not visible, and the information is not legible.

Author Contributions: Conceptualization, I.K.; methodology, I.K.; validation, I.K. and J.L.; formal analysis, I.K. and J.L.; investigation, I.K. and J.L.; resources, I.K.; writing—original draft preparation, I.K.; writing—review and editing, J.L.; visualization, I.K. and J.L.; supervision, I.K.

Acknowledgments: The anonymous reviewers were especially helpful; we would like to thank them for measurably improving the manuscript with penetrating observations and commonsense suggestions.

References

1. Kuhn, M.M. Optical time-domain eavesdropping risks of CRT displays. In Proceedings of the 2002 IEEE Symposium on Security and Privacy, Berkeley, CA, USA, 12–15 May 2002; pp. 3–18.
2. Kubiak, I. Video signal level (colour intensity) and effectiveness of electromagnetic infiltration. *Bull. Pol. Acad. Sci.* **2016**, *64*, 207–218. [CrossRef]
3. Ketenci, S.; Kayikcioglu, T.; Gangal, A. Recognition of sign language numbers via electromyography signals. In Proceedings of the Signal Processing and Communications Applications Conference (SIU), Malatya, Turkey, 16–19 May 2015; pp. 2593–2596.
4. Huzurbazar, S.; Kuang, D.; Lee, L. Landmark-based algorithms for group average and pattern recognition. *Pattern Recognit.* **2019**, *86*, 172–187. [CrossRef]
5. Toledo, I.; Carbonell, M.; Fornés, A.; Lladós, J. Information extraction from historical handwritten document images with a context-aware neural model. *Pattern Recognit.* **2019**, *86*, 27–36. [CrossRef]
6. Wasfy, W.; Zheng, H. Dual image processing algorithms and parameter optimization. In Proceedings of the Seventh International Conference on Natural Computation (ICNC), Shanghai, China, 26–28 July 2011; Volume 2, p. 946.

7. Goel, A.; Chandra, N. A technique for image encryption with combination of pixel rearrangement scheme based on sorting group wise of RGB values and explosive inter-pixel displacement. *Int. J. Image Graph. Signal Process.* **2012**, *4*, 16–22. [CrossRef]

8. Kubiak, I. LED printers and safe fonts as an effective protection against the formation of unwanted emission. *Turk. J. Electr. Eng. Comput. Sci.* **2017**, *25*, 4268–4279. [CrossRef]

9. Nguyen, K.; Fookes, C.; Raghavender, J.; Sridharan, S.; Ross, A. Long range iris recognition: A survey. *Pattern Recognit.* **2017**, *72*, 123–143. [CrossRef]

10. Xie, S.; Wang, T.; Hao, X.; Yang, M.; Zhu, Y.; Li, Y. Localization and Frequency Identification of Large-Range Wide-Band Electromagnetic Interference Sources in Electromagnetic Imaging System. *Electronics* **2019**, *8*, 499. [CrossRef]

11. Wu, C.; Gao, F.; Dai, H.; Wang, Z. A Topology-Based Approach to Improve Vehicle-Level Electromagnetic Radiation. *Electronics* **2019**, *8*, 364. [CrossRef]

12. Li, L.; Wang, Y.; Suen, Y.; Tang, Z.; Liu, D. A tree conditional random field model for panel detection in comic images. *Pattern Recognit.* **2015**, *48*, 2129–2140. [CrossRef]

13. Kuhn, M.M. Electromagnetic eavesdropping risks of flat-panel displays. In Proceedings of the 4th Workshop on Privacy Enhancing Technologies, Toronto, ON, Canada, 26–28 May 2004; LNCS 3424. p. 88.

14. Ulas, C.; Asik, U.; Karadeniz, C. Analysis and reconstruction of laser printer information leakages in the media of electromagnetic radiation, power, and signal lines. *Comput. Secur.* **2016**, *58*, 250–267. [CrossRef]

15. Yong, S.; Biao, L.; Wang, B.; Qi, Z.; Liu, J. Unsupervised Single-Image Super-Resolution with Multi-Gram Loss. *Electronics* **2019**, *8*, 833. [CrossRef]

16. Loughry, J.; Umphress, D.D. Information leakage from optical emanations. *Acm Trans. Inf. Syst. Secur.* **2002**, *5*, 262–289. [CrossRef]

17. Kim, J.; Park, N.; Kim, G.; Jin, S. CCTV Video Processing Metadata Security Scheme Using Character Order Preserving-Transformation in the Emerging Multimedia. *Electronics* **2019**, *8*, 412. [CrossRef]

18. Sadri, J.; Yeganehzad, M.M.; Saghi, J. A novel comprehensive database for offline Persian handwriting recognition. *Pattern Recognit.* **2016**, *60*, 378–393. [CrossRef]

19. Kubiak, I. Computer font resistant to electromagnetic infiltration process. *Prz. Elektrotechniczny* **2014**, *90*, 207–215.

20. Ketenci, S.; Gangal, A. Automatic reduction of periodic noise in images using adaptive Gaussian star filter. *Turk. J. Electr. Eng. Comput. Sci.* **2016**, *25*, 2336–2348. [CrossRef]

21. Song, T.T.; Jeong, Y.Y.; Yook, J.J. Modeling of leaked digital video signal and information recovery rate as a function of SNR. *IEEE Trans. Electromagn. Compat.* **2015**, *57*, 164–172. [CrossRef]

22. Kubiak, I.; Musial, S. Hardware raster generator as a tool that supports electromagnetic infiltration. *Telecommun. Rev. Telecommun. News* **2011**, *11*, 1601–1607.

23. Kubiak, I.; Przybysz, A. Printing technology and electromagnetic protection of information. *Prz. Elektrotechniczny* **2016**, *1*, 177–181.

24. Pishchulin, L.; Gass, T.; Dreuw, P.; Ney, H. Image warping for face recognition: From local optimality towards global optimization. *Pattern Recognit.* **2012**, *45*, 3131–3140. [CrossRef]

25. Jalilian, B.; Chalechale, A. Persian sign language recognition using radial distance and Fourier transform. *Int. J. Image Graph. Signal Process.* **2014**, *6*, 40–46. [CrossRef]

26. Kubiak, I. The influence of the structure of useful signal on the efficacy of sensitive emission of laser printers. *Measurement* **2018**, *119*, 63–76. [CrossRef]

27. Fortuna, L.; Arena, P.; Balya, D.; Zarandy, A. Cellular neural networks: A paradigm for nonlinear spatio-temporal processing. *IEEE Circuits Syst. Mag.* **2001**, *1*, 6–21. [CrossRef]

 *electronics*

Article

Bandwidth and Gain Enhancement of Endfire Radiating Open-Ended Waveguide Using Thin Surface Plasmon Structure

Abhishek Kandwal [1], Zedong Nie [1,*], Jingzhen Li [1], Yuhang Liu [1], Louis WY. Liu [2,*] and Ranjan Das [3]

[1] Shenzhen Institutes of Advanced Technology, Chinese Academy of Sciences, Shenzhen 518055, China; abhishek@siat.ac.cn (A.K.); lijz@siat.ac.cn (J.L.); yh.liu2@siat.ac.cn (Y.L.)
[2] EEIT Department, Vietnamese German University, Binh Duong 820000, Vietnam
[3] Indian Institute of Technology, Bombay 400076, India; ranjandas@ee.iitb.ac.in
* Correspondence: zd.nie@siat.ac.cn (Z.N.); liu.waiyip@vgu.edu.vn (L.W.L.)

Received: 24 April 2019; Accepted: 2 May 2019; Published: 7 May 2019

Abstract: This paper proposes a technique to enhance the bandwidth and gain of an endfire radiating open-ended waveguide using a thin slow-wave surface plasmon structure. Mounted in the E-plane of the stated waveguide, a thin corrugated slow-wave structure has been used in conjunction with a waveguide transition to generate an endfire electromagnetic beam. An efficient mode conversion from waveguide transition to the corrugate plate resulted in the improved performance of the design. An impedance bandwidth from 8 GHz to 18 GHz has been achieved along with a gain enhancement from 7 dBi to 14.8 dBi using the proposed hybrid design. Endfire radiations have been obtained with a beam width of less than 25° through the proposed hybrid design with an efficiency of about 96 percent.

Keywords: open-ended waveguide; slow-wave; endfire; radiation; dispersion; gain

1. Introduction

Rectangular waveguides have been in use for many decades for a range of communication applications. In principle, energy propagates along the waveguide in transverse electric modes (TE modes) with little energy loss. The lowest frequency in which a certain TE mode can propagate, which is known as the cutoff frequency of that mode, depends on the cross-sectional dimensions of the waveguide. The mode with the lowest cutoff frequency is known as the fundamental mode of the waveguide. Although other higher order modes do co-exist with the fundamental mode, the even order modes cannot propagate along the E-plane of the waveguide. Recently, the concept of spoof surface plasmon polaritons (SSPPs) has attracted a lot of interests from the microwave community, in part because of the following features and merits: groundlessness, high field confinement and feasibility of guiding an electromagnetic energy in surface modes. The effective exploitation of these features require a reliable evaluation of the important parameters of spoof surface plasmon (SSP)-based transmission lines (TLs). Surface plasmons (SPs) are highly localized electromagnetic (EM) waves with exponential field decay from the interface of metal and dielectric.

Recent applications of SSPs highlight the importance of a convenient and quick design procedure for the SSP structures. In a traveling-wave antenna, there are typically two modes of wave propagation: leaky modes and surface modes. A leaky mode propagates along the guiding body of the antenna with a group velocity greater than the speed of light, radiating electromagnetic energy in all directions other than the endfire direction. A surface mode propagates along the guiding body of the antenna with a group velocity less than the speed of light, with the electromagnetic field very confined on the surface or along an edge. It is for this reason why the device or medium supporting propagation of surface modes is collectively known as a slow-wave structure. The examples of slow wave structures include a dielectric material on top of a ground plane and a periodically corrugated structure. Slow wave antennas as well as other devices supporting propagation of an electromagnetic wave with a group velocity equal to the speed of light, i.e., V_g = c, are known to be able to generate an endfire radiation [1–7]. This paper focuses on a slow wave endfire antenna having a group velocity $V_g \leqslant$ c. Unlike other counterparts including those based on a transition from a microstrip line to spoof SPP waveguide [8–10] and those based on metamaterials [11], the proposed slow wave endfire antenna is to be mounted in the E-plane of a rectangular waveguide.

A variety of spoof SPP-based filters and wave splitters have been realized for SSPP communication systems [12–15]. On the other hand, surface plasmon waveguide based antenna designs, ultra wideband antennas and full-band waveguide differential phase shifters have been proposed recently for microwave applications [16–25]. In [20–25], the maximum bandwidth obtained is around 6–7 GHz with an average gain of <10 dBi.

Open-ended waveguides have been commercially marketed as a directional antenna for terahertz. These components are rarely used in microwave or millimeter wave frequencies in part because of their low gain and limited bandwidth. On the other hands, spoof surface plasmon antennas offer a larger bandwidth with limited gain. As explained in the sections which follow, these two can be combined together to achieve a wide bandwidth, a higher gain and better directional radiation characteristics. Furthermore, the effective aperture is reduced to less than half wavelength in a way to take advantage of the compactness of the design. In this paper, an efficient technique to enhance the bandwidth and gain of an open-ended waveguide using a slow-wave thin plate has been proposed. A simple and efficient endfire radiating hybrid waveguide spoof surface plasmon based design is investigated and analysed. An efficient mode conversion, and transition from waveguide to the proposed spoof surface plasmon design, have been achieved.

2. Principle

The Hansen Woodyard condition to achieve endfire radiation gives an optimum phase constant for maximizing the directivity of a lossless continuous endfire array of a particular length. For this, Hansen and Woodyard have proposed that the phase constant be increased from the usual endfire value β = k to an optimum value. For an endfire traveling-wave antenna with a constant phase velocity, there is an optimum velocity ratio $\dfrac{c}{V_{ph}}$ which results in maximum directivity as shown by Hansen Woodyard for very long antennas and Ehrenspeck-Poehler for an antenna of arbitrary length [1].

Optimum velocity ratio,

$$c/V_{ph} = 1 + 1/(2L/\lambda_0) \tag{1}$$

Using (1), the optimum velocity ratio calculated for the proposed endfire design is

$$\frac{c}{V_{ph}} = 1.1$$

which satisfies the condition for endfire travelling wave antennas.

The geometry of the proposed design is shown in Figure 1. The design is basically divided into two main parts. First is an open ended rectangular waveguide and the second is a spoof surface plasmon based slow wave thin plate (corrugated antenna). The spoof surface plasmon based corrugated antenna consists of periodically loaded unit cells/elements placed in a regular orientation along the forward y-axis. The design is tapered gradually at the end to obtain better reflection characteristic. The transition part of the antenna is also tapered at the beginning of the design which is connected to the open ended rectangular waveguide. This transition part has been optimised to provide better mode conversion from waveguide to corrugated antenna. As shown in Figure 1a, the length of the spoof surface plasmon antenna is 130 mm where the corrugated part is 110 mm long and the gradually tapered end is 35 mm long. The front tapered part of the antenna is inserted in the waveguide transition in such a way that it provides better mode conversion and mode matching. The effective aperture of the proposed design is less than half-wavelength i.e., 9 mm. The corrugated design is fabricated over a thin Roger Duroid substrate 5880 with a dielectric constant of 2.2 and thickness 0.1 mm. Figure 1b shows the geometry of the waveguide transition designed for the desired operating frequency band.

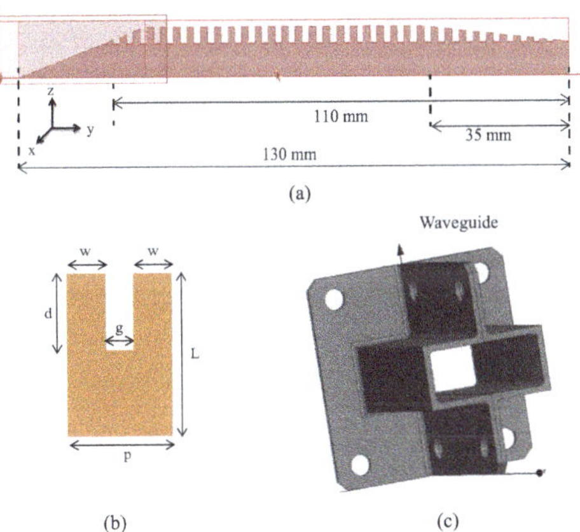

Figure 1. Geometry of the proposed design (**a**) Slow wave corrugated design with waveguide transition, (**b**) Unit element of the spoof surface plasmon (SSP) structure (**c**) waveguide transition.

Figure 2 shows the dispersion diagram of the periodic unit element of corrugated plate. The unit element has a period 'p' of 3.2 mm, groove depth 'd' of 3 mm, gap 'g' of 1 mm, 'w' of 1.1 mm and height 'L' of 9 mm. The first two modes are shown in the dispersion graph. The two modes are separated with an airline which defines the slow wave region and fast wave region. To satisfy the endfire condition according to [1], the operating dominant mode should be below the airline and the frequency of operation should be close to the airline. The proposed unit element satisfy the criteria for endfire radiations at frequency 11 GHz. Maximum gain and efficiency has been achieved at this frequency.

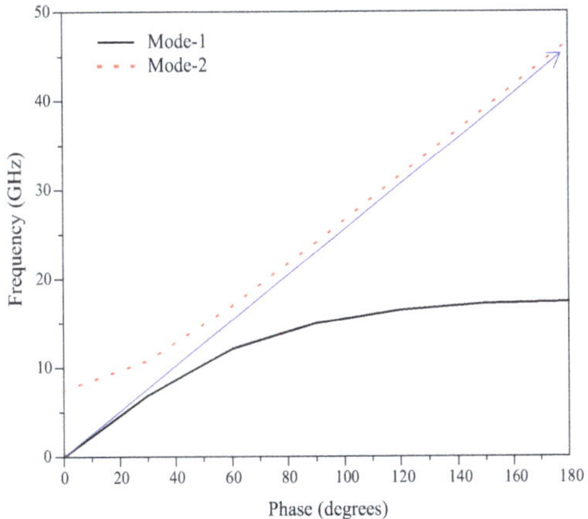

Figure 2. Dispersion curve for the proposed SSP Unit element.

Since the phase velocity remain below the free space velocity, the contributions of the various volume elements of the antenna to the radiation in the forward direction will add constructively only up to a certain length of the antenna. For antennas with a gradual taper the optimum length is approximately determined by the condition [2],

$$\int_0^L (K_x - K_0) = \pi \tag{2}$$

where, K_x is the phase constant of the guided wave and K_o is the free space wave number. If the length is exceeded, destructive interference will occur and a loss of directivity must be expected.

According to the theory, the optimum antenna length of the tapered antennas can be between 3λ to 8λ. Optimum length is approximately $4\lambda_0$ i.e., approx. 130 mm or less.

Increasing the cross-section of the antenna at the base will enlarge the bandwidth of the antenna and can be used as a method to reduce side lobe levels also (at the expense of a moderate gain reduction).

Furthermore, for long antennas with a gradual taper, it should be of a little consequence if the antenna termination at the far end is blunt or if the antenna is physically continued to its apex. The field strength within this tip section is small since most of its energy is now travelling in the air region.

For short antennas with a length of few wavelengths, tapering to a sharp tip has the advantage of a reduced reflection coefficient.

Here the length of the antenna and the tapering lengths have been optimised to achieve reflection <-10 dB, broadband and high efficiency (>90 percent). Optimum length has been obtained using the general equation of the tapered antennas which is providing a constructive interference to enhance the gain up to a certain limit.

3. Parametric Analysis

This section describes the parametric analysis and analyzes the effect of the length of corrugated plate, effect of transition part, effect of tapering end and peak gain variations with respect to appropriate variables. The simulations have been performed using 3D EM simulation software HFSS (High Frequency Structure Simulator).

Figure 3 shows the variation of S-parameters with the length of the proposed corrugated SSPP antenna. It can be seen from the graph that the length of the antenna is playing a vital role in achieving good reflection parameters. For length (120 mm), the reflection is not good at all and most of the power is reflected back. If the length is increased (140 mm), then the reflection is <-10 dB but still the antenna is not able to achieve a wideband. At the proposed length of 130 mm, the antenna has achieved a wide bandwidth of about 10 GHz.

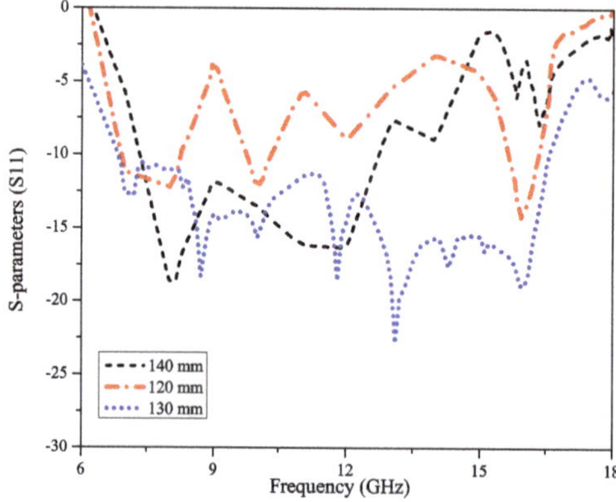

Figure 3. Variation of S-parameters with the length of the corrugated spoof surface plasmon polaritons (SSPP) antenna.

Figure 4 shows the effect of the tapering in the end part of the finite length antenna. Once the length has been optimized accordingly, the tapering part also needs to be optimized. As can be seen clearly in the graph, the proposed gradual taper (35 mm) is showing better reflection than a longer taper (40 mm) although it doesn't have much difference.

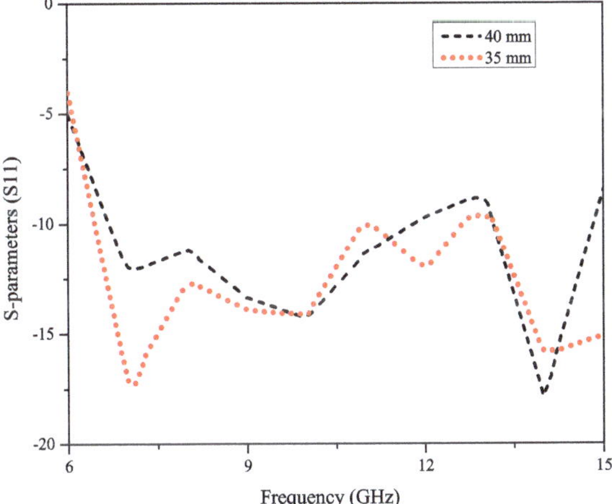

Figure 4. Effect of tapering on reflection parameter.

Figure 5 shows the effect of the transition part at the front of the antenna. Very short transition of 15 mm has not provided good matching with the waveguide transition whereas a longer transition 30 mm has achieved better matching. This may be due to the fact that the correct length of tapering results in better mode matching and energy conversion from waveguide transition to the corrugated slow wave antenna design.

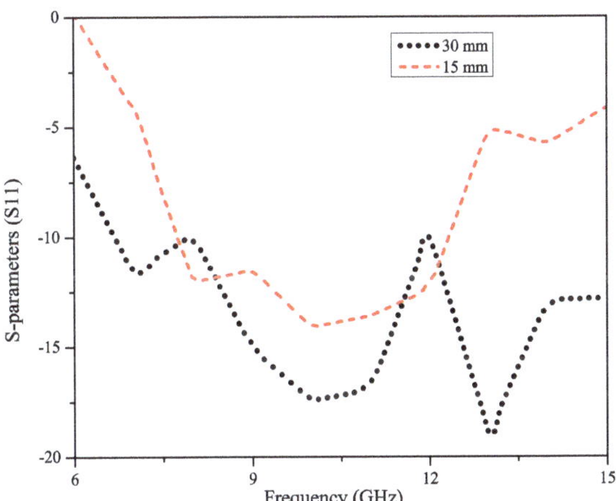

Figure 5. Effect of transition inside waveguide on matching.

Figure 6 shows the gain variations with the length of the proposed design. The peak gain varies a lot with the length of the antenna. This is due to the optimum length, better mode matching and better reflection. The peak gain for the optimum length of 130 mm is around 14.8 dBi whereas the gain is less than this peak value for other lengths due to the constructive and destructive interference phenomenon.

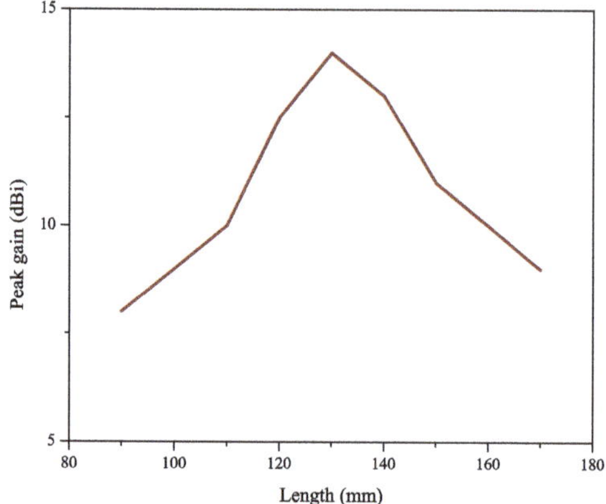

Figure 6. Formulated variation of peak gain with length of SSPP.

Figure 7 shows 2D and 3D-radiation patterns of the simple waveguide and proposed corrugated SSPP at 11 GHz frequency. For comparison purpose, the patterns are shown at 12 GHz also as it is also providing nice endfire radiation with little lower gain. At different frequencies in the operating frequency band, the proposed hybrid waveguide SSP antenna design is showing the endfire or near-endfire radiations. The simulated peak gain of the simple rectangular waveguide is around 7 dB and for proposed design is 14.8 dB.

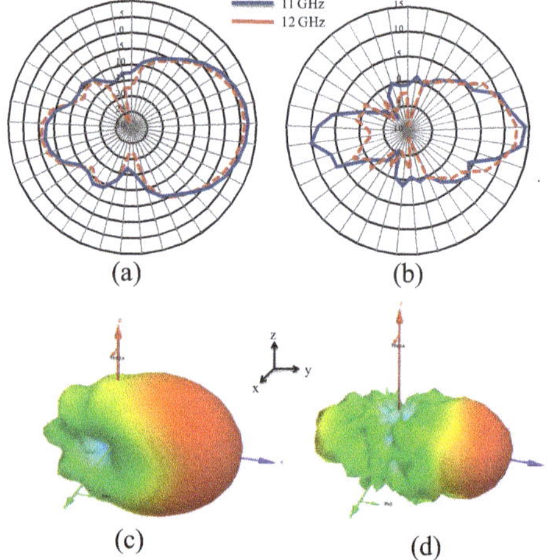

Figure 7. 2D and 3D Radiation Patterns: (**a**) Waveguide 2D radiation pattern, (**b**) Proposed waveguide-SSPP 2D radiation pattern, (**c**) Waveguide 3D radiation pattern, and (**d**) Proposed waveguide-SSPP 3D radiation pattern.

Figure 8 shows the E-field distribution for the proposed hybrid waveguide-endfire SSP design along the axis. As can be seen from the plot, the field is highly confined to the corrugated design and has achieved a very good mode matching or energy conversion from the waveguide region to the corrugated surface.

Despite increase in the gain, more back radiation has been observed. Generally, a reflected surface wave spoils the pattern and bandwidth of the antenna.

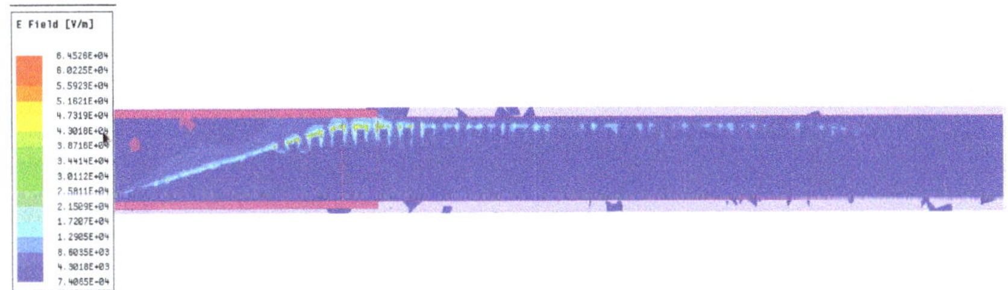

Figure 8. Field distribution: E-Field along the axis.

The optimum formulas derived for the proposed hybrid SSPP design are:

Peak gain ∝ Length of the structure (up to a certain limit as criteria)

For normal endfire antennas and broad bandwidth case, directivity (D) in dBi can be written as [2],

$$D = 10\log(\frac{4L}{\lambda_0}) \tag{3}$$

In this design, it can be calculated from (3) as:

$$D = 10\log(4\frac{130}{30}) = 12.4 \text{ dBi}$$

For tapered antennas (or rods), the directivity gain (G) is primarily determined by the antenna length also.

Gain initially increases with L according to (4)

$$G = p\frac{L}{\lambda_0} \text{ dBi} \tag{4}$$

From (4), it will be

$$G = 3.2\frac{130}{30} = 13.9 \text{ dBi}$$

Higher gain has been obtained in the simulation because of gradually tapered end which reduced the reflections to a good extent thereby reducing the back radiations significantly which further leads to gain enhancement.

Similarly half power beamwidth can be calculated as,

$$\theta = 55\sqrt{\frac{\lambda_0}{L}} = 25 \text{ degree}$$

From the results, it can be seen that the calculated beamwidth is consistent with the measured one and varies from 20 degree to 30 degree.

4. Experimental Validation

The design is fully analyzed using a 3D electromagnetic (EM) simulation software HFSS and measured using an Agilent vector network analyzer.

Figure 9 shows the fabricated prototype. Figure 9a is the slow wave corrugated design. Figure 9b is the waveguide transition and Figure 9c is the combined final design of waveguide and SSPP antenna.

Figure 10 shows the measured S-parameters for the proposed hybrid waveguide-endfire SSP design. From the S-parameters, it is observed that the proposed design has good impedance matching below −10 dB in the region from 8 GHz to 18 GHz.

Figure 11 shows the experimental setup for the proposed design in the anechoic chamber.

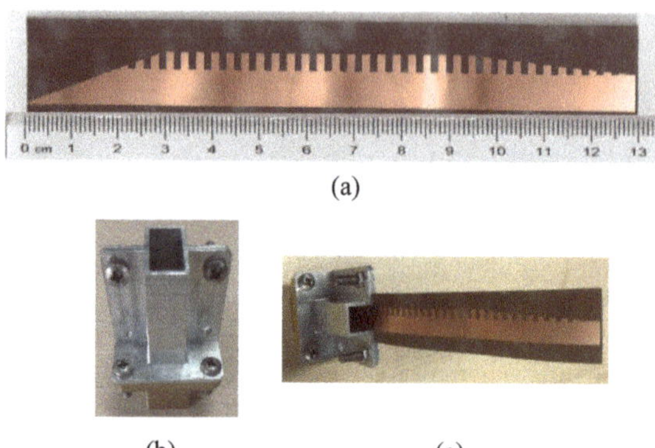

(a)

(b)   (c)

Figure 9. Prototype of the proposed hybrid waveguide-SSPP (**a**) corrugated SSPP design, (**b**) waveguide transition, and (**c**) proposed final design.

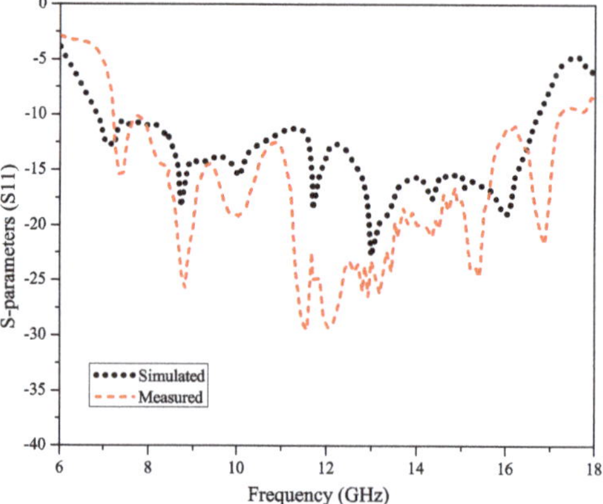

Figure 10. S-parameters of the proposed hybrid waveguide-SSPP design.

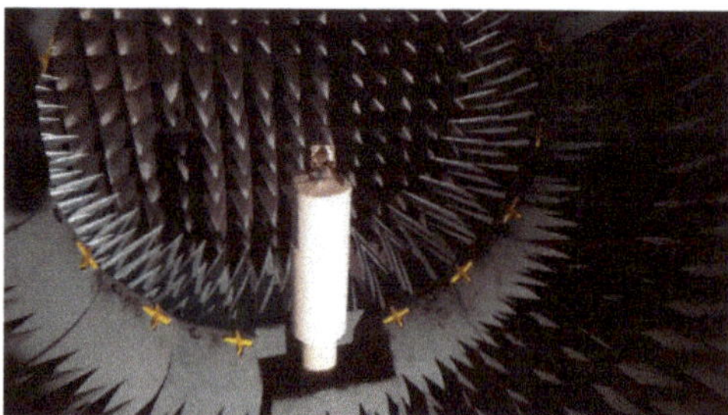

Figure 11. Experimental setup in the anechoic chamber.

Figure 12 shows the radiation patterns (in dB) at different frequencies from 8 GHz to 15 GHz. It has been observed from the radiation patterns that in the whole operating frequency band, endfire and near endfire radiations have been observed.

Figure 13 shows the comparison of simulated and measured gain for the proposed hybrid waveguide-endfire SSP design. The peak gain for the proposed hybrid design in the operating frequency band obtained is around 14.8 dB. The total gain varies from 7 dB to 14.8 dB in operating frequency band. The measured gain is around 13 dB which is less than simulated one but is close to the calculated one. This may be due to the small experimental problems such as connections between waveguide transition and the corrugated design.

Figure 14 shows the efficiency for the proposed hybrid waveguide-endfire SSP design. The radiation efficiency for this design is quite high around 96 percent.

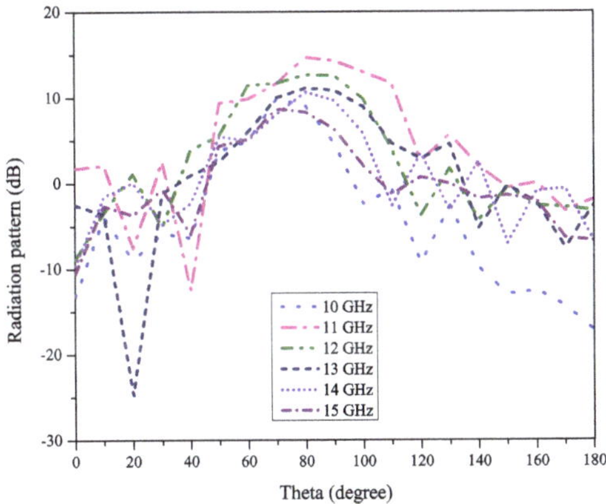

Figure 12. Radiation pattern of the proposed waveguide-SSPP design.

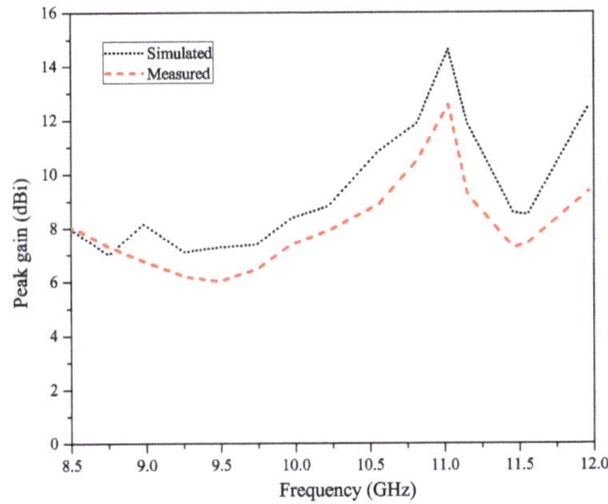

Figure 13. Measured peak gain of the proposed hybrid waveguide-SSPP design.

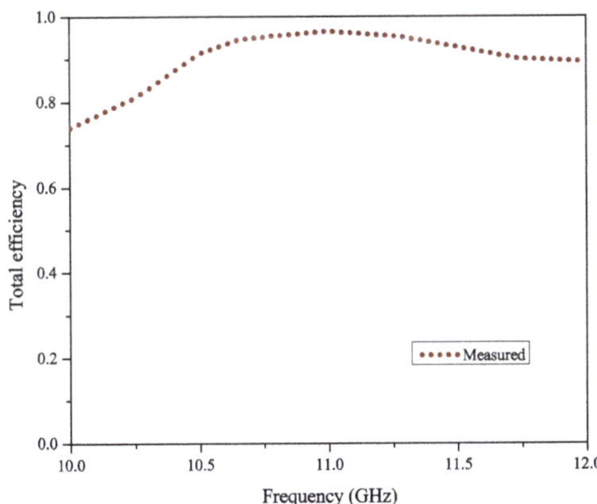

Figure 14. Efficiency of the proposed hybrid waveguide-SSPP design.

5. Conclusions

A hybrid waveguide design using slow-wave thin corrugated structure has been proposed in this paper. The proposed design has obtained endfire radiations along with an enhanced impedance bandwidth and peak gain. It provides a simple and efficient technique to obtain endfire radiations with improved antenna characteristics using waveguide transition to SSP structure. Peak gain is increased from 7 dBi to 14.8 dBi and a high efficiency of about 96 percent has been obtained.

Author Contributions: A.K. conceived the idea and performed the experiments, Z.N. supervised the work, J.L. and Y.L. made revisions and corrections, L.W.L. helped with simulations and experiments and R.D. performed numerical modeling.

Funding: This work was supported by the National Natural Science Foundation of China under Grant No. 61871375, National Key R&D Program of China under Grant No. 2018YFC2001002, and Shenzhen Basic Research Project under Grant No. JCYJ20180507182231907.

Conflicts of Interest: The authors declare no conflict of interest.

Abbreviations

The following abbreviations are used in this manuscript:

SSPP	Spoof surface plasmon polaritons
SPs	Surface Plasmons
TLs	Transmission lines
EM	Electromagnetic
CPW	Coplanar waveguide
D	Directivity

References

1. Hansen, W.W. *Radiating Electromagnetic Wave Guide, in Google Patent 2 402 622*; Wiley: New York, NY, USA, 1946.
2. Zucker, F.J. Surface- and leaky-wave antennas. In *Antenna Engineering Handbook*; Jasik, H., Ed.; McGraw-Hill: New York, NY, USA, 1961.
3. Kianinejad, A.; Chen, Z.N.; Qiu, C.W. Design and Modeling of Spoof Surface Plasmon Modes-Based Microwave Slow-Wave Transmission Line. *IEEE Trans. Microw. Theory Tech.* **2015**, *63*, 1817–1825. [CrossRef]
4. Yin, J.Y.; Ren, J.; Zhang, H.; Pan, B.C. Broadband frequency-selective spoof surface plasmon polaritons on ultrathin metallic structure. *Sci. Rep.* **2015**, *5*, 483–486. [CrossRef] [PubMed]
5. Yi, H.; Qu, S.W.; Bai, X. Antenna Array Excited by Spoof Planar Plasmonic Waveguide. *IEEE Antennas Wirel. Propag. Lett.* **2014**, *13*, 1227–1230.
6. Cao, W.; Zhang, B.; Liu, A.; Yu, T.; Guo, D.; Wei, Y. Gain Enhancement for Broadband Periodic Endfire Antenna by Using Split-Ring Resonator Structures. *IEEE Trans. Antennas Propag.* **2012**, *60*, 3513–3516. [CrossRef]
7. Cao, W. Broadband High-Gain Periodic Endfire Antenna by Using I-Shaped Resonator (ISR) Structures. *IEEE Antennas Wirel. Propag. Lett.* **2012**, *11*, 1470–1473.
8. Liao, Z.; Zhao, J.; Pan, B.C.; Shen, X.P.; Cui, T.J. Broadband transition between microstrip line and conformal surface plasmon waveguide. *J. Phys. Appl. Phys.* **2014**, *47*, 315103.
9. Zhang, H.C.; Liu, S.; Shen, X.; Chen, L.H.; Li, L.; Cui, T.J. Broadband amplification of spoof surface plasmon polaritons at microwave frequencies. *Laser Photonics Rev.* **2015**, *9*, 83–90. [CrossRef]
10. Pan, B.C.; Liao, Z.; Zhao, J.; Cui, T.J. Controlling rejections of spoof surface plasmon polaritons using metamaterial particles. *Opt. Express* **2014**, *22*, 13940–13950. [CrossRef]
11. Yin, J.Y.; Bao, D.; Ren, J.; Zhang, H.C.; Pan, B.C.; Fan, Y.; Cui, T.J. Endfire Radiations of Spoof Surface Plasmon Polaritons. *IEEE Antennas Wirel. Propag. Lett.* **2017**, *16*, 597–600. [CrossRef]
12. Wu, J.J.; Hou, D.J.; Yang, T.J.; Hsieh, I.J.; Kao, Y.H.; Lin, H.E. Bandpass filter based on low frequency spoof surface plasmon polaritons. *Electron. Lett.* **2012**, *48*, 269–270. [CrossRef]

13. Gao, X.; Zhou, L.; Liao, Z.; Ma, H.F.; Cui, T.J. An ultra-wideband surface plasmonic filter in microwave frequency. *Appl. Phys. Lett.* **2014**, *104*, 191603. [CrossRef]
14. Chen, Z.; Hu, Y.; Zheng, X.; Yang, A.; Li, P.; Zhu, L.; Liu, M.; Cheng, L.; Zhou, Q.; Hu, X. Slot-array antenna devising for surface microwave discharge of surface plasmon polaritons. In Proceedings of the 2011 Cross Strait Quad-Regional Radio Science and Wireless Technology Conference, Harbin, China, 26–30 July 2011; Volume 1, pp. 363–366.
15. Wu, J.J.; Hou, D.J.; Chiueh, H.L.; Shen, J.Q.; Wu, C.J.; Kao, Y.H.; Lo, W.C.; Yang, T.J.; Hsieh, I.J. Kind of high directivity scanning radiation based on subwavelength periodic metal structure. *Electron. Lett.* **2014**, *50*, 1611–1613. [CrossRef]
16. Xu, J.J.; Zhang, H.C.; Zhang, Q.; Cui, T.J. Efficient conversion of surface-plasmon-like modes to spatial radiated modes. *Appl. Phys. Lett.* **2015**, *106*, 021102. [CrossRef]
17. Yin, J.Y.; Zhang, H.C.; Fan, Y.; Cui, T.J. Direct Radiations of Surface Plasmon Polariton Waves by Gradient Groove Depth and Flaring Metal Structure. *IEEE Antennas Wirel. Propag. Lett.* **2016**, *15*, 865–868. [CrossRef]
18. Shibayama, J.; Yamauchi, J.; Nakano, H. Metal disc-type splitter with radially placed gratings for terahertz surface waves. *Electron. Lett.* **2015**, *51*, 352–353. [CrossRef]
19. Shen, X.; Moreno, G.; Chahadih, A.; Akalin, T.; Cui, T.J. Spoof surface plasmonic devices and circuits in THz frequency. In Proceedings of the 39th International Conference on Infrared, Millimeter, and Terahertz Waves (IRMMW-THz), Tucson, AZ, USA, 14–19 September 2014; p. 1.
20. Guo, Y.J.; Xu, K.D.; Liu, Y.; Tang, X. Novel Surface Plasmon Polariton Waveguides with Enhanced Field Confinement for Microwave-Frequency Ultra-Wideband Bandpass Filters. *IEEE Access* **2018**, *6*, 10249–10256. [CrossRef]
21. Zhang, D.; Zhang, K.; Wu, Q.; Ding, X.; Sha, X. High-efficiency surface plasmonic polariton waveguides with enhanced low-frequency performance in microwave frequencies. *Opt. Express* **2017**, *25*, 2121–2129. [CrossRef] [PubMed]
22. Gao, X.; Che, W.; Feng, W. Novel non-periodic spoof surface plasmon polaritons with H-shaped cells and its application to high selectivity wideband bandpass filter. *Sci. Rep.* **2018**, *8*, 2456. [CrossRef] [PubMed]
23. Cano, J.L.; Mediavilla, A.; Tribak, A. Parametric Design of a Class of Full-Band Waveguide Differential Phase Shifters. *Electronics* **2019**, *8*, 346. [CrossRef]
24. Li, K.; Dong, T.; Xia, Z. Wideband Printed Wide-Slot Antenna with Fork-Shaped Stub. *Electronics* **2019**, *8*, 347. [CrossRef]
25. Rahman, M.; NagshvarianJahromi, M.; Mirjavadi, S.S.; Hamouda, A.M. Compact UWB Band-Notched Antenna with Integrated Bluetooth for Personal Wireless Communication and UWB Applications. *Electronics* **2019**, *8*, 158. [CrossRef]

Article

Analysis and Design of an Energy Verification System for SC200 Proton Therapy Facility

Pengyu Wang [1,2,3], Jinxing Zheng [1,*], Yuntao Song [1], Wuquan Zhang [1] and Ming Wang [1,2]

[1] Institute of Plasma Physics, Hefei Institutes of Physical Science, Chinese Academy of Sciences, Hefei 230031, China; pywang@mail.ustc.edu.cn (P.W.); songyt@ipp.ac.cn (Y.S.); zhangwuquan@ipp.ac.cn (W.Z.); wm102518@mail.ustc.edu.cn (M.W.)

[2] Science Island Branch of Graduate School, University of Science and Technology of China, Hefei 230026, China

[3] College of Mechanical Engineering, Anhui University of Science and Technology, Huainan 232001, China

* Correspondence: jxzheng@ipp.ac.cn; Tel.: +86-0551-65669355

Received: 23 April 2019; Accepted: 10 May 2019; Published: 13 May 2019

Abstract: The purpose of this study is to provide an energy verification method for the nozzle of the SC200 proton therapy facility to ensure safe redundancy of treatment. This paper first introduces the composition of the energy selection system of the SC200 proton therapy facility. Secondly, according to IEC60601 standard, the energy verification requirement that correspond to 1 mm error in water is presented. The allowable difference between the measured magnetic field and the reference are calculated based on the energy verification requirements to select the field resolution of the Hall probe. To ensure accuracy and stability, two Hall probes are mounted on the dipole to monitor the magnetic field strength to verify the proton beam energy in real time. In addition, the test results of the residual field of the dipole show that the probe system meets the accuracy requirements of energy verification. Furthermore, the maximum width of the slit of the energy selection system in accordance with the IEC standard at the corresponding energy is calculated and compared with the actual position of the movable slit to verify the momentum divergence of the proton beam. Finally, we present an energy verification method.

Keywords: energy verification; Hall probe; dipole; movable slit

---

1. Introduction

Proton beams have received increasing attention for cancer treatment due to their high dose localization and high biological effect at the Bragg peak [1,2]. Therefore, proton therapy has been successfully carried out at various facilities around the world [3–6]. Since 2016, the Joint Institute for Nuclear Research (JINR) and the Institute of Plasma Physics of the Chinese Academy of Sciences (ASIPP) have collaborated in the development of proton therapy facility in Hefei, China [7]. The proton therapy facility consists mainly of a 200 MeV superconducting cyclotron (SC200), energy selection system (ESS), beam transport line, gantry treatment room and fixed beam room. A typical ESS often includes a degrader, quadrupole magnets, bending dipoles, steering magnets, collimators, slits, beam diagnostic equipment and so on [8].

The energy selection system prototype has been designed to work in the energy range between 70 MeV and 200 MeV allowing to select a beam with an energy spread ranging from 0.193% to 1.93% according to the transmission efficiency that one wants to achieve [9]. The proton beam extracted from a superconducting cyclotron is focused at the degrader, by passing it through four quadrupole magnets (QM01-QM04). The beam energy is modulated by the wedge graphite degrader. As the proton beam passes through the degrader, energy spread increases significantly due to multiple scatterings inside the graphite block. Therefore, in order to efficiently transmit the beam, a double-bend achromatic system is used to eliminate the dispersion. It is better to design a vertical beam waist lactating in the bending magnet as much as possible for the purpose of reducing the air gap. There are four quadrupole magnets (QM07-QM10) and a movable slit between the two bending dipoles. The first dipole will create a large horizontal dispersion to the beam which will be enlarged by using a defocusing quadrupole magnet (QM07). Next, a quadrupole magnet (QM08) is used to focus the beam and make the derivative of dispersion equal to zero ($R_{26} = 0$). Then, the movable slit is used to select the momentum spread according to clinical requirements. The schematic of the part of the ESS in SC200 as shown in Figure 1.

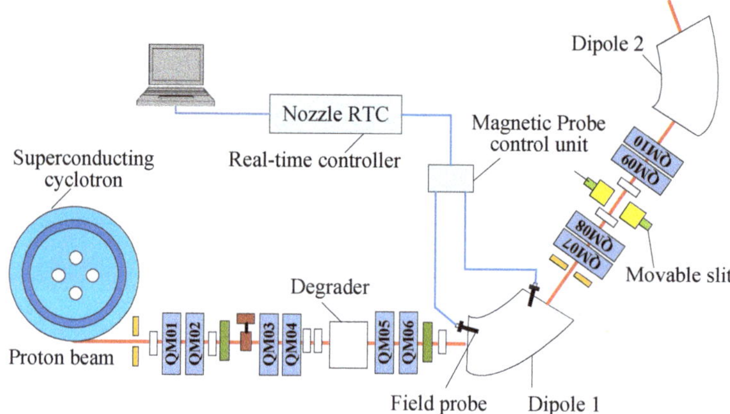

Figure 1. Schematic of the part of the energy selection system (ESS) in 200 MeV superconducting cyclotron (SC200).

As shown in Figure 2, the PBS nozzle is composed of scanning magnets, helium chamber and three Ionization chambers (IC), primarily used to control and monitor the beam [10,11]. The tumor is divided into multiple irradiated slices and is irradiated from deep to shallow by changing the energy. In order to ensure safety redundancy and meet the requirements of the threshold of 1mm particle range (in water), Light Ion Therapy standard IEC60601-2-64:2014 requires independent validation of beam energy. It is required that energy verification can be performed during beam delivery to the patient without destructive measurements.

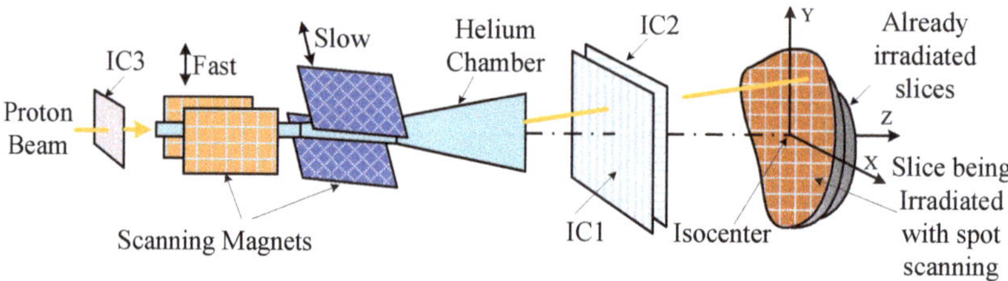

Figure 2. Nozzle layout principle.

In this paper, the component configuration of the energy selection system is introduced. The energy validation requirement and momentum analysis requirement are given based on IEC60601 standard, and then the design scheme of the energy validation system is discussed. Furthermore, the momentum spread, slit horizontal half width, the allowable difference between the measured magnetic field and the reference are calculated. Finally, test results of the probe system will be given.

2. Analysis for Energy Verification System

2.1. Energy Validation Requirement

In our project, the cyclotron produces a proton beam with a fixed energy of 200 MeV. Therefore, an energy selection system (ESS) is employed to change energy to match the particle range requirements during treatment. The range-energy relationship can be calculated with Equation (1).

$$R = \alpha E^p, \tag{1}$$

where, R is the particle range (unit: cm), E is the kinetic energy (unit: MeV), $\alpha \approx 2.2 \times 10^{-3}$ and $p \approx 1.77$ for protons in water [12].

The difference of energy ΔE between the sample and the reference can be expressed as:

$$\Delta E = (E_r^p + \frac{R_s - R_r}{\alpha})^{p^{-1}} - E_r, \tag{2}$$

where, R_r is the range of reference particle, E_r is the kinetic energy of reference particle, R_s is the range of sample particle. Since $R_s - R_r \leq 1$ mm, the maximum difference of energy ΔE_{max} is given as:

$$\Delta E_{max} = (E_r^p + \frac{0.1}{\alpha})^{p^{-1}} - E_r. \tag{3}$$

According to Equation (3), the threshold of 1 mm range with different energies can be calculated. Taking a full energy of 200 MeV as an example, the energy verification system must detect if a beam with proton energy outside 200 MeV ±0.43 MeV can reach the patient. In addition, we use different range errors to calculate the energies difference of the kinetic energy, from 70 MeV to 200 MeV. The energy difference under different range errors in water is shown in Figure 3.

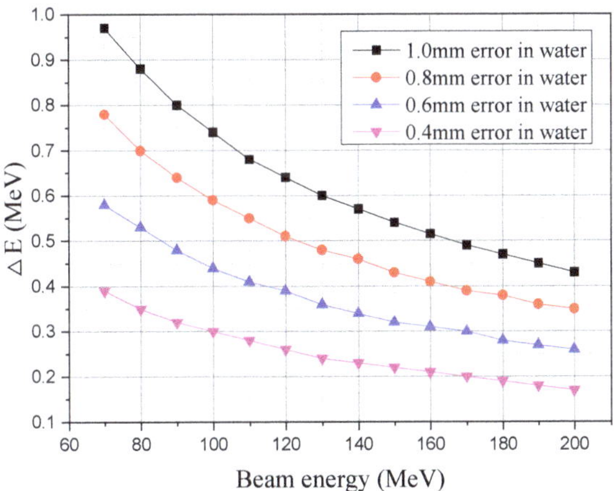

Figure 3. Energies difference under different range errors in water.

2.2. Momentum Analysis Requirement

When the difference of energy between sample particle and reference particle are known, the momentum spread (1sigma) can be expressed as:

$$\delta = \frac{\Delta p}{p_r} = \frac{p_s - p_r}{p_r} = \frac{\gamma_s \beta_s - \gamma_r \beta_r}{\gamma_r \beta_r},$$
$$\gamma_s = \frac{\gamma_r \varepsilon_0 + \Delta E}{\varepsilon_0}, \gamma_r = \frac{\varepsilon_0 + E_r}{\varepsilon_0}, \beta_s = \sqrt{1 - 1/\gamma_s^2}, \beta_r = \sqrt{1 - 1/\gamma_r^2}, \tag{4}$$

where, ε_0 is the rest energy of proton (unit: MeV, $\varepsilon_0 = 938$ MeV for protons), γ_s is the relativistic factor of sample particle, γ_r is the relativistic factor of reference particle, β_s is the relative velocity of sample particle, β_r is the relative velocity of reference particle. According to Equation (4), the maximum allowed momentum spread δ_{max} can be obtained. Therefore, when the energy is selected by the ESS, the momentum spread needs to be less than the maximum allowed momentum spread δ_{max}. The maximum allowed momentum spread δ_{max} under different energies is shown in Figure 4. Obviously, at lower beam energy, the measurement requirement for a momentum spectrometer is relatively easier and the full energy of 200 MeV is the most demanding case.

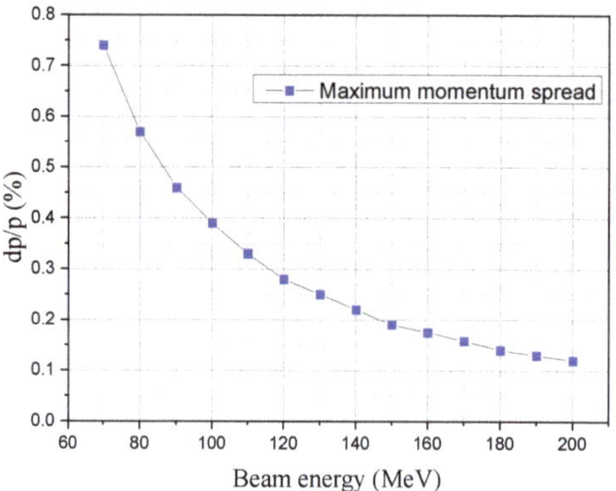

Figure 4. The maximum allowed momentum spread under different energies.

2.3. Principle of Energy Verification

The proton beam reference energy are determined by the response of the field measurement and deflection radius of the dipole, which can be calculated according to Equation (5) [13].

$$E_r = m_0 c^2 \left[\sqrt{1 + \left(\frac{qB\rho}{m_0 c}\right)^2} - 1 \right], \tag{5}$$

where, B is the magnetic strength of the measurement by field probe (unit: Telsa), ρ is the deflection radius of the beam (unit: meter), c is the speed of light (unit: meter/sec), m_0 is the mass of proton (unit: MeV/c^2, $m_0 = 938$ MeV/c^2), q is the charge of proton (unit: Coulombs, $q = 1.6 \times 10^{-19}$ Coulombs). The energy spread is determined by the dispersion and beam transverse position constraint in the limiting slit, to the limits that correspond to 1 mm error in water.

2.4. Energy Verification Calculation

When passing the energy selector dipole, the proton beam deflects under Lorentz force. The magnetic field strength of the dipole is measured by two Hall probes. The deflection radius of the dipole is 1.637 m. Due to the non-uniformity of the dipole magnetic field and the positional limitation of the Hall probe, it is necessary to add a compensation to be equivalent to equal the center field strength of the dipole. The reference beam kinetic energy can be calculated as:

$$E_r = \varepsilon_0 \left[\sqrt{1 + \left(\frac{q(B + B_c)\rho c}{\varepsilon_0}\right)^2} - 1 \right], \tag{6}$$

where, ε_0 is the rest energy of proton ($\varepsilon_0 = m_0 c^2 = 938$ MeV), B_c is the compensated magnetic field. The magnitude of the compensated magnetic field depends on the position of the Hall probe and the magnitude of the dipole field strength.

3. Magnetic Measurements

3.1. Dipole Magnet

In our project, the maximum magnetic field strength of the dipole magnet is 1.32×10^4 Gauss, the deflection radius is 1.637 m, the deflection angle is 63 degrees, and the effective length is 1.809 m. The lateral field uniformity and the integration field uniformity of 0.7×10^4 to 1.32×10^4 Gauss are better than $\pm 5 \times 10^{-4}$, that is, the field strength difference between the edge trajectory and the central trajectory is 3.5 Gauss to 6.6 Gauss. In order to improve precision and accuracy, two Hall probes are mounted on a dipole magnet to monitor the magnetic field strength in real time, as shown in Figure 5.

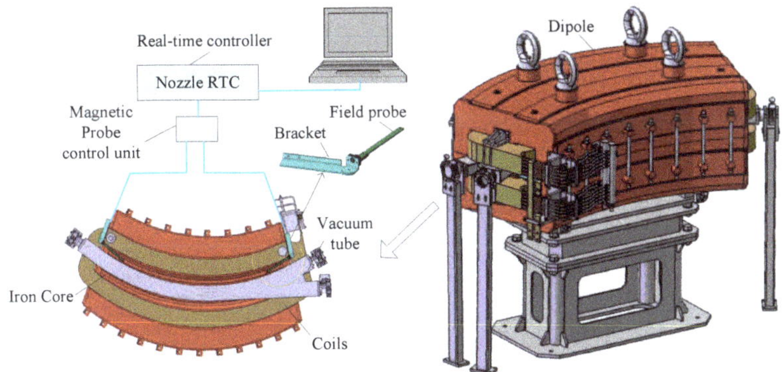

Figure 5. Schematic diagram of Hall probe and dipole magnet.

3.2. Magnetic Field Probe

The energy verification requirement needs to meet the precision limit that corresponds to 1 mm error in water. The precision of the Hall probe system is an important factor affecting the accuracy of energy verification. The maximum allowable difference between the measured field and the reference is given as:

$$\Delta B_{\max} = \frac{\sqrt{[2\varepsilon_0 + (E_r + \Delta E_{\max})](E_r + \Delta E_{\max})} - \sqrt{(2\varepsilon_0 + E_r)E_r}}{299.8\rho}. \tag{7}$$

We calculate the magnetic field difference according to the above formula. The comparison of the magnetic field difference under different proton energies is listed shown in Table 1. With the increase of the proton energy, the measurement error of the magnetic field has a great influence on the precision of energy validation. Therefore, we have chosen a higher precision Hall probe system for magnetic field measurement. The field range of the Hall probe system is required to be greater than $\pm 1.5 \times 10^4$ Gauss, and the field resolution is better than 2 Gauss.

Table 1. Comparison of the magnetic field allowable difference under different proton energies.

Energy E_r (MeV)	Energy Validation Requirement $\Delta E_{\max}$ (MeV)	Magnetic Field Strength B (Gauss)	Allowable Difference $\Delta B_{\max}$ (Gauss)
70	0.97	7521	54
90	0.80	8572	40
110	0.68	9525	31
130	0.60	10407	26
150	0.54	11234	22
170	0.49	12019	19
190	0.45	12768	17
200	0.43	13131	16

3.3. Test Hall Probe Stability and Accuracy

Unfortunately, the mounting bracket for the Hall probe has not been manufactured. Therefore, we measured the remnant field in a dipole magnet to verify the accuracy and stability of the Hall probe system, as shown in Figure 6.

Figure 6. The remnant field of the dipole magnet is measured by Hall probe system.

The test result is shown in Figure 7. The maximum spread in the data is under 0.5 Gauss, which meets the stated stability requirement of 2 Gauss. There were no anomalous readings. At full energy of 200 MeV, the momentum spread of the energy selection system must be less than 0.12% to meet the IEC standard of 1 mm water maximum error. Meanwhile, the reference magnet field is 13131 Gauss at full energy of 200 MeV, this would correspond to a field resolution of 16 Gauss, which is easily met.

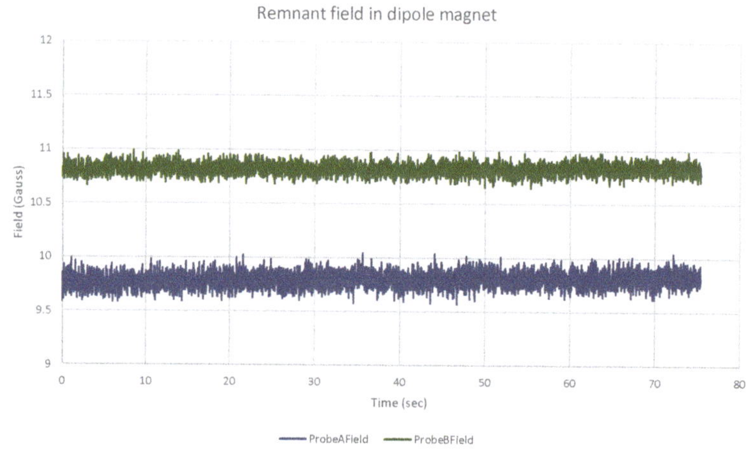

Figure 7. The test result of Hall probe system.

4. Design Scheme of the Energy Verification System

4.1. Momentum Analysis at the Limiting Slit

In the energy selection system of the SC200, the momentum spread is controlled by a movable slit. The movable slit is located in the middle of the two horizontal focusing quadrupoles (QM08 and

QM09). As the beam passes through the bending magnets and the quadrupoles, momentum dispersion effect causes center shift of phase ellipse, the dispersion effect is shown in Figure 8.

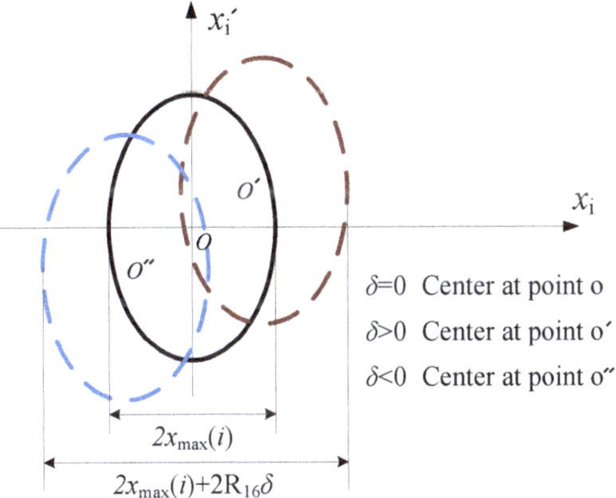

Figure 8. Momentum dispersion effect.

For beams with momentum p_r and $p_r + \Delta p$, the center distance of such beams can be calculated with Equation (8). The necessary condition for separating them is $\Delta x \geq x_{max}(i)$.

$$\Delta x = R_{16}\delta. \tag{8}$$

According to the final design, the momentum spread should not exceed 0.1% to 1% depending on the energy. At the movable slit, the correlation between the momentum spread and the horizontal distance(x) was calculated by the computer code TURTLE and TRANSPORT. The position dispersion coefficient R_{16} is −27.54 mm/% [9].

4.2. Energy Verification Scheme

The schematic of the energy verification system for the nozzle in SC200 is shown in Figure 9. Two precision Hall probes are added to the energy selector magnet (dipole 1) at different locations to improve precision and accuracy for energy validation. The stability and accuracy of the Hall probe depends upon temperature, so the Hall probe includes a separate temperature sensor in close thermal proximity to the Hall device, and the compensation is carried out in real time using coefficients stored by the magnetic probe control unit. The magnetic field is sampled, typically at 1 kHz sampling rate. A movable slit is used to restrict the transverse beam position. The real-time Controller (RTC) is intended to provide convenient, high-performance connection of field probe devices via fiber optic loops and real-time processing of the data. Connection to a host computer system is via a standard Ethernet interface. A set of interlock relays allow the RTC to gate external processes based on the incoming data and the state of health of the RTC itself. The proton beam energy is determined by the response of the field measurement plus dispersion and beam transverse position constraint in the ESS.

A flowchart of the process in the energy verification system is shown in Figure 10. Firstly, the beam energies are determined according to the magnetic field tested by the Hall probe. Then, the maximum difference of energy ΔE and the maximum allowed momentum spread δ_{max} are calculated by the clinical requirements that correspond to 1 mm error in water. The energy verification system compares the actual slit position with the target position, if the horizontal half width of the slit is less than or equal to the target position $2R_{16}\delta_{max}$, irradiation is continued. Otherwise, a signal is sent to the interlock system and the irradiation stops.

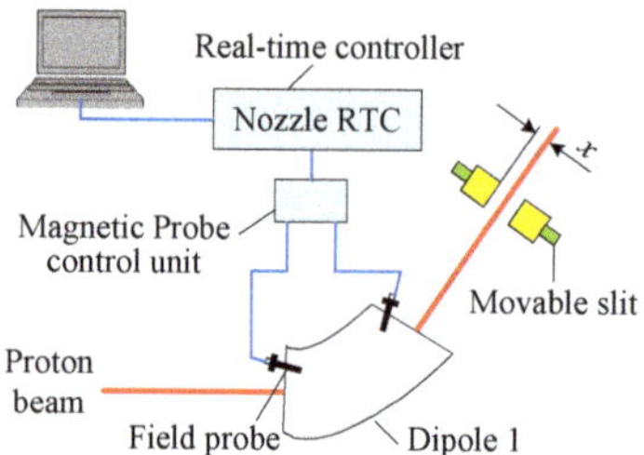

Figure 9. Schematic of the energy verification system for the nozzle in SC200.

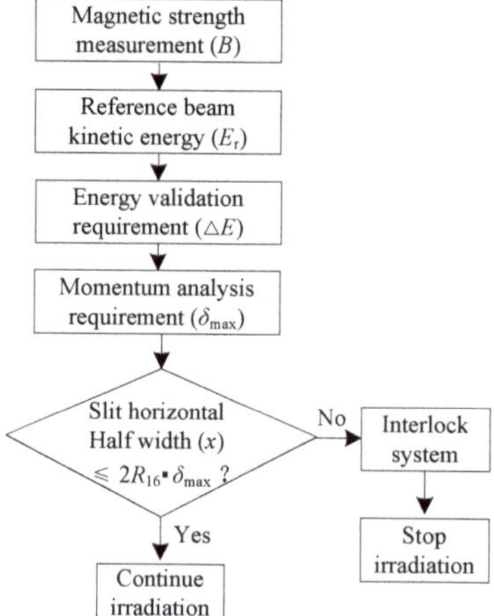

Figure 10. A flowchart of the process in the energy verification system.

The correspondence between the slit maximum positions and the energies selected is shown in Figure 11. The energy selection system must resolve at least δ_{max} 0.12% to meet the IEC standard of 1 mm water maximum error, with a corresponding slit horizontal half width of less than 6.61 mm. During the treatment, the nozzle system needs to obtain the actual position information of the movable slit in real time, and then perform energy verification compared with the maximum allowable position of the slit.

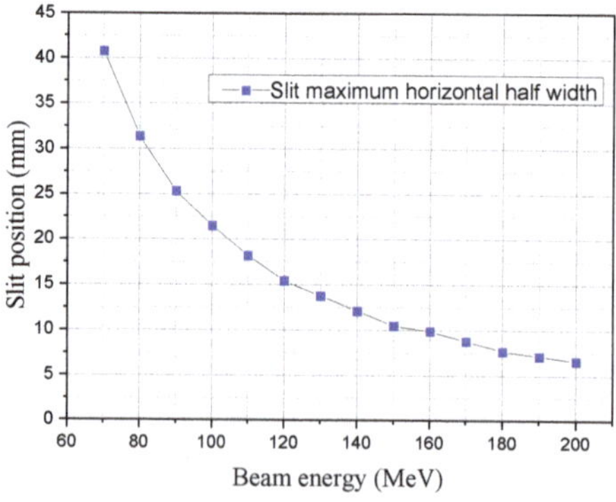

Figure 11. The analytical calculation of the correspondence between the slit maximum positions and the energies selected.

5. Discussion and Conclusions

We calculated the maximum allowed momentum spread that satisfies the energy verification requirements of a proton therapy system, according to IEC60601 standard. The energy verification scheme for the nozzle of the SC200 proton therapy is designed. In order to improve precision and accuracy, two Hall probes are mounted on a dipole magnet, and the beam energies are determined based on the magnetic field monitored by the Hall probe. Meanwhile, the momentum dispersions at the corresponding energies are verified by combining the dispersion and the movable slit horizontal width. Furthermore, the Hall probe system is tested, and the results meet the requirements. Finally, we calculate the target slit positions at the corresponding energies and give an energy verification method.

Author Contributions: Conceptualization, P.W., J.Z. and Y.S.; investigation, W.Z and M.W.; Formal analysis, P.W.; writing—original draft preparation, P.W. and M.W.; writing—review and editing, P.W.

Funding: This research was supported in part by the Major Science and Technology Projects of Anhui under Grant 17030801003 and in part by the Natural Science Research Project of Anhui University under Grant KJ2016A199.

Acknowledgments: The authors are grateful for the support from Hefei CAS Ion Medical and Technical Devices Co., Ltd.

Conflicts of Interest: The authors declare no conflict of interest.

References

1. Lomax, A. Intensity modulation methods for proton radiotherapy. *Phys. Med. Biol.* **1999**, *44*, 185–205. [CrossRef] [PubMed]
2. Shtemberg, A.S.; Kokhan, V.S.; Kudrin, V.S.; Matveeva, M.I.; Bazyan, A.S. The effect of high-energy protons in the Bragg Peak on the behavior of rats and the exchange of monoamines in some brain structures. *Neurochem. J.* **2015**, *9*, 66–72. [CrossRef]
3. Smith, A.R. Proton therapy. *Phys. Med. Biol.* **2006**, *51*, 491–504. [CrossRef] [PubMed]
4. Bonnett, D.E. Current developments in proton therapy: A review. *Phys. Med. Biol.* **1993**, *38*, 1371–1392. [CrossRef]
5. Smith, A.; Gillin, M.; Bues, M.; Zhu, X.R.; Suzuki, K.; Mohan, R.; Woo, S.; Lee, A.; Komaki, R.U.; Cox, J.D.; et al. The MD Anderson proton therapy system. *Med. Phys.* **2009**, *36*, 4068–4083. [CrossRef] [PubMed]
6. Matsuda, K.; Itami, H.; Chiba, D.; Kazuyoshi Saito, E. World-first Proton Pencil Beam Scanning System with FDA Clearance. *Hitachi Rev.* **2009**, *58*, 225–231.
7. Karamyshev, O.V.; Bi, Y.F.; Chen, G.; Ding, K.Z.; Karamysheva, G.A.; Morozov, N.A.; Samsonov, E.; Shirkov, G.; Shirkov, S.; Song, Y.T. Beam tracking simulation for SC200 superconducting cyclotron. In Proceedings of the 7th International Particle Accelerator Conference, Busan, Korea, 8–13 May 2016.
8. Jiang, F.; Song, Y.T.; Zheng, J.X.; Zeng, X.H.; Wang, P.Y.; Zhang, J.S.; Zhang, W.Q. Energy loss of degrader in SC200 proton therapy facility. *Nucl. Sci. Tech.* **2019**, *30*, 4. [CrossRef]
9. Zeng, X.H.; Song, Y.T.; Zheng, J.X.; Jiang, F.; Li, M.; Zhang, J.S.; Zhang, W.Q.; Zhu, L. Beam optics study for energy selection system of SC200 superconducting proton cyclotron. *Nucl. Sci. Tech.* **2018**, *29*, 134. [CrossRef]
10. Kanai, T.; Kawachi, K.; Matsuzawa, H.; Inada, T. Three-dimensional beam scanning for proton therapy. *Nucl. Instrum. Methods Phys. Res.* **1983**, *214*, 491–496. [CrossRef]
11. March, B.; Prieels, D.; Bauvir, B.; Sépulchre, R.; Gérard, M. IBA proton pencil beam scanning: An innovative solution for cancer treatment. In Proceedings of the EPAC, Vienna, Austria, 26–30 January 2000.
12. Bortfeld, T. An analytical approximation of the Bragg curve for therapeutic proton beams. *Med. Phys.* **1997**, *24*, 2024–2033. [CrossRef]
13. Deng, J.J.; He, G.R.; Ding, B.N.; Cheng, N.A.; Yao, L.B. Energy spectrum measurement of 3.3 MeV LIA IPEB. *High Power Laser Part. Beams* **1993**, *5*, 353–358.

Article

Analysis and Experimental Test of Electrical Characteristics on Bonding Wire

Wenchao Tian, Hao Cui * and Wenbo Yu

School of Electro-Mechanical Engineering, Xidian University, Number 2 Taibai South Road, Xi'an 710071, China; wctian@xidian.edu.cn (W.T.); wenbo.yu.albert@outlook.com (W.Y.)
* Correspondence: hcui@stu.xidian.edu.cn; Tel.: +86-29-8820-2954

Received: 20 February 2019; Accepted: 23 March 2019; Published: 26 March 2019

Abstract: In this paper, electrical characteristic analysis and corresponding experimental tests on gold bonding wire are presented. Firstly, according to EIA (Electronic Industries Association)/JEDEC97 standards, this paper establishes the electromagnetic structure model of gold bonding wire. The parameters, including flat length ratio, diameter, span and bonding height, were analyzed. In addition, the influence of three kinds of loops of bonding wire is discussed in relation to the S parameters. An equivalent circuit model of bonding wire is proposed. The effect of bonding wire on signal transmission was analyzed by eye diagram as well. Secondly, gold bonding wire design and measurement experiments were implemented based on radio frequency (RF) circuit theory analysis and test methods. Meanwhile, the original measurement data was compared with the simulation model data and the error was analyzed. At last, the data of five frequency points were processed to eliminate the fixture error as much as possible based on port embedding theory. The measurement results using port extension method were compared with the original measurement data and electromagnetic field simulation data, which proved the correctness of the simulation results and design rules.

Keywords: bonding wire; S parameters; electromagnetic simulation; port embedding

---

1. Introduction

Electronic packaging is the connection between the chip and the external pin. It is an important part for maintaining the electrical, thermal and mechanical properties of the device. Electronic components are developing towards being small volume, high power, high frequency and highly reliable, which requires higher packaging techniques. However, packaging technology has gradually become a bottleneck in the development of semiconductor industry. Although there are advanced package forms such as flip-chip, tape automated bonding, and wafer level package, more than 90% of the device packages are still using wire bonding (WB). Therefore, WB is the dominant form of electronic packaging for its mature technology, low cost and high reliability [1]. Precious metal bonding wires are key materials in electronic packaging including gold bonding wire, copper bonding wire, sliver bonding wire and composite metal bonding wire. Currently, semiconductor packaging is using gold wire bonding as the main connection from the chip to the lead frame or substrate. The high price of gold wire has led to the development of bonding wire made from copper silver and palladium-coated-copper as lower cost alternatives. So far, gold wire accounts for the highest proportion of bonding wire in high-end electronic products because of its stable chemical properties, good ductility and excellent weldability. According to authoritative prediction and a realistic encapsulation of the industry situation, WB will still be the main interconnection method of electronic products until 2020 and will continue developing [2].

In the package structures, bonding wire can be used as signal transmission line, power, grounding and so on. With the improvement of the transmission signal frequency (currently the system operating frequency up to GHz), bonding wire is no longer a simple transmission line. Moreover, it performs like radio frequency (RF) in a variety of ways, such as crosstalk, coupling, distortion, parasitism, ground bounce, interference and other electromagnetic phenomena [3]. Therefore, how to solve the electrical problems and improve the signal transmission quality of the gold bonding wire in circuit design, analysis and testing have become research hotspots. At present, research on the electrical properties in WB packaging is mainly focused on the following aspects: (1) the process and geometric parameters of bonding wire such as bonding process, bonding wire loop parameters, materials and so forth [4–7]; (2) the wiring form of bonding wire including wire spacing, signal pin distribution, interconnection etc., [8–11]; (3) the overall package design such as grounding copper cover, through silicon vias (TSV) technology, and the redistribution layer (RDL) of flip chip or the like [12–15]. Zhang analyzed the influence of bonding wire on parasitic parameters under different span and arch height. The larger the span and arch height, the greater the influence on the circuit characteristics by simulation [16]. Lu investigated the relationship between materials with different dielectric constants and return losses, but the geometric parameters of bonding wire were not taken into consideration, unfortunately [17]. Liang studied the height arch of single gold bonding wire and the results showed that the lower the arch height of bonding gold wire the better in the case of single gold and the same span. But the value of high arches in the simulation cannot guarantee the welding stability of flat bonding gold because of the characteristics of bonding process [18]. Owing to the diversity of the IC products and package types, there are still no satisfactory results on effects such as wire loops, geometric parameters, electromagnetic models, overall package design and other factors on comprehensive IC electromagnetic properties.

In this paper, the electromagnetic properties of bonding wire are studied. The transmission performance of bonding wire is analyzed based on the electromagnetic field model. The design rule is derived from the simulation result. Furthermore, the WB and measurement experiments are designed. The experimental data and simulation data are compared. The error is also analyzed. Finally, the correctness of the simulation is proved by comparison with other experiments, which provides the theoretical basis for the bonding wire selection and design in high frequency.

2. Bonding Wire Model

Figure 1 shows the package structure of a ball grid array (BGA) composed of a BGA solder ball, PCB board, lead frame, chip, gold bonding wires, epoxy resin encapsulation and so on. In addition, bonding wire can be used as signal transmission line, power, and grounding. In low frequency signal, the bonding wire is equivalent to a line without resistance, capacitor, or inductive transmission. However, in high frequency, the electrical properties of bonding wire have an important influence on the performance of the whole structure and even the circuit. On the one hand, the influence of the parasitic parameters of the bonding wire on the circuit cannot be ignored. At the same time, the inductance, resistance and capacitance (RLC) parameters of bonding wire are constantly changing because of the skin effect with frequency. On the other hand, the cumulative effect of many bonding wires greatly influences the signal transmission capacity, especially in high frequency system. Once the poor WB transmission performance affects the signal transmission and cause large signal loss, it may easily lead to signal crosstalk and other serious signal integrity problems.

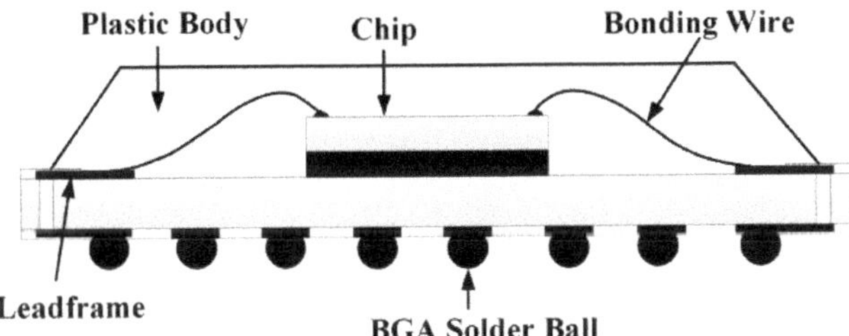

Figure 1. Wire bonding (WB) package structure.

2.1. Geometric Parameters

S-loop wire is the most common bonding wire. S-loop wire that has a longer flat distance then breaks is mainly used in two conditions: when the actual loop length of wire is too long, or the welding wire arc height should be under control in some devices. S-loop can prevent the sweep and sag problem when epoxy compound flows during the transfer injection molding process because of its strong stability and stiffness. The geometric parameters of the S-loop wire are shown in Figure 2 (d is the diameter, L is the span, L1 is the flat distance of S-loop wire, H1 is the bonding height of the arch and H2 is the connection height).

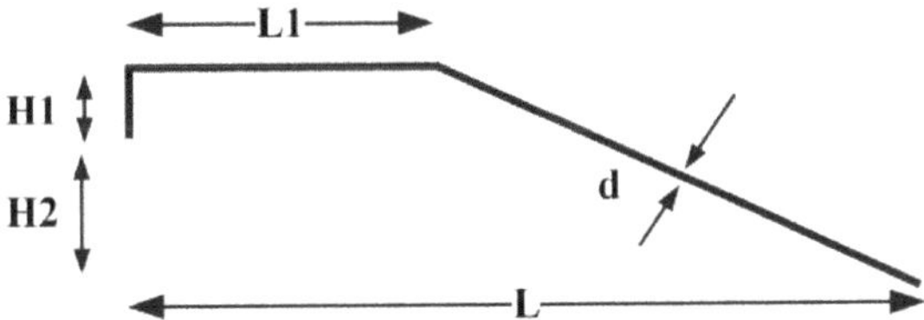

Figure 2. S-loop wire geometric parameters.

In this section, the influence of the geometrical parameters on the scattering parameters is discussed. The four geometric parameters factors are the ratio of the L1/L, the diameter d, the span L and the bonding height H1. Finally, by comparing the scattering parameters, the transmission characteristics of Q-loop, S-loop and M-loop are analyzed and derived.

2.2. Simulation Model of Single Gold Bonding Wire

The simulation model of single bonding wire (as shown in Figure 3 was composed using S-loop wire, microstrip lines, chips, substrate and ground. In practice, the bonding wire was embedded in a plastic body. The plastic material was not taken into consideration in order to simplify this model. In actual work conditions, bonding wire usually connects the chip I/O pin to the other chip pins or the external PCB board. Therefore, the two ends of the bonding wires were connected to different materials. This meant that the dielectric constant and the dielectric thickness were different. As the simulation frequency is up to GHz level, Rogers RO4350B was chosen as the substrate in order to reduce substrate dielectric loss. The materials of each part in the simulation model are shown in Table 1. The ambient temperature can be maintained at a constant and the loss factor of materials ignored as well.

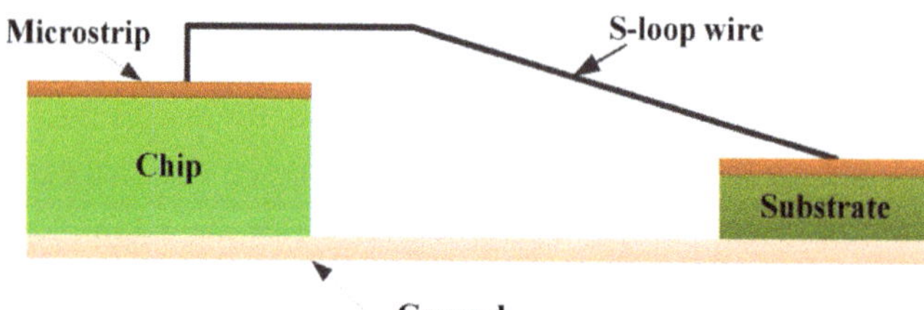

Figure 3. Single bonding wire simulation model.

Table 1. Model material parameters.

Name	Material	Dielectric Constant
S-loop wire	Gold	0.99996
Microstrip	Copper	0.99991
Chip	Silicon	11.9
Substrate	Rogers RO4350B	3.66
Ground	Copper	0.99991

2.3. Solve Setting

The bonding wire models with different geometric parameters were established by using HFSS (High Frequency Structure Simulator) electromagnetic field simulation software. After adding the radiation boundary condition and the port excitation, the corresponding analysis was set to solve the frequency of the simulation. The return loss (S11) and insertion loss (S21) of different geometric parameters models were obtained after simulation. The HFSS simulation model is shown in Figure 4. The total port excitations were set respectively on the two sides of the bonding wire. The simulation frequency was from 1MHz to 20 GHz and the logarithmic scale was used to solve the step. In this paper, skin effect can be ignored as a secondary factor. The electrical performance of the bonding wire was obtained and analyzed by using the terminal-driven solution.

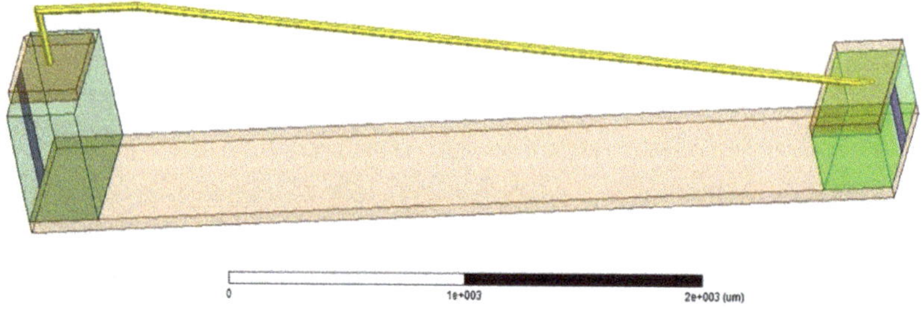

Figure 4. HFSS simulation model.

3. Frequency Domain Simulation

3.1. Influence of Flat Length Ratio on S Parameter of Bonding Wire

S-loop wire is the most common bonding wire. Due to its flat length, S-loop wire can withstand the impact from the epoxy compound in plastic molding process and prevent the gold bonding wire from having a horizontal sweep and vertical sag problem. However, too long a length of flat length wire will bring serious signal crosstalk and coupling problems. According to the Kulicke and Soffa package model library in Cadence APD, WL_LH3_SPXX series of models were selected for modeling and simulation. The value after SP is the ratio of the flat length with span, that is, the value of L1/L (as shown in Figure 5). According to the standard, there are six sizes in total: 00, 12, 20, 40, 60 and 70.

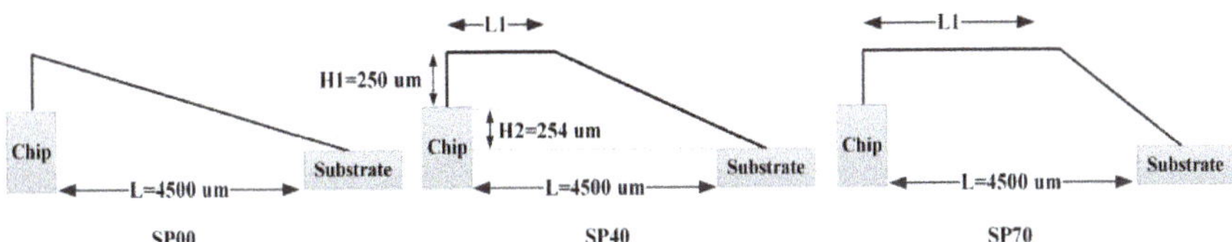

Figure 5. Three S-loop wires with different flat lengths.

According to the geometric relationship shown in Figure 5, the parameters of the model were selected (L = 4500 μm, H1 = 250 μm, H2 = 254 μm, d = 25 μm). The ratios of the flat length to the span L are 0, 12%, 20%, 40%, 60% and 70%, respectively through changing the length of L1, while sweep frequency ranges from 1 MHz to 20 GHz. The results of S11 and S21 with the ratio and frequency of the bonding wire are shown in Figure 6a,b respectively.

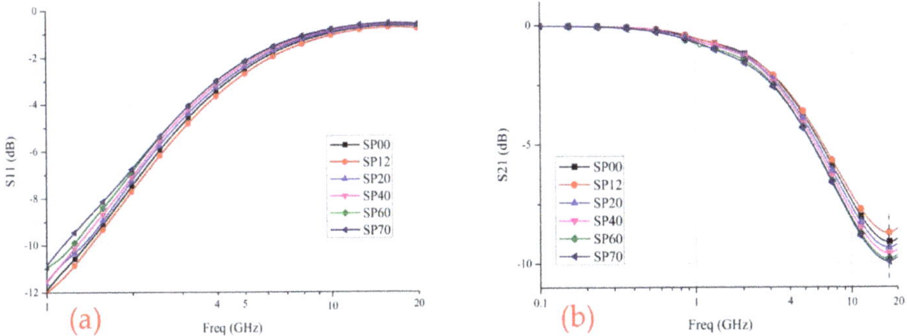

Figure 6. (**a**) S11 of bonding wire with different flat length; (**b**) S21 of bonding wire with different flat length.

In order to get a clear graph and accurate data, the S11 log coordinate was set from 1 to 20 GHz. The return loss describes the signal reflection performance of the transmission line. The smaller return loss means less signal loss and better signal integrity performance. From Figure 6, it can be seen that the loss increases with the frequency. In addition, the SP12 bonding wire has a higher cutoff frequency, which demonstrates that it can work in a higher frequency band. Compared with SP70 bonding wire, the working frequency increases by 16% (0.64 GHz). Figure 7 shows S21 (insertion loss) parameters. The abscissa is set from 0.1 to 20 GHz. It can be found that the insertion loss also increases with frequency, which suggests the ratio of the signal power delivered to the terminal prior to insertion versus the signal power delivered to the terminal after insertion is constantly decreasing. The value of the minimum point of S21 and its corresponding frequency under the condition of different flat length are selected in Table 2. The minimum point indicates the signal transmission is the worst when the ratio of the passing and input signal reaches the minimum. It is easy to get from Table 2 that the minimum values of S21 in SP12 are larger than that in other cases. At 17.38 GHz, the modulus of S21 reaches 8.71, which increases by 13.9% compared with SP70. According to the definition of S21, the ratio of passing signal with incoming signal is 36.69% after calculation.

Table 2. Comparison of S21 extremum under different flat length.

Ratio of Flat Length to Span	SP00	SP12	SP20	SP40	SP60	SP70
Min S21 (Insertion Loss)/dB	−9.10	−8.71	−9.34	−9.58	−9.82	−9.92
Corresponding frequency/GHz	17.63	17.38	17.58	17.58	17.48	17.48
Passing signal/incoming signal/%	35.07	36.69	34.12	33.19	32.28	31.92

3.2. Influence of Diameter on S Parameter of Bonding Wire

Considering the density of the pins connection, the diameter of the bonding wire is an important design parameter. Six kinds of nominal diameters of bonding wire were selected for simulation: 15 μm, 20 μm, 25 μm, 30 m, 35 μm and 40 μm. Taking the SP12 bonding wire as a model, the span of L was 4500 μm and the connection height of H1 was 250 μm. The other simulation parameters were kept constant, while the diameter of d was chosen as the variable. The sweep frequency ranges from 1MHz to 20 GHz. The results of the S11 and S21 with the diameter and frequency of the bonding wire are shown in Figure 7a,b.

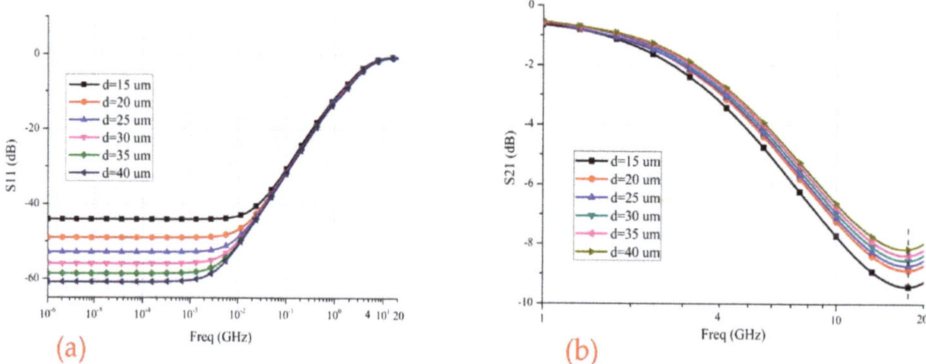

Figure 7. (a) S11 of bonding wire with different diameter; (b) S21 of bonding wire with different diameter.

Figure 7a is the logarithmic coordinate graph of S11. It can be easily seen that the return loss grows with increase of frequency. For example, the bonding wire with a diameter of 40 μm can operate at the near 5 GHz frequency. Compared with the diameter of 15 μm, the working frequency band increased by 17.5% (0.73 GHz,). Figure 7b is the logarithmic coordinate graph of S21. With the increase of frequency, the insertion loss rises, while the signal transmission performance decreases. The S21 extreme points of different diameters were extracted as shown in Table 3. When the diameter was 40 μm (17.18 GHz), S21 had the minimum value of −8.16 dB, and the ratio of passing with incoming signal was 39.06%. Compared with the diameter of 15 μm, the modulus of S21 went up by 1.26 dB (about 15.4%). Also, the signal transmission quality increased by 5.26%.

Table 3. Comparison of S21 extremum under different diameter.

Diameter/μm	15	20	25	30	35	40
Min S21 (Insertion Loss)/dB	−9.42	−8.87	−8.71	−8.55	−8.37	−8.16
Corresponding frequency/GHz	17.53	17.43	17.38	17.38	17.33	17.18
Passing signal/incoming signal/%	33.80	36.01	36.67	37.38	38.16	39.06

3.3. Influence of Span on S Parameter of Bonding Wire

With the development of the narrow spacing (fine-pinch) and the stacked package, the ratio of low radian and long span bonding wire are also rising. The span L is the horizontal distance of interconnection, which is the most important geometric parameter in the bonding wire. While keeping the other parameters unchanged, the bonding wire span L was selected as a variable. The span scanning ranged from 2000 μm to 4500 μm. In addition, the interval was 500 μm: 2000 μm, 2500 μm, 3000 μm, 3500 μm, 4000 μm and 4500 μm. The sweep frequency ranged from 1 MHz to 20 GHz.

The results of S11 and S21 with the span and frequency of the bonding wire are shown in Figure 8a,b ranging from 1MHz to 20 GHz.

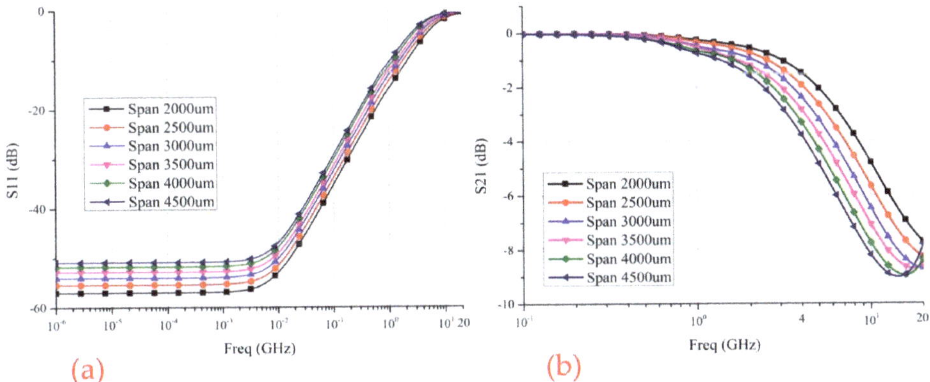

Figure 8. (**a**) S11 of bonding wire with different span; (**b**) S21 of bonding wire with different span.

According to the trends of return loss and graph of insertion loss, the transmission quality of the signal decreases with the rise of the span because of the increase of the span and the length of the bonding wire. Figure 8a is the logarithmic graph of return loss S11. Taking half the power point as an example, when the longitudinal coordinate was −3 dB, the horizontal coordinate of span 2500 µm bonding wire stood at 7 GHz, while the coordinate of span 5000 µm was 3.65 GHz. The working frequency also increased by 91%. In the insertion loss graph, the working band of the bonding wire with a span of 2500 µm was higher than 3.36GHz in the −3 dB (101%). Figure 8b is the insertion loss S21 logarithmic coordinate graph. With the increase of frequency, the insertion loss grows and the signal transmission performance drops. The S21 extreme points of different spans were extracted (as shown in Table 4). When the span was 2500 µm at 25.12 GHz, the S21 had the minimum value of −8.01 dB, and the ratio of passing with incoming signal was 39.76%. Compared with the case of the 5000 µm, the modulus of S21 increased by 0.98 dB (about 12.2%). The signal transmission quality grew by 4.4%, and the corresponding frequency of the extreme point increased by 10.79 GHz.

Table 4. Comparison of S21 extremum under different span.

Span/µm	2500	3000	3500	4000	4500	5000
Min S21 (Insertion Loss)/dB	−8.01	−8.37	−8.69	−8.71	−8.96	−9.03
Corresponding frequency/GHz	25.12	22.54	19.82	17.33	15.67	14.33
Passing signal/incoming signal/%	39.76	38.15	36.77	36.69	35.65	35.36

3.4. Influence of Bonding Height on S Parameter of Bonding Wire

The height of the bonding wire and the arch has a great influence on the electrical properties of the bonding wire. The signal transmission path also increases because of the rise of the height, which exacerbates the signal transmission loss. Keeping the other parameters unchanged, the bonding height of H1 was chosen as the variable. The bonding height scanning ranged from 210 µm to 310 µm, and the interval was 20 µm: 210 µm, 230 µm, 250 µm, 270 µm, 290 µm and 310 µm. The sweep frequency ranged from 1 MHz to 20 GHz. The results of S11 and S21 with the bonding height and frequency of the bonding wire are shown in Figure 9a,b ranging from 1 MHz to 20 GHz.

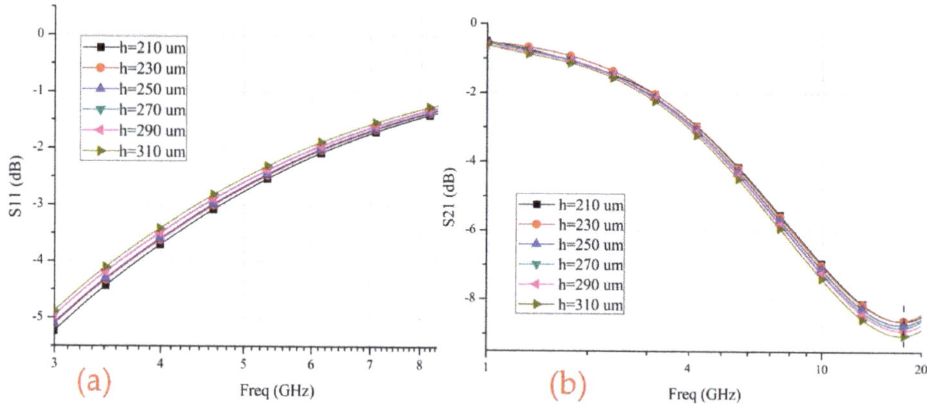

Figure 9. (**a**) S11 of bonding wire with different bonding height; (**b**) S21 of bonding wire with bonding height.

In order to get clear graph and accurate data, the S11 log coordinate was intercepted from three to 10 GHz in Figure 9a. With the increase of frequency, the return loss also increased. Furthermore, the wire with a smaller bonding height had better electrical performance. For example, when the longitudinal coordinate was −3 dB, the corresponding abscissa of bonding wire with height 210 µm was 4.69 GHz, and the abscissa of bonding wire with 310 µm is 4.40 GHz. Figure 9b is the logarithmic coordinate graph of insertion loss S21. The S21 extreme points of different bonding height are extracted in Table 5. When the bonding height was 210 µm, the S21 had the minimum value of −8.61 dB at 17.58 GHz, and the ratio of passing with incoming signal was 37.11%. Compared with the case of 310 µm, the modulus of S21 increased by 0.41 dB (about 4.8%). The signal transmission quality also rises by 1.71%. Although the increase of bonding height will cause the rise of the whole wire length and a decrease in signal quality, the connection height cannot be over reduced. A low height connection will increase stress of bonding wire, making the solder joint unstable and causing other mechanical problems.

Table 5. Comparison of S21 extremum under different bonding height.

Bonding Height/µm	210	230	250	270	290	310
Min S21 (Insertion Loss)/dB	−8.61	−8.59	−8.71	−8.79	−8.88	−9.02
Corresponding frequency/GHz	17.58	17.33	17.38	17.28	17.43	17.38
Passing signal/incoming signal/%	37.11	37.20	36.69	36.35	35.97	35.40

3.5. Influence Of Three Loops On S Parameter Of Bonding Wire

Q-loop, S-loop and M-loop bond modes are shown in Figure 10. The Q-loop is the most universal bonding wire form. Compared with the Q-loop, the S-loop has a longer flat distance. The M-loop bond has the most kinking points. Through the finite element analysis and experimental verification, the result shows that the M-loop is 13–75% as good as the S-loop [19], which suggests that the sweep resistance is better than the S-loop relying on bond span and height. This is because the M-loop bond has five kinking nodes. During the transfer molding process, the M-loop bond has higher stiffness to resist the deformation when epoxy compound flows. Keeping the diameter, span and bonding height unchanged, the S parameters of three kinds of loop wires in 1 MHz–30 GHz are simulated and compared. The results of the 1 MHz–20 GHz scan are shown in Figure 11a,b.

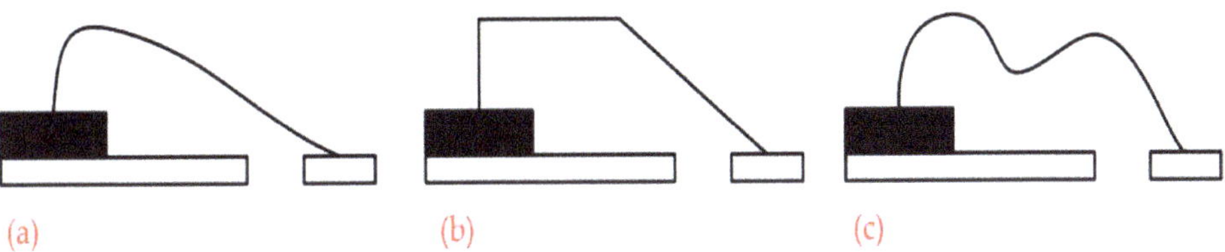

Figure 10. Bonding wire loop modes: (**a**) Q-loop bonding wire; (**b**) S-loop bonding wire; (**c**) M-loop bonding wire.

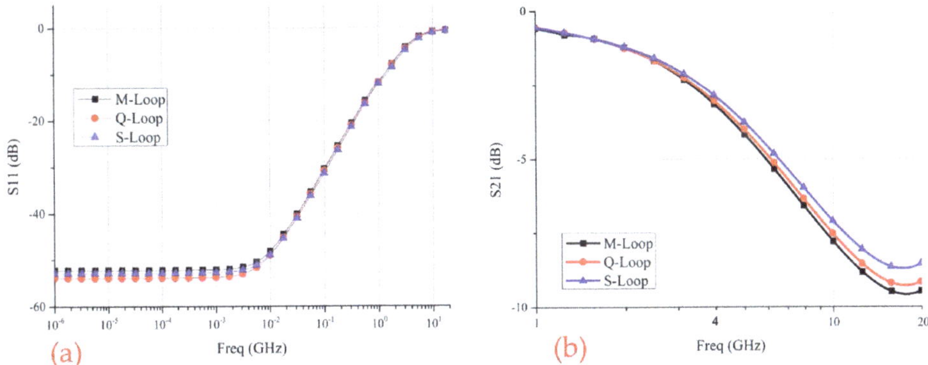

Figure 11. (**a**) S11 of bonding wire with different loop modes; (**b**) S21 of bonding wire with loop modes.

The results of the return loss S11 from 1MHz to 1GHz are shown in Figure 11a. As shown in Figure 11a, the return loss S11 of three different loop modes has little difference. Figure 11b is the logarithmic coordinate graph of the insertion loss S21. It can be seen that the electrical performance of the S-loop wire is the best, while the Q type is slightly worse than that and the M type is the worst in Figure 11b. For example, the working frequency of S type wire increases by 0.32 GHz (about 8%) compared with M-loop wire at −3 dB. The S21 extreme points of different bonding height were extracted in Table 6. For the S type bonding wire, S21 has the minimum value of −8.71 dB at 17.38 GHz, and the ratio of passing and incoming signal is 36.69%. Compared with the case of the M type wire, the modulus of S21 rises by 0.88 dB (about 10.1%), and the signal transmission quality increases by 3.54%. Therefore, reasonable bonding wire modes should be made according to the process and mechanical reliability requirements comprehensively.

Table 6. Comparison of S21 extremum under different loop modes.

Bonding Wire Loop	M	Q	S
Min S21 (Insertion Loss)/dB	−9.59	−9.29	−8.71
Corresponding frequency/GHz	17.73	17.68	17.38
Passing signal/incoming signal/%	33.15	34.32	36.69

3.6. Design Rules

With the increase of the frequency, the bonding parameters have a great influence on the characteristics of microwave transmission. Therefore, it is important to analyze, optimize and design the parasitic characteristics of high-power bonding wires under high frequency. The appropriate design parameters should be selected according to different working frequency bands. When the working frequency is too high, the transmission quality of the signal may be very poor. Combined with the practical project application, in order to reduce the parasitic characteristics while improving the frequency characteristics and electromagnetic properties of bonding wires, the optimization measures are obtained from the following five aspects (as shown in Table 7).

Table 7. The influence of WB parameters on electrical properties.

Factors	Measurement	Restriction
Flat Length (S-loop wire)	Flat length changes nonlinearly with return loss and insertion loss. The flat length only stands in a specific value, and its signal transmission performance is good	Through the comparison of the data, it is found that signal quality is the best when the flat ratio of length and span is 12%, maybe the best L1/L is 11% or 13% in real operation.
Diameter	The bigger diameter is, the better electrical properties are. Insertion loss and equivalent inductance decline with increasing diameter.	Excessive increase in wire diameter: 1. Aggravate the cost of materials 2. Decline in toughness
Span	The shorter loop span is, the better electrical properties are. Shorter loop can reduce signal loss and improve the quality of signal transmission.	Too shorter span: 1. Increase line tension, 2. Make welding spot unstable 3. Cause wire sag and sweep problem
Bonding Height	The lower wire height is, the better electrical properties are.	Too lower height: 1. Stress concentration 2. Phenomenon of expanding under hot condition and shrinking under cold condition in bonded component.

3.7. SPICE Equivalent Circuit Model Of Bonding Wire

As is shown in Figure 12, the SPICE (simulation program with integrated circuit emphasis) equivalent circuit model of bonding wire consists of resistance R, inductance L and two different capacitors C_1 and C_2. The resistance R and the inductance L are related to the parameters of the bonding wire itself. The two ends of bonding wires are connected to different materials. Therefore, the equivalent capacitance C at the ends of the bonding wire is not the same.

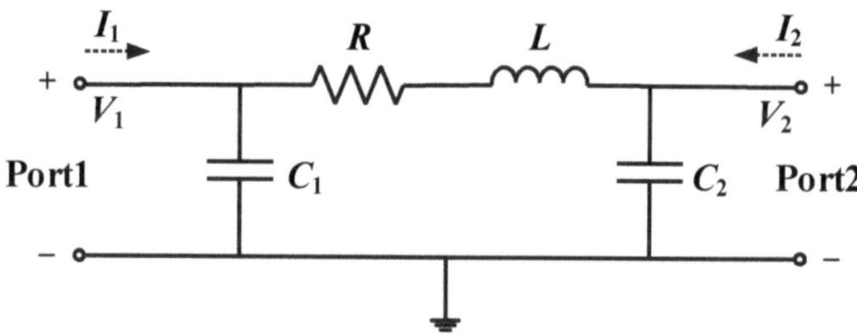

Figure 12. The SPICE equivalent circuit model of bonding wire.

In Figure 12, There are four variables; V_1, V_2, I_1 and I_2. Based on Two Port Network Theory, the following equations can be obtained:

$$\begin{cases} V_1 = I_1 \cdot Z_{11} + I_2 \cdot Z_{12} \\ V_2 = I_1 \cdot Z_{21} + I_2 \cdot Z_{22} \end{cases} \tag{1}$$

In this case the relationship between the port currents, port voltages and the Z-parameter is given by:

$$Z_{11} = \left.\frac{V_1}{I_1}\right|_{I_2=0}, Z_{12} = \left.\frac{V_1}{I_2}\right|_{I_1=0} \\ Z_{21} = \left.\frac{V_2}{I_1}\right|_{I_2=0}, Z_{22} = \left.\frac{V_2}{I_2}\right|_{I_1=0} \tag{2}$$

Figure 13 shows the equivalent impedance model of Z_{11}. Taking Z_{11} as an example, the following equations can be obtained through math deduction.

$$Z_{11}(\omega) = Z_{11}(s)|_{s=j\omega} = \frac{RC_2\omega + j(LC_2\omega^2 - 1)}{\omega(-LC_1C_2\omega^2 + C_1 + C_2 + jRC_1C_2\omega)} \tag{3}$$

Similarly, other parameters can be derived as follows:

$$Z_{22}(\omega) = Z_{22}(s)|_{s=j\omega} = \frac{RC_1\omega + j(LC_1\omega^2 - 1)}{\omega(-LC_1C_2\omega^2 + C_1 + C_2 + jRC_1C_2\omega)} \tag{4}$$

$$Z_{12}(\omega) = Z_{21}(\omega) = Z_{12}(s)|_{s=j\omega} = \frac{1}{\omega[-jLC_1C_2\omega^2 - RC_1C_2\omega + j(C_1 + C_2)]} \tag{5}$$

Based on the principle of RF circuit, the two-port S parameters can be obtained from the equivalent two-port Z parameters by means of the following expressions:

$$\begin{aligned} S_{11}(\omega) &= \frac{[Z_{11}(\omega)-Z_0]\cdot[Z_{22}(\omega)+Z_0]-Z_{12}(\omega)\cdot Z_{21}(\omega)}{[Z_{11}(\omega)+Z_0]\cdot[Z_{22}(\omega)+Z_0]-Z_{12}(\omega)\cdot Z_{21}(\omega)} \\ &= \frac{2\cdot[R+Z_0(1-LC_2\omega^2)+j(RC_2+L)\omega]}{-RC_1C_2Z_0^2\omega^2-L(C_1+C_2)Z_0\omega^2+2Z_0+R+j(-LC_1C_2Z_0^2\omega^2+RC_1Z_0+RC_2Z_0+C_1Z_0^2+C_2Z_0^2+L)\omega} - 1 \end{aligned} \tag{6}$$

$$S_{21}(\omega) = \frac{2Z_{21}(\omega)\cdot Z_0}{[Z_{11}(\omega)+Z_0]\cdot[Z_{22}(\omega)+Z_0]-Z_{12}(\omega)\cdot Z_{21}(\omega)} \tag{7}$$

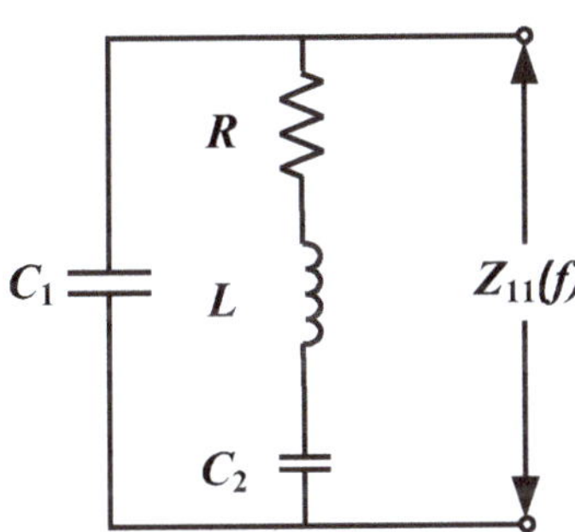

Figure 13. The equivalent impedance model of Z11.

In order to compare the data accurately and eliminate the inconsistency of the software, the simulation tools of ADS software were adopted. The diameter, flat length ratio, span, height and different loops of bonding wire will certainly affect the resistance R, inductance L and two different capacitors C1 and C2 in SPICE equivalent circuit model. As previously mentioned, the S-loop wire is the best loop modes. Meanwhile, based on the design rules, SP12 bonding wire is appropriate for the model and the parameters of this model were selected (L = 4500 µm, H1 = 250 µm, H2 = 254 µm, d = 25 µm). Compared with the result file solved by the HFSS electromagnetic field simulation software and S parameter curve obtained by the ADS frequency domain simulation solver, the results of the S11 and S21 are shown in Figure 14. As is shown below, the two curves are basically well fitted, which can verify the correctness of the SPICE equivalent circuit model.

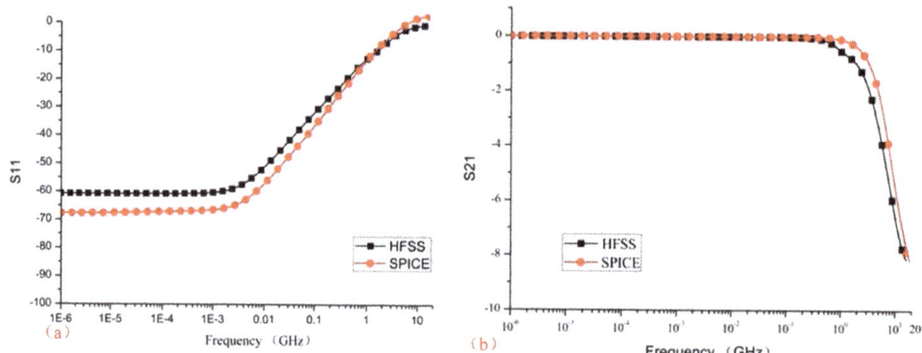

Figure 14. Comparison between SPICE and HFSS (**a**) S_{11}; (**b**) S_{21}.

3.8. Eye Diagram Analysis

The eye diagram is a very successful and effective way of showing the quality and parametric information in digital transmission. Careful analysis of this visual display can give the user a first-order approximation of signal-to-noise, clock timing jitter and skew.

SP12 bonding wire (H1 = 250 µm, H2 = 254 µm, d = 25 µm) were selected as an illustration. In order to control variables, H2 is constant in this part. The span scanning ranged from 2000 µm to 4500 µm. In addition, the interval was 500 µm: 2000 µm, 2500 µm, 3000 µm, 3500 µm, 4000 µm and 4500 µm. Meanwhile, L1 varied from different span According to the schematic in Figure 15, the excitation signal was added at the input port as follows, the high level was 1 V, the low level was 0 V, the rising and falling time was set to 50 ps, and the bit rate of the signal was 10 Gbps. In this part, it recognizes a high level to be a logic 1 and a low level as a logic 0. The voltage values of logic 0 and logic 1, eye height and eye width in the eye diagram are shown in Table 8.

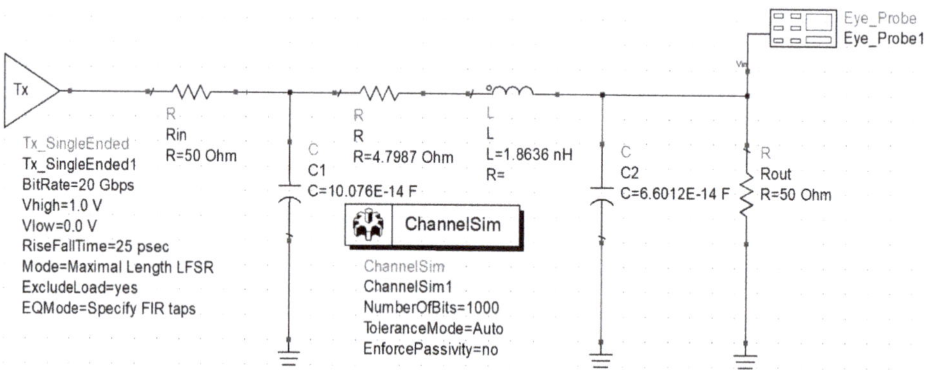

Figure 15. Eye diagram analysis schematic.

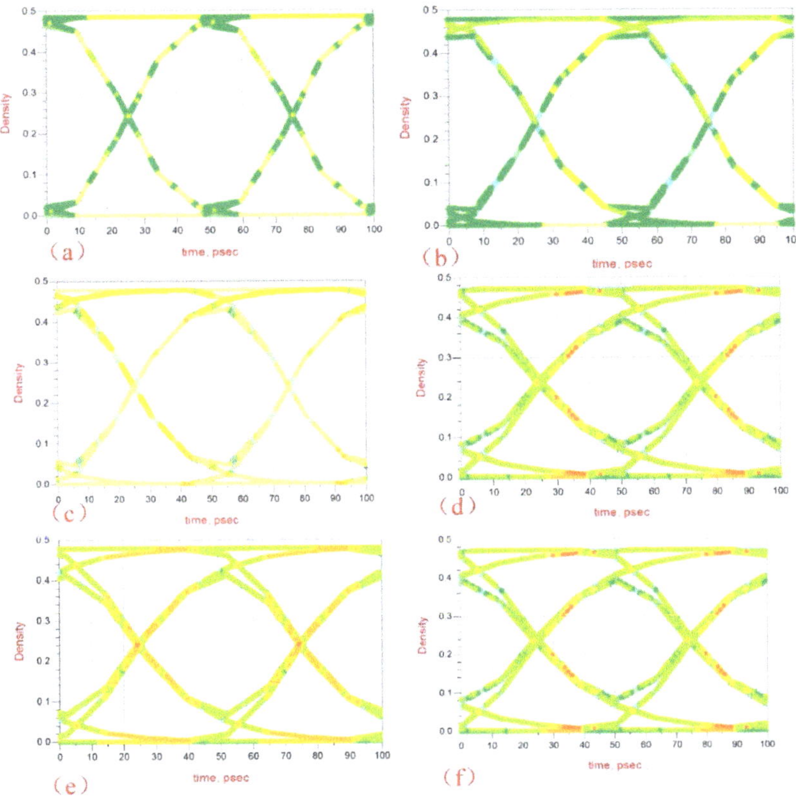

Figure 16. Eye diagram with different span. (**a**) 2000 μm (**b**) 2500 μm (**c**) 3000 μm (**d**) 3500 μm (**e**) 4000 μm (**f**) 4500 μm.

As illustrated in Table 8, with the increase of span, logic 1 voltage decreases, logic 0 voltage increases, and this will cause signal inconsistency between the input port and the output port and the increase of the bit error rate. Furthermore, as shown in Figure 16, the problem of unclear trace was aggravated with the increase of span, which will led to inter-symbol interference.

Table 8. Eye diagram characteristics under different span.

Number	Span/μm	Logical 1 Voltage/V	Logical 0 Voltage/V	Eye Height/V	Eye Width/ps
a	2000	0.473	0.012	0.424	50
b	2500	0.458	0.021	0.392	50
c	3000	0.450	0.028	0.351	49.5
d	3500	0.441	0.032	0.320	49.25
e	4000	0.440	0.039	0.294	48.75
f	4500	0.425	0.046	0.261	48

4. Bonding Wire Design and Measurement Experiment

In order to improve the correctness of the bonding wire simulation model, the bonding wire measurement experiments were designed in this section. The TPT HB05 type manual bonding machine was used to bond the gold wire in the test board. In addition, the vector network analyzer (VNA) was used to measure the bonding wire S parameter in the experiment. At the same time, based on the network port extension and port embedding theory, the S parameters of the device under test (DUT) were derived. With the error analysis, the results were compared with the original measurement data, which proves the correctness of the simulation results and design rules.

4.1. Circuit Board Design

In order to reduce the measurement error, it is necessary to shorten the length of the microstrip line and make the impedance matching design. Since the measurement frequency band had reached the GHz level, a Rogers RO4350B high-speed plate was chosen as substrate in accordance with simulation model. The test board length and width were 3.5 cm × 5.0 cm.

By using the empirical formula of the characteristic impedance for microstrip line, the approximate calculation of characteristic impedance is [20]:

$$Z_0 = \frac{87}{\sqrt{\varepsilon_r + 1.41}} \cdot \ln\left(\frac{5.98H}{0.8W + T}\right) \tag{8}$$

In Equation (1), Z_0 is the characteristic impedance of the microstrip line. As shown in Equation (1), ε_r is the dielectric constant of the substrate, H is the thickness of the substrate, W is the width of the microstrip line, and the T is the thickness of the microstrip line. The test board was connected with the VNA using the SMA connector. In order to weld with the SMA joint and shorten the microstrip line, the microstrip line length L was designed as 8 mm. The plate thickness H was 0.508 mm. A Rogers RO4350B was selected as double-layer plate with a dielectric constant of 3.48. Keeping the Z_0 as constant 50 Ω based on Equation (1) and Cadence two-dimensional field solver impedance design tool, the rest of design parameters are obtained after adjustment: microstrip width W is 0.85 mm, thickness T is 0.05 mm. The specific microstrip line design parameters are as shown in Figure 17.

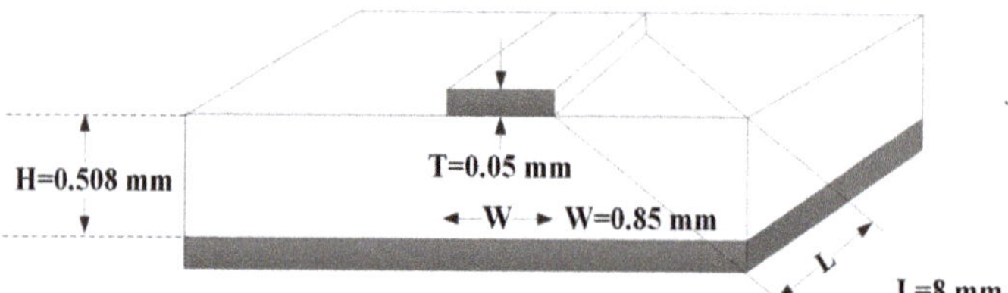

Figure 17. Microstrip line design parameters.

A TPT HB05 type semi-automatic bonding machine works on the basis of thermal compress bonding method. High pressure and high temperature caused metal plastic deformation in the bonding region, which realized the connection between the pad and the lead. Therefore, the hot pressing bonding method needs the hot pressing area to adopt the same metallurgy material, so that the hot pressing bonding process can be completed. The conventional PCB technique is not suitable for the bonding. This means that depositing Sn on copper pads as tin does not make the wire pressing successful. Considering about the bonding machine hot press temperature, hot press reliability and cost, bare gold process is the final choice for the microstrip line. There is no solder layer on the line, while the back of circuit board is deposited with gold as well as ground plane. Finally, the front and back of the test board are shown in Figure 18a,b.

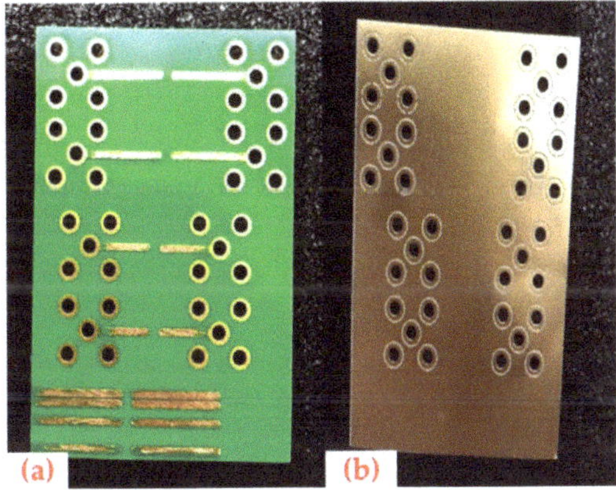

Figure 18. Test circuit board: (**a**) front (**b**) back.

4.2. Bonding Wire Shape

The semi-automatic bonding machine (TPT HB05 type) was used to bond on the surface of a gold plated microstrip line. After adjusting the bonding parameters, the bonding wire under the requirements of the ball pad was finally obtained. The TPT HB05 bonding machine and bonding parameter settings are shown in Figure 19a,b.

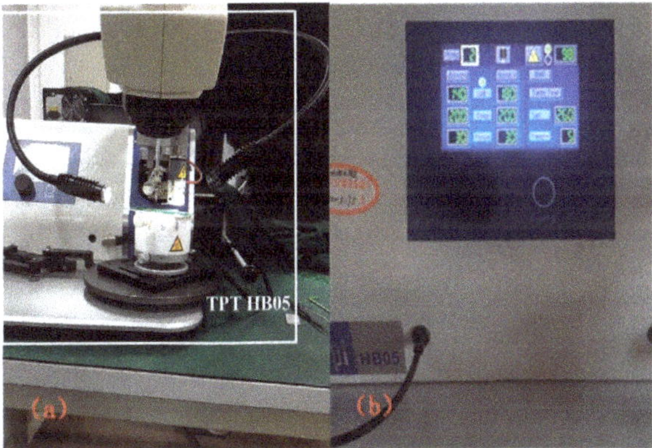

Figure 19. (**a**)TPT HB05 bonding machine (**b**) Bonding parameter settings.

The observation of bonding wire under video meter system (VMS) is shown in Figure 20a. The span was 4480 µm by measurement. To make sure there was no false welding, the first and second bonding point were checked through a microscope. The first and second bonding point of the bonding wire under the metallographic microscope is shown from Figure 20b–d.

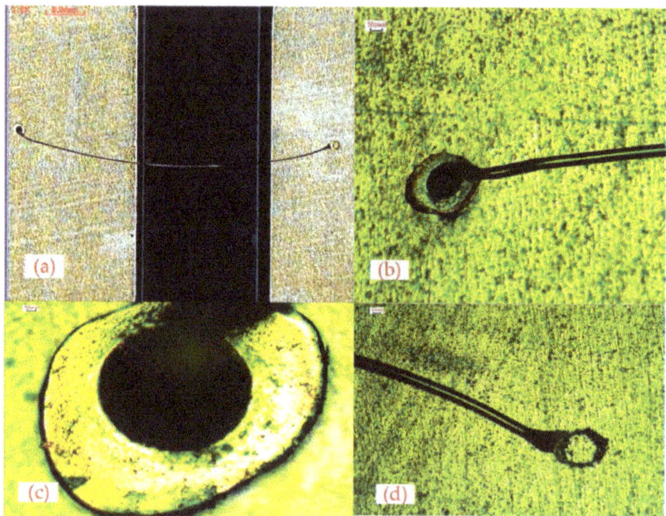

Figure 20. (**a**) Bonding wire under VMS; (**b**) First bonding point under 10 times microscope; (**c**) First bonding point under 40 times microscope; (**d**) Second bonding point under 20 times microscope.

4.3. S Parameter Measurement And Error Analysis

The measurement band is determined from 0.3 MHz to 10 GHz using VNA. The test board was connected to the instrument through SMA. Also, the S parameters of the device were measured after calibration. The S11, ϑ11, S21 and ϑ21 measured graphs of bonding wire are shown from Figure 21a–d. The measured data were compared with the simulation data (as shown in Figure 22).

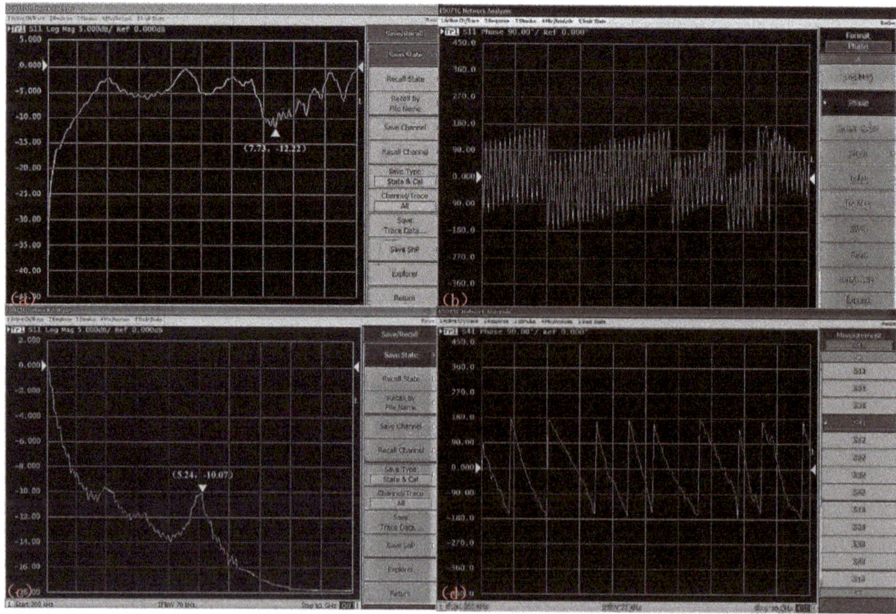

Figure 21. (**a**) Measured S11 of circuit-under-test; (**b**) Measured ϑ11 of circuit-under-test; (**c**) Measured S21 of circuit-under-test; (**d**) Measured ϑ21 of circuit-under-test.

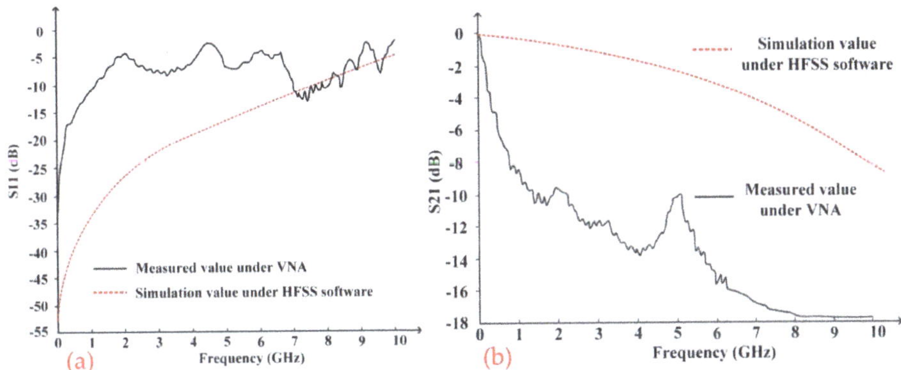

Figure 22. Comparison between simulated and measured values: (**a**) S11; (**b**) S21.

It can be seen from Figure 22 that the trend of simulation curve and measured curve are consistent, that is, the S11 increases and S21 decreases with the rise of the frequency. This means that the signal reflection increases and the passing signal drops. The peak value of the test curve may appear because of the resonance problem of the whole test system. However, there is a certain gap between the measured value and the simulation value. The maximum error was about −20 dB in S11 curve, while −7 dB in S21 curve. The error comes mainly from the following three aspects:

1. Systematic error of measurement experiment

Due to the use of the SMA connector, the error of the measurement system is introduced to complete the measurement such as SMA conversion error, reflection/transmission measurement circuit error and fixture embedded error, as well as the random error in measurement.

2. Model error

For example, the simulation model is not precise enough resulting in less accurate modeling of the first and second bonding point; unable to establish completely accurate bonding wire model of actual experiment wire; the excitation port of stimulation is inconsistent with the actual measurement port.

3. Manufacturing error

The dielectric constant of the test plate is not constant resulting in dielectric loss; the actual impedance is inconsistent with calculation causing impedance mismatch and reflection loss problem.

4.4. Port Extension

Based on network analysis theory, the measured S parameters will inevitably have a fixture effect because the measured parameter is a joint cascading result with DUT and the fixture. The port extension is to move the position of the calibration reference and compensate the transmission line loss by measuring the electrical delay and loss for each single port. That is to say, the fixture is equivalent to a lossless transmission line (or lossy transmission line) under certain conditions. The measurement method is also used to measure the loss and phase shift of the transmission line. Finally, the reference surface is extended to the actual device to reduce the influence of the fixture on the final measuring results. The extension reference and actual reference in test are shown in Figure 23.

Figure 23. Extension reference and actual reference in test.

Based on the RF network theory:

$$\begin{bmatrix} b_1 \\ b_2 \end{bmatrix} = [S]_{origin} \cdot \begin{bmatrix} a_1 \\ a_2 \end{bmatrix} = \begin{bmatrix} S_{11} & S_{12} \\ S_{21} & S_{22} \end{bmatrix} \cdot \begin{bmatrix} a_1 \\ a_2 \end{bmatrix} \tag{9}$$

$$\begin{bmatrix} b'_1 \\ b'_2 \end{bmatrix} = [S]_{extension} \cdot \begin{bmatrix} a'_1 \\ a'_2 \end{bmatrix} = \begin{bmatrix} S'_{11} & S'_{12} \\ S'_{21} & S'_{22} \end{bmatrix} \cdot \begin{bmatrix} a'_1 \\ a'_2 \end{bmatrix} \tag{10}$$

From the signal flow graph in Figure 24, the relationship between normalized reflection wave and incident wave is as follows:

$$\begin{bmatrix} a_1 \\ a_2 \end{bmatrix} = \begin{bmatrix} a'_1 e^{-j\theta_1} \\ a'_2 e^{-j\theta_2} \end{bmatrix} = \begin{bmatrix} e^{-j\theta_1} & 0 \\ 0 & e^{-j\theta_2} \end{bmatrix} \cdot \begin{bmatrix} a'_1 \\ a'_2 \end{bmatrix} \tag{11}$$

$$\begin{bmatrix} b'_1 \\ b'_2 \end{bmatrix} = \begin{bmatrix} b_1 e^{-j\theta_1} \\ b_2 e^{-j\theta_2} \end{bmatrix} = \begin{bmatrix} e^{-j\theta_1} & 0 \\ 0 & e^{-j\theta_2} \end{bmatrix} \cdot \begin{bmatrix} b_1 \\ b_2 \end{bmatrix} \tag{12}$$

According to the relationship among the parameters, the scattering matrix of extended surface [S] extension can be obtained after the calculation:

$$\begin{aligned} \begin{bmatrix} b'_1 \\ b'_2 \end{bmatrix} &= \begin{bmatrix} b_1 e^{-j\theta_1} \\ b_2 e^{-j\theta_2} \end{bmatrix} = \begin{bmatrix} e^{-j\theta_1} & 0 \\ 0 & e^{-j\theta_2} \end{bmatrix} \cdot \begin{bmatrix} b_1 \\ b_2 \end{bmatrix} = \begin{bmatrix} e^{-j\theta_1} & 0 \\ 0 & e^{-j\theta_2} \end{bmatrix} \cdot \begin{bmatrix} S_{11} & S_{12} \\ S_{21} & S_{22} \end{bmatrix} \cdot \begin{bmatrix} a_1 \\ a_2 \end{bmatrix} \\ &= \begin{bmatrix} e^{-j\theta_1} & 0 \\ 0 & e^{-j\theta_2} \end{bmatrix} \cdot \begin{bmatrix} S_{11} & S_{12} \\ S_{21} & S_{22} \end{bmatrix} \cdot \begin{bmatrix} e^{-j\theta_1} & 0 \\ 0 & e^{-j\theta_2} \end{bmatrix} \cdot \begin{bmatrix} a'_1 \\ a'_2 \end{bmatrix} \end{aligned} \tag{13}$$

$$[S]_{extention} = \begin{bmatrix} e^{-j\theta_1} & 0 \\ 0 & e^{-j\theta_2} \end{bmatrix} \cdot \begin{bmatrix} S_{11} & S_{12} \\ S_{21} & S_{22} \end{bmatrix} \cdot \begin{bmatrix} e^{-j\theta_1} & 0 \\ 0 & e^{-j\theta_2} \end{bmatrix} = \begin{bmatrix} S_{11}e^{-j2\theta_1} & S_{12}e^{-j(\theta_1+\theta_2)} \\ S_{21}e^{-j(\theta_1+\theta_2)} & S_{22}e^{-j2\theta_2} \end{bmatrix} \tag{14}$$

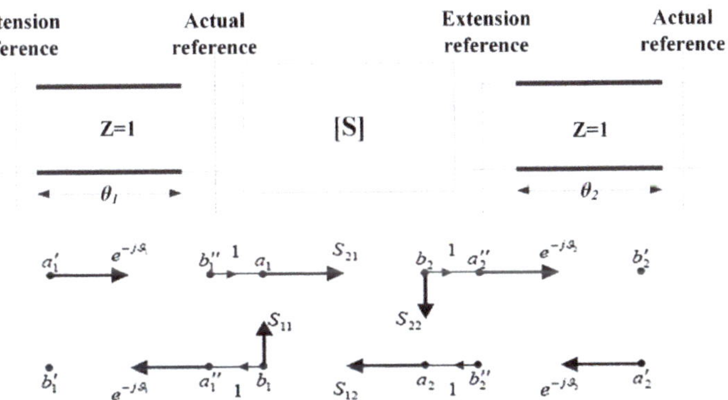

Figure 24. Port extension outward and signal flow graph.

Taking the bonding wire with 4480 μm span as the example, the port extension and data analysis were performed at 2.5 GHz, 3.0 GHz, 3.5 GHz, 4.0 GHz and 4.5 GHz frequency points. After calibration of the VNA, the test circuit board was connected to Port 1, while Port 2 was kept empty. Then the reflection coefficient file was obtained. Furthermore, according to the cascade algorithm and signal flow chart, the port extension file was written in MATLAB. The S parameters of the final bonding wire are shown in the Table 9 after calculation.

Table 9. Bonding wire S parameter results file DUT_WireBond.TXT.

Frequency/GHz	$|S_{11}|$	ϑ_{11}	$|S_{21}|$	ϑ_{21}
2.5	6.67×10^{-1}	-2.51×10	5.51×10^{-2}	3.27×10
3.0	6.94×10^{-1}	5.87×10	5.03×10^{-2}	-4.62×10
3.5	7.03×10^{-1}	2.60×10^2	4.16×10^{-2}	-8.54×10
4.0	7.10×10^{-1}	-4.21×10	4.21×10^{-2}	-1.08×10
4.5	7.71×10^{-1}	8.59×10	4.05×10^{-2}	1.85×10^2

Frequency/GHz	$|S_{12}|$	ϑ_{12}	$|S_{22}|$	ϑ_{22}
2.5	4.65×10^{-2}	3.24×10	6.89×10^{-1}	-2.69×10
3.0	4.47×10^{-2}	-2.89×10	7.23×10^{-1}	4.17×10
3.5	4.32×10^{-2}	-5.69×10	7.42×10^{-1}	3.28×10^2
4.0	4.02×10^{-2}	-6.78×10^2	7.38×10^{-1}	-8.26×10
4.5	3.65×10^{-2}	1.95×10^2	7.94×10^{-1}	1.22×10

The data of 5 fixed frequency points are calculated by extending the port, and the calculated data were compared with the result and original test data without error elimination using HFSS three-dimensional electromagnetic field simulation. The comparison graph of the three cases above is shown in Figure 25.

From the experimental data, system error can be eliminated to some extent by the port extension method making the measurement result more accurate. From the comparison between the simulation and the experimental data, the trends of S11 and S21 parameters were consistent. There was a deviation in some specific values, and the S parameter error contrast value was extracted in Table 10 among the five fixed points.

Due to limited experimental method and inability to establish the precise and accurate bonding wire loop model of experiment wire, even after port extension method and data processing, there was still some deviation between the simulation and measurement values. However, the experimental results have confirmed the results of S parameters simulation for the bonding wire and design rules. The error is also within the reasonable range. Therefore, the correctness of the simulation is proved.

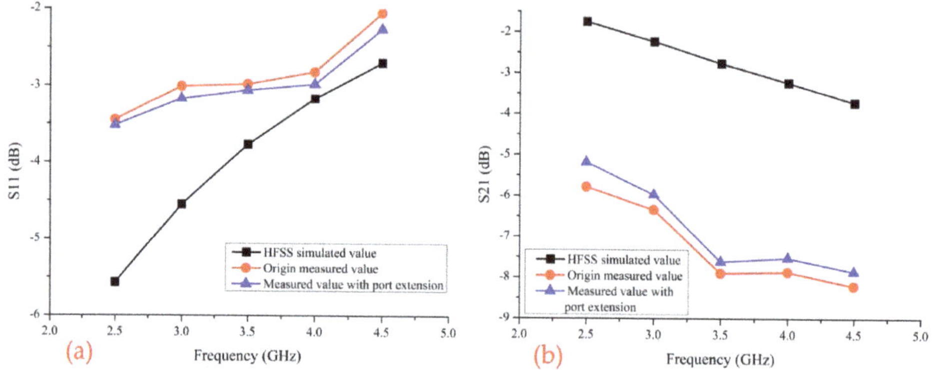

Figure 25. Comparison of S parameters: (**a**) S11; (**b**) S21.

Table 10. Comparison between simulated and measured value with port extension.

S Parameter	Max Relative Error	Frequency	Min Relative Error	Frequency
S11	33%	2.5 GHz	11%	4.0 GHz
S21	69%	3.5 GHz	39%	2.5 GHz

5. Conclusions

This article adopted the method of electromagnetic field simulation analysis and actual experiment test. In addition, the electrical properties of gold bonding wire in WB package were studied. According to EIA/JEDEC97 standard, this paper establishes the electromagnetic structure model of gold bonding wire. The equivalent circuit model of bonding wire was proposed and discussed. The effect of bonding wire on signal transmission was analyzed by an eye diagram as well. Although it was difficult to obtain accurate mathematical relations between the electrical parameters and geometries and materials of bonding wire, the reasonable parameter select method and general rules can be obtained after qualitative analysis in this paper. The platform length, diameter, span, bonding height parameters and the influence of three kinds of loops on the S parameters of bonding wire were discussed. The results showed that: (1) the transmission performance of the bonding wire with high diameter, short span and arch height is good; (2) the flat length ratio only in a specific value, its signal transmission performance is good; (3) S-type arc signal transmission performance is the best, Q-type the second, M-type the worst. Furthermore, the design rules were derived from the simulation data. Meanwhile, based on RF circuit theory analysis and test method, gold bonding wire design and measurement experiments were implemented. The original measurement data was compared with the simulation model data and the error was analyzed. At last, the data of five frequency points were processed to eliminate the fixture error as much as possible based on port embedding theory. The measurement results using port extension method were compared with the original measurement data and electromagnetic field simulation data, which proves the correctness of the simulation results and design rules.

Author Contributions: W.T. guided other authors to complete this paper. He also supplied the latest supporting material. H.C. wrote the paper. W.Y. helped to revise the grammar of literature.

Acknowledgments: This work was supported by the National Natural Science Foundation of China (61176130), the Natural Science Foundation of Ningbo city of China (2016A610030).

References

1.   Fischer, A.C.; Korvink, J.G.; Roxhed, N. Unconventional applications of wire bonding create opportunities for microsystem integration. *J. Micromechan. Microeng.* **2013**, *23*, 905–923. [CrossRef]

2. Chao, Y.Q.; Yang, Z.J.; Qiao, H.L. Progress on Technology of Wire Bonding. *J. Electron. Process Technol.* **2007**, *4*, 205–210.

3. Tian, W. *Electronic Packaging, Microelectromechanical and Micro System*; Xidian University Press: Xi'an, China, 2012; Volume 1, pp. 59–66.

4. Casper, T. Electrothermal simulation of bonding wire degradation under uncertain geometries. In Proceedings of the Automation & Test in Europe Conference & Exhibition, Dresden, Germany, 14–18 March 2016.

5. Manoharan, S.; Patel, C.; Mccluskey, P.; Pecht, M. Effective decapsulation of copper wire-bonded microelectronic devices for reliability assessment. *Microelectron. Reliab.* **2018**, *84*, 197–207. [CrossRef]

6. Ke, M.; Li, D.; Dai, X.; Jiang, H.; Deviny, I.; Luo, H.; Liu, G. Improved surge current capability of power diode with copper metallization and heavy copper wire bonding. In Proceedings of the IEEE European Conference on Power Electronics & Applications, Karlsruhe, Germany, 5–9 September 2016.

7. Lorenz, G.; Naumann, F.; Mittag, M.; Petzold, M. In Mechanical characterization of bond wire materials in electronic devices at elevated temperatures. In Proceedings of the IEEE 5th Electronics System-integration Technology Conference (ESTC), Helsinki, Finland, 16–18 September 2014; pp. 1–6.

8. Zhong, Z.W. Overview of wire bonding using copper wire or insulated wire. *Microelectron. Reliab.* **2011**, *51*, 4–12. [CrossRef]

9. Liu, P.; Tong, L.; Wang, J.; Shi, L.; Tang, H. Challenges and developments of copper wire bonding technology. *Microelectron. Reliab.* **2012**, *52*, 1092–1098. [CrossRef]

10. Ndip, I.; Huhn, M.; Brandenburger, F.; Ehrhardt, C.; Schneider-Ramelow, M.; Reichl, H.; Lang, K.D.; Henke, H. Experimental verification and analysis of analytical model of the shape of bond wire antennas. *Electron. Lett.* **2017**, *53*, 906–908. [CrossRef]

11. Zuo, P.; Wang, M.; Li, H.; Song, T.; Liu, J.; Li, E.P. Modeling and analysis of transmission performance of bonding wire interconnection. In Proceedings of the IEEE MTT-S International Conference on Numerical Electromagnetic & Multiphysics Modeling & Optimization, Beijing, China, 27–29 July 2016.

12. Yu, X.Q.; Bai, Y.D.; Zhou, Y.; Bai, W.; Yang, L.; Min, J.X. In EMI study of high-speed IC package based on pin map. In Proceedings of the IEEE Asia-Pacific International Symposium on Electromagnetic Compatibility, New York, NY, USA, 21–24 May 2012; pp. 305–308.

13. Huang, N.K.H.; Jiang, L.J.; Yu, H.C.; Li, G.; Xu, S.; Ren, H.S. Fundamental components of the IC packaging electromagnetic interference (EMI) analysis. In Proceedings of the IEEE 21st Conference on Electrical Performance of Electronic Packaging and Systems, New York, NY, USA, 21–24 October 2012; pp. 141–144.

14. Lim, J.H.; Kwon, D.H.; Rieh, J.S.; Kim, S.W.; Hwang, S.W. RF characterization and modeling of various wire bond transitions. *IEEE Trans. Adv. Packag.* **2005**, *28*, 772–778.

15. Radojcic, R.; Chandrasekaran, A.; Lane, R. Hybrid Package Construction with Wire Bond and through Silicon Vias. U.S. Patent No. 8,803,305, 12 August 2014.

16. Zhang, Y. Study on Influence of Grounding Vias Substrate on Performance of Millimeter Wave Products. *J. Radio Eng.* **2017**, *47*, 60–63.

17. Lu, F.; Cao, Y.; Lian, B. Study on the electrical performance of one differential pair bonding fingers on different layers in wire bonding package. In Proceedings of the IEEE International Conference on Electronic Packaging Technology, Wuhan, China, 16–19 August 2016.

18. Liang, Y.; Huang, C.; Wang, W. Modeling and characterization of the bonding-wire interconnection for microwave. In Proceedings of the IEEE MCM International Conference on Electronic Packaging Technology & High Density Packaging, Xi'an, China, 16–19 August 2010.

19. Kung, H.K.; Chen, H.S.; Lu, M.C. The wire sag problem in wire bonding technology for semiconductor packaging. *Microelectron. Reliab.* **2013**, *53*, 288–296. [CrossRef]

20. Liu, G.M. Dielectric Resonator Antenna Research. Master's Thesis, Zhejiang University, Hangzhou, China, 2005.

MDPI
St. Alban-Anlage 66
4052 Basel
Switzerland
Tel. +41 61 683 77 34
Fax +41 61 302 89 18
www.mdpi.com

Electronics Editorial Office
E-mail: electronics@mdpi.com
www.mdpi.com/journal/electronics